This book is a comprehensive introduction to electron–atom collisions, covering both theory and experiment. The interaction of electrons with atoms is the field that most deeply probes both the structure and reaction dynamics of a many-body system. The book begins with a short account of experimental techniques of cross-section measurement. It then introduces the essential quantum mechanics background needed. The following chapters cover one-electron problems (from the classic particle in a box to a relativistic electron in a central potential), the theory of atomic bound states, formal scattering theory, calculation of scattering amplitudes, spin-independent and spin-dependent scattering observables, ionisation and electron momentum spectroscopy. The connections between experimental and theoretical developments are emphasised throughout.

Graduate students and researchers in atomic, molecular, and chemical physics will find this text a valuable introduction to a subject of central importance.

Cambridge Monographs on Atomic, Molecular,
and Chemical Physics 5

General editors: A. Dalgarno, P. L. Knight, F. H. Read, R. N. Zare

ELECTRON–ATOM COLLISIONS

Cambridge Monographs on
Atomic, Molecular, and Chemical Physics

1. R. Schinke: *Photodissociation Dynamics*
2. L. Frommhold: *Collision-induced Absorption in Gases*
3. T. F. Gallagher: *Rydberg Atoms*
4. M. Anzinsh and R. Ferber: *Optical Polarization of Molecules*
5. I. E. McCarthy and E. Weigold: *Electron–atom Collisions*

ELECTRON–ATOM COLLISIONS

IAN E. McCARTHY

The Flinders University of South Australia

ERICH WEIGOLD

Australian National University

CAMBRIDGE
UNIVERSITY PRESS

CAMBRIDGE UNIVERSITY PRESS
Cambridge, New York, Melbourne, Madrid, Cape Town, Singapore, São Paulo

Cambridge University Press
The Edinburgh Building, Cambridge CB2 2RU, UK

Published in the United States of America by Cambridge University Press, New York

www.cambridge.org
Information on this title: www.cambridge.org/9780521413596

First published 1995
This digitally printed first paperback version 2005

A catalogue record for this publication is available from the British Library

Library of Congress Cataloguing in Publication data
McCarthy, I. E. (Ian Ellery), 1930–
Electron–atom collisions / Ian E. McCarthy, Erich Weigold.
p.   cm. – (Cambridge monographs on atomic, molecular, and
chemical physics; 5)
Includes bibliographical references.
ISBN 0 521 41359 1
1. Electron–atom collisions.   I. Weigold, Erich.   II. Title.
III Series.
QC793.5.E628M3   1995
539.7′57 – dc20   94-16810 CIP

ISBN-13 978-0-521-41359-6 hardback
ISBN-10 0-521-41359-1 hardback

ISBN-13 978-0-521-01968-2 paperback
ISBN-10 0-521-01968-0 paperback

Contents

Preface   xi

1   Introduction   1

2   Experimental techniques for cross-section measurements   4
2.1   Concept of cross sections   5
2.2   Measurement of total and integral cross sections   8
2.3   Measurement of differential cross sections   14
2.4   Ionisation   22
2.5   Polarised electrons   32
2.6   Polarised atom sources   39
2.7   Electron–photon correlation experiments   45

3   Background quantum mechanics in the atomic context   50
3.1   Basic mathematical constructions   50
3.2   Physical interpretation   58
3.3   Angular momentum   61
3.4   The Pauli exclusion principle   71
3.5   The Dirac equation   77

4   One-electron problems   81
4.1   Particle in a cubic box   81
4.2   The Schrödinger equation for a local, central potential   82
4.3   Bound states in a local, central potential   82
4.4   Potential scattering   87
4.5   Integral equations for scattering   98
4.6   Resonances   104
4.7   Relativistic electron in a local, central potential   111

5	**Theory of atomic bound states**	115
5.1	The Hartree—Fock problem	116
5.2	Numerically-specified orbitals	120
5.3	Analytic orbitals	123
5.4	Frozen-core Hartree—Fock calculations	125
5.5	Multiconfiguration Hartree—Fock	126
5.6	Configuration interaction	128
5.7	Perturbation theory	133
5.8	Comparison with spectroscopic data	135
6	**Formal scattering theory**	139
6.1	Formulation of the problem	139
6.2	Box-normalised wave-packet states	142
6.3	Integral equation for the box-normalised collision state	143
6.4	The physical limiting procedure : normalisation	144
6.5	Transition rate and differential cross section	145
6.6	The optical theorem	146
6.7	Differential cross section for scattering	147
6.8	Differential cross section for ionisation	148
6.9	The continuum limit : Lippmann—Schwinger equation	149
6.10	The distorted-wave transformation	152
7	**Calculation of scattering amplitudes**	156
7.1	Antisymmetrisation	157
7.2	Reduced Lippmann—Schwinger equations	164
7.3	Potential matrix elements	168
7.4	The complete set of target states	178
7.5	The optical potential	179
7.6	Alternative methods for restricted energy ranges	190
8	**Spin-independent scattering observables**	199
8.1	Collisional alignment and orientation	200
8.2	Hydrogen	212
8.3	Sodium	225
8.4	Two-electron atoms	230
9	**Spin-dependent scattering observables**	235
9.1	Origin of spin-dependent effects	236
9.2	Combined effects of several polarisation mechanisms	241
9.3	One-electron atoms	247
9.4	Closed-shell atoms	252

10   Ionisation   261
10.1   Formulation of the three-body ionisation problem   263
10.2   Inner-shell ionisation   274
10.3   Ionisation near threshold   275
10.4   Excitation of autoionising resonances   279
10.5   Integrated cross sections   283
10.6   Total ionisation asymmetry   288

11   Electron momentum spectroscopy   289
11.1   Basic theory   289
11.2   Examples of structure information   300
11.3   Excited and oriented target states   307

References   310
Index   321

Preface

The advancement of knowledge of electron–atom collisions depends on an iterative interaction of experiment and theory. Experimentalists need an understanding of theory at the level that will enable them to design experiments that contribute to the overall understanding of the subject. They must also be able to distinguish critically between approximations. Theorists need to know what is likely to be experimentally possible and how to assess the accuracy of experimental techniques and the assumptions behind them. We have aimed to give this understanding to students who have completed a program of undergraduate laboratory, mechanics, electromagnetic theory and quantum mechanics courses.

Furthermore we have attempted to give experimentalists sufficient detail to enable them to set up a significant experiment. With the development of position-sensitive detectors, high-resolution analysers and monochromators, fast-pulse techniques, tuneable high-resolution lasers, and sources of polarised electrons and atoms, experimental techniques have made enormous advances in recent years. They have become sophisticated and flexible allowing complete measurements to be made. Therefore particular emphasis is given to experiments in which the kinematics is completely determined. When more than one particle is emitted in the collision process, such measurements involve coincidence techniques. These are discussed in detail for electron–electron and electron–photon detection in the final state. The production of polarised beams of electrons and atoms is also discussed, since such beams are needed for studying spin-dependent scattering parameters. Overall our aim is to give a sufficient understanding of these techniques to enable the motivated reader to design and set up suitable experiments.

Theorists have been given enough detail to set up a calculation that can be expected to give a realistic description of an experiment. For scattering this level of detail is only given for methods that take into account the whole space of reaction channels, which is necessary in general to describe experiments. These are mainly the recently-developed momentum-space methods, based on the solution of coupled integral equations, which have given an excellent account of experiments in a sufficient variety of cases to support the belief that they are generally valid and computationally feasible. Calculations of ionisation have not in general reached this stage of development, although kinematic regions are known where reactions are fully understood within experimental error. These reactions are extremely sensitive to the details of atomic structure and constitute a structure probe of unprecedented scope and sensitivity.

While reactions are the subject of the book, an essential ingredient in their understanding is the calculation of atomic structure. There is a chapter describing many-body structure methods and how to obtain the results in a form suitable for input to reaction calculations. This aims at an understanding of the methods without giving sufficient detail to set up calculations.

There is a chapter summarising background quantum mechanics from the undergraduate level and developing aspects such as angular momentum, second quantisation and relativistic techniques that are not normally taught at that level.

For certain mathematical functions and operations it is necessary for the physicist to know their context, definition and mathematical properties, which we treat in the book. He does not need to know how to calculate them or to control their calculation. Numerical values of functions such as $\sin x$ have traditionally been taken from table books or slide rules. Modern computational facilities have enabled us to extend this concept, for example, to Coulomb functions, associated Legendre polynomials, Clebsch–Gordan and related coefficients, matrix inversion and diagonalisation and Gaussian quadratures. The subroutine library has replaced the table book. We give references to suitable library subroutines.

We would like to acknowledge the help of Dmitry Fursa, Jim Mitroy and Andris Stelbovics in reading and criticising parts of the manuscript, Igor Bray and Yiajun Zhou for special calculations, and particularly Win Inskip for her patience, good humour and expertise in typing.

Ian McCarthy
Erich Weigold

1

Introduction

The detailed study of the motion of electrons in the field of a nucleus has been made possible by quite recent developments in experimental and calculational techniques. Historically it is one of the newest of sciences. Yet conceptually and logically it is very close to the earliest beginnings of physics. Its fascination lies in the fact that it is possible to probe deeper into the dynamics of this system than of any other because there are no serious difficulties in the observation of sufficiently-resolved quantum states or in the understanding of the elementary two-body interaction.

The utility of the study is twofold. First the understanding of the collisions of electrons with single-nucleus electronic systems is essential to the understanding of many astrophysical and terrestrial systems, among the latter being the upper atmosphere, lasers and plasmas. Perhaps more important is its use for developing and sharpening experimental and calculational techniques which do not require much further development for the study of the electronic properties of multinucleus systems in the fields of molecular chemistry and biology and of condensed-matter physics.

For many years after Galileo's discovery of the basic kinematic law of conservation of momentum, and his understanding of the interconversion of kinetic and potential energy in some simple terrestrial systems, there was only one system in which the dynamical details were understood. This was the gravitational two-body system, whose understanding depended on Newton's discovery of the $1/r$ law governing the potential energy. By understanding the dynamics we mean keeping track of all the relevant energy and momentum changes in the system and being able to predict them accurately.

For the next 250 years Newton's dynamics of force was applied with incomplete success to many incompletely-observed systems. At the same time an understanding of the relationship of momentum, energy, space and time was developed by Maupertuis, Euler and Lagrange. The understanding of processes involving the production and absorption of bosons began with Maxwell's equations, although their significance in this sense was not realised until Einstein's development of the photon concept. Atomic and nuclear physics were born at the same instant, the discovery of the nucleus by Rutherford (1911).

The early understanding of atomic systems came almost exclusively from the observation of photons. In this sense atomic physics was like astronomy. The physicist had to be content with the information the system (with some encouragement from random excitation processes) chose to give. The first development of dynamical probes, involving the observation of internal energy changes in colliding systems, came in the study of systems of nucleons where the huge energy changes (on the scale of the energies of electronic matter) are easily observed. However in nuclear physics one cannot hope to calculate dynamical details that agree with experiment to better than about 10% from assumed two-body forces because the creation of new strongly-interacting particles, particularly pions, is not far away energetically in most observations.

The field of electron–atom collisions involves the observation of energy, momentum and spin changes in colliding systems, governed for smaller nuclei by the elementary electrostatic potential $1/r$ and for larger nuclei by additional relatively-small magnetic potentials understood in terms of Maxwell's equations and special relativity. It involves one electron in the initial state and one, two or (very occasionally) three electrons in the final state. The observation is considerably aided by photon emission and

Table 1.1. *Atomic units ($\hbar = m = e = 1$) and constants in terms of laboratory units (Cohen and Taylor, 1987). The error in the final significant figures is given in parentheses*

Quantity	Significance	Magnitude
charge	$e\sqrt{4\pi/\mu_0 c^2}$: electron charge	$-1.602\,177\,33(49) \times 10^{-19}$C
speed	c : speed of light	$2.997\,924\,58 \times 10^{10}$cm s^{-1}
time	lab. unit : 1 nanosecond	10^{-9}s
energy	lab. unit : 1 eV	$1.602\,177\,33(49) \times 10^{-19}$J
mass	mc^2 : mass of one electron $\times c^2$	$0.510\,999\,06(15) \times 10^6$eV
action	$\hbar$: Dirac–Planck constant	$6.582\,122\,0(20) \times 10^{-16}$eV s
–	$e^2/\hbar c$: fine-structure constant	$1/137.035\,989\,5(61)$
energy	optical unit : 1 cm$^{-1} \times 2\pi\hbar c$	$1.239\,842\,44(38) \times 10^{-4}$eV
length	$a_0 = \hbar^2/me^2$: 1 bohr (a.u.)	$0.529\,177\,249(24) \times 10^{-8}$cm
energy	$me^4/\hbar^2$: 1 hartree (a.u.)	$27.211\,396(81)$ eV
momentum	$\hbar/a_0$: 1 inverse bohr (a.u.)	$3.728\,940\,6(11) \times 10^3$eV/$c$

absorption processes but they do not play a part in the collision dynamics, since photons interact extremely weakly with free electrons.

The orders of magnitude of quantities involved in electron–atom collisions are seen in table 1.1. The units that are used for the description of experiments and of the individual atomic processes are chosen so that the order of magnitude of numbers to be discussed is not very far from 1. In particular, distances characteristic of atoms are far too small to observe directly, but energies and momenta are of roughly the same order of magnitude as the corresponding laboratory units. It is natural therefore to consider collisions in momentum space.

Experiments involve the counting of electrons and photons with particular energy, momentum and spin characteristics. The resulting probability distributions are described by quantum mechanics, which has been the accepted language of physics for sixty years. Yet it is perhaps not fully realised that for systems involving more than two elementary bodies there has been very little serious testing of dynamical calculations. Electron–atom collisions provide an ideal field for such testing because it is possible to keep track of energy and momentum changes where the potential creation of interacting particles is completely irrelevant. The elementary two-body interactions require only electrostatic and occasionally magnetic potential energy laws. In spite of this conceptual simplicity, a numerically-convergent calculation of the dynamics of even the three-body, electron–hydrogen, system was first implemented as late as 1993, and this calculation is not consistent with all observations. The field is very much alive.

The development of experimental techniques goes hand in hand with the development of dynamical approximations and calculational techniques. New types of observation require new approximations. The approximations are useful for deeper understanding only if they can be extended outside the range of validity relevant to the original experiments. This requires further experiments. The whole process is the subject of the book.

2
Experimental techniques for cross-section measurements

Quantitative studies of the scattering of electrons by atoms began in 1921 with Ramsauer's measurements of total collision cross sections. Ramsauer (1921) with his single-collision beam technique showed that electron–atom collision cross sections for noble gas targets pass through maxima and minima as the electron energy is varied, and can have very low minima at low electron energies. The marked transparency of rare gases over a small energy range to low energy (~ 1 eV) electrons was also noted by Townsend and Bailey (1922) in swarm experiments. This result was in total disagreement with the classical theory of scattering, which predicts a monotonic increase in the total collision cross section with decreasing energy. The Ramsauer–Townsend effect provided a powerful impetus to the development of quantum collision theory.

Although the history of electron impact cross-section measurements is quite long, the instrumentation and the experimental techniques used have continued to evolve, and have improved significantly in recent years. Part of the motivation for this progress has been the need for electron collision data in such fields as laser physics and development, astrophysics, plasma devices, upper atmospheric processes and radiation physics. The development of electron–atom collision studies has also been strongly motivated by the need of data for testing and developing suitable theories of the scattering and collision processes, and for providing a tool for obtaining detailed information on the structure of the target atoms and molecules and final collision products. It has been aided by advances in vacuum techniques, sources of charged and neutral targets, progress in electron energy analysis and detection, progress in the development of electron sources (in particular the development of suitable sources of polarised electrons), the development of tuneable lasers, and the use of computers for online control of experiments and for data handling and analysis. Refinements in the experimental techniques have made it possible to study individual processes which have had to be averaged over in previous measurements.

We cannot in this chapter make a comprehensive coverage of all of the experimental procedures presently being utilised. We will give a brief overview of the major modern techniques. The emphasis will be on differential cross-section measurements using single-collision beam—beam scattering geometry, since these are the most widely used techniques. They are versatile and also demonstrate most of the basic techniques involved in cross-section measurements.

2.1 Concept of cross sections

The time-independent probability for the occurrence of a particular collision process is represented by the corresponding scattering cross section. It characterises the scattering process and is well defined in most scattering experiments. There are some situations when a time-dependent probability must be considered and the normal definition of a scattering cross section is not applicable.

2.1.1 *Differential cross section*

The effective interaction between an electron and an atom depends strongly on the electron velocity as well as the scattering angle and the nature of the process. The cross section, which measures the probability that a given type of reaction will occur, will therefore in general depend on both the incoming and outgoing energies and angles. Such cross sections are usually called doubly-differential cross sections.

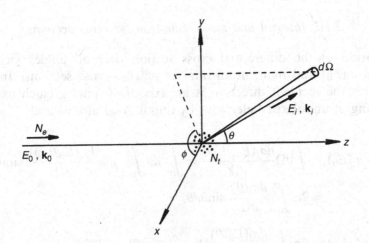

Fig. 2.1.   Schematic diagram of a scattering experiment.

In order to develop the concept of a cross section let us consider a beam of monoenergetic electrons, of energy and momentum E_0 and $\mathbf{k}_0$ respectively, approaching a number of target particles as in fig. 2.1. The uniform incident beam of N_e particles/(a.u.)^2.s is directed along the z axis and the N_t target particles are at the origin of the coordinate system. We assume in this discussion that the target particles can be considered to be at rest and that they are so massive relative to the incident electrons that the laboratory and centre-of-mass systems are identical. Then the number of particles N_i with energy between E_i and $E_i + dE_i$ scattered per second into the solid angle element $d\Omega$ making an angle (θ, ϕ) defined by the direction $\mathbf{k}_i$ relative to the incident direction is given by

$$N_i dE_i d\Omega = d^3\sigma_i(E_0, E_i, \Omega)N_e N_t, \tag{2.1}$$

where the subscript i denotes the particular excitation process leading to final state i. For a discrete process it is usually possible to integrate over the energy-loss profile, which reduces the double-differential cross section $d^3\sigma_i/dE_i d\Omega$ to the (single) differential (in solid angle) cross section, usually written as $d\sigma_i/d\Omega$. Since the target may be degenerate, one usually measures cross sections averaged over these degeneracies. Target degeneracies can sometimes be removed by using either laser techniques or inhomogeneous magnets to provide spin-selected targets (i.e. target atoms in specific magnetic substates). Similarly the spin direction of the initial and final electrons must in general be averaged or respectively summed over. The use of spin-polarised electron beams and electron spin analysers can remove the degeneracies.

2.1.2 Integral and momentum-transfer cross sections

Integration of the differential cross section over all angles yields the integral $\sigma_i(E_0)$ and momentum-transfer $\sigma_i^M(E_0)$ cross sections. In many situations the incident direction is an axis of symmetry (such as in the scattering of unpolarised electrons by unpolarised atoms) and

$$\sigma_i(E_0) = \int d\Omega \frac{d\sigma_i(E_0, \theta, \phi)}{d\Omega} = \int_0^\pi d\theta \int_0^{2\pi} d\phi \frac{d\sigma_i(E_i, \theta, \phi)}{d\Omega} \sin\theta$$

$$= 2\pi \int_0^\pi \frac{d\sigma_i(\theta)}{d\Omega} \sin\theta d\theta, \tag{2.2}$$

$$\sigma_i^M(E_0) = 2\pi \int_0^\pi \frac{d\sigma_i(E_0, \theta)}{d\Omega}[1 - \frac{k_i}{k_0} \cos\theta] \sin\theta d\theta. \tag{2.3}$$

2.1.3 Total cross section

The total cross section is obtained by summing and integrating over all cross sections

$$\sigma_T(E_0) = \sum_i \int \int \frac{d^3\sigma_i(E_0, E_i, \Omega)}{dE_i d\Omega} d\Omega dE_i. \tag{2.4}$$

The sum notation includes continuum states i.

If there is more than one particle emitted in the scattering process, and one or more of these (electron, photon, etc.) are detected in coincidence with the 'scattered' electron, then the cross sections are also differential with respect to the energy and angular distributions of these 'secondary' particles. Such measurements are called correlation measurements, since the cross section depends on the correlation between say the angles (or energies, or spin directions) of the final particles.

2.1.4 Polyenergetic incident particles

Up to the present we have assumed that our incident beam is monoenergetic. Since the cross section often depends sensitively on the incident beam energy, particularly near threshold or near a resonance, and the incident beam is never strictly monoenergetic, we must extend our treatment to cover the general case where we may not assume that the cross section is that given by the mean corresponding to the mean of the energy distribution.

Let $n(E_0)$ be the number of projectiles of energy E_0 per unit volume per unit energy interval, so that $n(E_0)dE_0$ is the number per unit volume in the range E_0 to $E_0 + dE_0$. The total incident flux is

$$\Phi = \int_0^\infty n(E_0)v_0 dE_0 = \int_0^\infty \Phi(E_0)dE_0. \tag{2.5}$$

The effective collision cross section is given by the reaction rate for stationary targets by

$$\int_0^\infty \frac{d^3\sigma(E_0, E_i, \Omega_i)}{dE_i d\Omega} n(E_0)v_0 dE_0 = \int_0^\infty \frac{d^3\sigma}{dE_i d\Omega} \Phi(E_0)dE_0$$

$$= \overline{\frac{d^3\sigma}{dE_i d\Omega_i}}\Phi, \tag{2.6}$$

where

$$\overline{\frac{d^3\sigma}{dE_i d\Omega_i}} = \frac{\int_0^\infty \frac{d^3\sigma(E_0, E_i, \Omega)}{dE_i d\Omega} n(E_0)v_0 dE_0}{\int_0^\infty n(E_0)v_0 dE_0}. \tag{2.7}$$

One can similarly define average integral, momentum-transfer and total cross sections.

2.2 Measurement of total and integral cross sections

2.2.1 *Total cross section*

When a beam of electrons of current I enters a collision chamber containing a target gas of uniform density n, the fractional loss of intensity is given in terms of the total cross section σ_T by

$$dI = -\sigma_T I n dz, \tag{2.8}$$

which yields the well known Lambert–Beer law

$$\sigma_T = (n\ell)^{-1}\ln(I_0/I_c), \tag{2.9}$$

where ℓ is the effective length of the electron path through the gas and I_0/I_c is the ratio of the beam intensity in front of and behind the absorption cell respectively. From this it is also obvious that σ_T has the dimensions of area. Fig. 2.2 shows a schematic diagram of a linearised version of the Ramsauer technique, which does not use magnetic fields, used by Wagenaar and de Heer (1985) to obtain total cross sections for electron scattering from Ar, Kr and Xe. They used a short collision cell with small entrance and exit apertures and of adjustable length, which permitted the use of higher target pressures, which in turn allowed a more accurate determination of pressure. Total cross sections have been measured using this technique since the measurements of Ramsauer (1921).

Care must be taken in designing the collision apparatus and electron detector (normally a Faraday cup or retarding field analyser) to ensure that equation (2.9) is applicable and small-angle scattering is eliminated (Wagenaar and de Heer, 1985; Bederson and Kieffer, 1971). To allow for

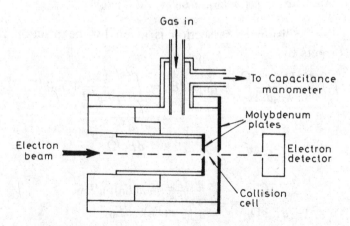

Fig. 2.2.   Schematic of the apparatus used by Wagenaar and de Heer (1985) to obtain total cross sections for electron scattering from noble gases.

scattering from the edges of apertures one usually measures the primary beam current before (I_0) and behind (I_c) the scattering cell. The total cross section is then derived by comparing the ratio I_c/I_0 with and without gas in the collision cell.

$$(I_c/I_0)_{gas}/(I_c/I_0)_{vac} = \exp(-n\ell\sigma_T). \tag{2.10}$$

It is obvious that n and ℓ have to be determined accurately. n is usually determined by measuring the pressure, for example with a capacitance manometer such as a baratron, taking care to allow for temperature differences between the gas cell and the measuring region (Blaauw *et al.*, 1980). Usually one has an effusive molecular flow through the entrance and exit apertures, which leads to large density gradients in the gas, and the product $n\ell$ in the exponent of (2.10) has to be replaced by

$$(n\ell)_{eff} = \int_g^d n(z)dz = \alpha n\ell, \tag{2.11}$$

where the integral is from the electron beam source to detector and α is a correction factor which can in principle be calculated (Blaauw *et al.*, 1980; Wagenaar and de Heer, 1985). The quantity α depends critically on the length, and hence Wagenaar and de Heer investigated α as a function of ℓ making certain they took data in the region where α was essentially equal to unity.

If the energy distribution of the primary beam is not narrow, one has to allow for it in the manner discussed in section 2.1.4. Using this method it is possible to measure σ_T as a function of E_0 with an accuracy of better than 5%.

Instead of measuring the transmission of electrons one can measure the transmission of the target particles in a crossed-beam experiment, where a beam of electrons intersects a beam of target particles (Bederson, 1968). The recoil imparted to the target atom or molecule deflects it out of the original beam and the decreased beam intensity is a measure of the total cross section. Usually one uses a d.c. molecular beam and a modulated electron beam and associated phase-sensitive detection of the transmitted molecular beam. If the two beams are rectangular with a common dimension h, then (Bederson, 1968)

$$\sigma_T = \frac{I_s}{I_A} \frac{h\overline{V}}{I_e}, \tag{2.12}$$

where $\overline{V}$ is the average velocity of the target particles, I_s and I_A the total scattered and incident atom beam fluxes and I_e the total electron beam flux. Care must be taken in allowing for geometrical and flux distribution considerations.

The recoil technique is capable of yielding angular distributions with velocity-selected targets. In addition it is particularly useful in studying the change of spin states in a scattering process, since it is easier to analyse the spin states of scattered atoms than those of electrons. Such measurements are not discussed here but the reader is referred to the articles by Bederson (1968) and Bederson and Kieffer (1971).

Crossed-beam techniques must also be used when the targets are chemically unstable systems, such as hydrogen atoms. These techniques are now being used in many laboratories and have become indispensable in atomic collision physics.

The time-of-flight technique has been used in transmission experiments to obtain total cross sections with extremely high energy resolution at low energies (Land and Raith, 1974; Ferch *et al.*, 1985; Raith, 1976). For the low and intermediate energy range Kennerly and Bonham (1978) developed a nice technique which allowed total cross sections to be derived simultaneously for all impact energies below 50 eV. A schematic representation of their time-of-flight apparatus is shown in fig. 2.3. An electron pulse of 2000 eV with a 100 ps time width is directed at a graphite target. Secondary electrons emitted at 90° are transmitted through the gas cell to an electron detector, namely a microchannel plate electron multiplier.

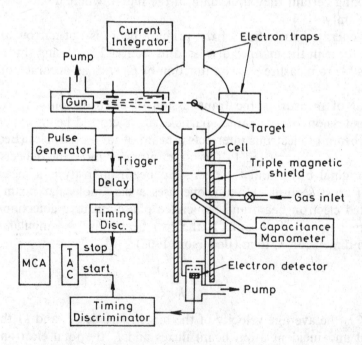

Fig. 2.3. Outline of the time-of-flight apparatus used by Kennerly and Bonham (1978) for total cross section measurements.

The cell serves as both a gas cell and a field-free flight tube. The gas cell had a length of 38 cm whereas the total flight path had a length of 45 cm. The fast pulse from the electron detector was used as the start pulse for the time-to-amplitude converter (TAC) and a delayed trigger pulse from the pulser used to sweep the 2000 eV electron beam across a small aperture in front of the solid target was used as the stop pulse. A time-to-amplitude converter produces an output analogue voltage pulse whose height is proportional to the time difference between the start and stop pulses, the pulse height being inversely proportional to the velocity of the electron. A multichannel analyser (MCA) was used to record the detected electron count rate as a function of electron velocity (or energy) with and without target gas in the cell. The use of differential pumping of the electron source and detector region as well as the low pressures used (a few millitorr) meant that the effects of changing gas pressure on the electron gun and detector are negligible. The low effective current in the gas cell minimises any space charge and surface charging effects. Careful calibration of the capacitance manometer resulted in a claimed accuracy of about 3% in the measured total cross sections.

2.2.2 Integral cross sections

Integral cross sections for selected electron-impact excitation and ionisation processes have been largely obtained by measuring optical excitation functions. These need to be corrected to a varying degree of accuracy for effects such as cascade contributions and photon polarisation. The details of the experimental procedures, sources of errors and data evaluation have been discussed by Heddle and Keesing (1968).

The optical excitation function is the cross section, measured as a function of energy, for observing a photon of energy $\epsilon_i - \epsilon_j$, which is emitted in the decay of the state i to the state j, which is normally the ground state. Here ϵ_i is the energy eigenvalue of the state i. Such a photon is emitted after the state i has been directly excited by the electron collision. If this were the only means of producing the state the optical excitation function would be equal to the integral cross section σ_i (2.2). However the state i may also be produced by the decay of higher-energy states i' to i. These processes are cascades. If i' can decay only to i the cross section $\sigma_{i'}$ must be added to σ_i. If it decays to i with a probability (branching ratio) r then we add $r\sigma_{i'}$.

The most useful measurements of optical excitation functions are made for the first dipole excitation of an atom, which has an allowed photon transition to the ground state. This excitation normally has a large cross section σ_1. Integral cross sections for other states may be determined relative to σ_1.

The cross section σ_1 is often put on an absolute scale by considering the generalised oscillator strength

$$f(K) = \tfrac{1}{2}(\epsilon_1 - \epsilon_0)K^2(k_0/k_1)\sigma_1(K), \qquad (2.13)$$

where $\mathbf{K} = \mathbf{k}_1 - \mathbf{k}_0$ is the momentum transfer and $\sigma_1(K)$ is the differential cross section. In the Born approximation for $\sigma_1(K)$, which is given by the first-order term of the Born series (6.74) and whose claims to validity improve with increasing incident energy, $f(0)$ is the optical oscillator strength. This measures the rate of the photon decay and its value is well determined by photon spectroscopy. Lassetre, Skeberle and Dillon (1969) were able to show that $f(K)$ converges to the optical limit even in cases where the Born approximation does not hold.

In practice one either measures the angular distribution of the scattered electrons and plots $f(K)$ as a function of K^2, and extrapolates the resulting curve to $K = 0$ and fits this curve to the known optical oscillator strength $f(0)$, or one carries out the experiment at high impact energy and low (preferably zero) scattering angle and assumes that the optical limit has been achieved.

Angular distribution measurements for selected excitation processes can be integrated over all angles to give integral cross sections. Such measurements will be discussed in detail in the next section. The finite angular resolving power, the angular dependence of the collision region down to zero degrees, as well as the absolute calibration must all be known in determining the integral cross section from $d\sigma_i/d\Omega$ by use of (2.2).

2.2.3 Momentum-transfer cross section

Momentum-transfer cross sections are normally determined by the electron swarm technique. A detailed discussion of the drift and diffusion of electrons in gases under the influence of electric and magnetic fields is beyond the scope of this book and only a brief summary will be given. The book by Huxley and Crompton (1974) should be consulted for a full description of the experimental methods and analysis procedures.

The technique involves high precision measurements of characteristic transport properties, the transport coefficients, of an ensemble or swarm of electrons as they drift and diffuse through a gas at pressure ranging from a few torr to many atmospheres. The most commonly measured transport coefficients are the drift velocity $W_\parallel$, which is defined as the velocity of the centroid of the swarm in the direction of the applied uniform electric field $\mathbf{E}$, the ratio D_T/μ (where D_T is the diffusion coefficient perpendicular to the electric field and μ is the electron mobility, defined as $W_\parallel E$) and, when a magnetic field $\mathbf{B}$ transverse to the electric field $\mathbf{E}$ is present, the ratio $W_\perp/W_\parallel$ (where $W_\perp$ is the drift velocity at right angles to $\mathbf{E}$ and $\mathbf{B}$). For a given gas all these coefficients are functions only of the ratio E/N (where N is the gas number density), the gas temperature T, and when a magnetic field is present, of B/N. The coefficients are related to the momentum-transfer cross section and the relevant inelastic-scattering cross sections by complex integral expressions involving the electron energy distribution function. It is these expressions, coupled with a solution of the Boltzmann equation to determine the distribution function, which form the basis of the iterative procedure to derive cross section information.

A typical swarm experiment, to measure D_T/μ in this case, is shown in fig. 2.4. Electrons from a suitable source enter a diffusion chamber through a small hole (typically 1 mm diameter) and drift and diffuse in a uniform electric field to a collector consisting of a central disk A_1 and surrounding annulus A_2. The ratio of the currents received by A_1 and A_2 is measured and D_T/μ found for the particular value of E/N and T from a solution of the diffusion equation with appropriate boundary conditions.

The distribution of the positions and velocities of electrons at a given instant can be found by solving the Boltzmann equation

$$\frac{\partial f}{\partial t} + \mathbf{v} \cdot \nabla_r f + \mathbf{a} \cdot \nabla_v f = \left(\frac{\partial f}{\partial t}\right)_{\text{coll}}, \qquad (2.14)$$

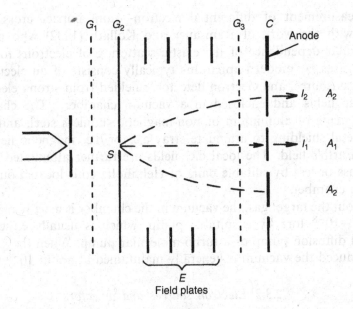

Fig. 2.4.   Schematic diagram of a swarm experiment to measure the ratio D_T/μ. G_1 is an electrode containing a small central hole and C is a collector containing a central disk A_1 and a surrounding annulus A_2.

where $f(\mathbf{r}, \mathbf{v}, t)$ is the distribution function, $\mathbf{a}$ is the acceleration due to external forces and $(\partial f / \partial t)_{coll}$ is the collision-induced rate of change of the number of electrons per unit volume of phase space. The solution normally involves the assumptions that the fields are time independent, that spatial gradient terms can be omitted and that the distribution in velocity space is only slightly disturbed from spherical, i.e. the distribution can be expanded in spherical harmonics with only the first two terms retained.

The analysis and experimental procedures are very demanding in obtaining accurate cross-section data and there are particular problems in obtaining unique cross sections (see Huxley and Crompton (1974) and Kumar (1984) for details). If only the elastic channel is open the momentum-transfer cross section can be obtained reliably and accurately in some cases (e.g. He, 2%). With the addition of inelastic channels the uncertainty in the derived cross sections due to lack of uniqueness increases.

The transport coefficients depend on a balance between the rates of acquiring energy from acceleration in the field and losing it in collisions. Since this balance can be made very close the swarm technique is particularly suited for providing cross-section data at low electron energies.

2.3 Measurement of differential cross sections

2.3.1 Instrumentation

The measurement of differential electron–atom impact cross sections began with the work of Ramsauer and Kollath (1932) who measured the angular dependence of the elastic scattering of electrons for several simple gases. A modern apparatus typically consists of an electron gun, target gas source, and electron detector, shielded from strong electric and magnetic fields and enclosed in a vacuum chamber. The chamber is usually made of aluminium or non-magnetic stainless steel, and is lined by μ-metal shielding to attenuate stray a.c. or d.c. magnetic fields, such as the Earth's field. The local d.c. fields are further attenuated to a few milligauss or less by suitable pairs of Helmholtz coils located outside the vacuum chamber.

Without the target gas, the vacuum in the chamber is usually maintained at 10^{-7}–10^{-8} torr by a suitable pump, which is usually either a well trapped diffusion pump or a turbo-molecular pump. When the gas target is introduced the vacuum is generally maintained at about 10^{-6} torr.

2.3.2 Electron sources and detectors

The apparatus of Ramsauer and Kollath (1932) is primitive by today's standards. Excellent energy resolution and satisfactory beam intensity can be achieved by utilizing energy-dispersing and direction-focussing properties of electrostatic fields of various geometries and by taking care in the construction and operation of the apparatus. Vastly-improved angular resolution was achieved by using detectors of great sensitivity for the scattered electrons. Electron multipliers, including channeltrons and microchannel plates, when operating in the pulse-counting mode, are capable of measuring currents as low as 10^{-19}A, a factor of 10^4 lower than that detectable by the most sensitive electrometers.

The most frequently-used lenses in electron optics are aperture lenses or tube (cylindrical) lenses. There is a substantial literature on the design and characteristics of electron lenses and on charged-particle optics. Harting and Read (1976), Hawkes and Kasper (1988) and Wollnik (1987) are especially useful.

Any system whose transmission depends on the electron energy can be used as a monochromator, and it is therefore not surprising that a large variety of energy selectors are used (e.g. retarding field filters, Wien filters, Mollenstedt analyser, trochoidal, toroidal, parallel-plate, cylindrical, and hemispherical analysers, etc.). At present the most widely-used analyser is probably the 180° hemispherical analyser (Kuyatt and Simpson, 1967), although some other electrostatic analysers are capable of collecting data simultaneously at several angles (Toffoletto, Leckey and Riley, 1985; Schnetz and Sandner, 1992). Analysers have also been designed to simultaneously record data with good energy resolution over a wide range of energies (e.g. Cook *et al.*, 1984), and more recently over a range of both energy and angle (Storer *et al.*, 1994). Excellent reviews of many of the different energy analyser designs currently in use have been given by Read *et al.* (1974), Sevior (1972) and Leckey (1987).

The most common source of electrons is a thermionic emitter, although photoionisation sources are sometimes used. Typical thermionic emitters include directly-heated thoriated tungsten or thorium-coated iridium filaments and indirectly-heated oxide-coated cathodes. Electrons are extracted from the source and formed into a well-collimated beam by a lens system. They can then pass through a monochromator for energy selection before being focussed and accelerated to the desired input energy E_0 with another lens system incorporating a zoom lens. A similar zoom lens and electron optics system prepare the electrons emitted from the collision region for the scattered electron energy analyser. After transmission through the analyser the electrons are detected by a suitable electron multiplier.

As an example a spectrometer with 12 meV resolution constructed by Linder's group in Kaiserslautern (Weyhreter *et al.*, 1988) is shown schematically in fig. 2.5. This instrument, which can operate down to 50 meV incident electron energy, is constructed of stainless steel and ceramics. It employs double tandem hemispheres for energy selection and analysis. A real aperture is placed between the two hemispheres to eliminate the unwanted part of the electron flux after the initial dispersion. Apertures at the entrance of the detector define the collecting volume. The acceptance solid angle is defined by these apertures and the collision volume, which is determined by the intersection of the incident electron beam and the target gas beam. The effect of an extended scattering volume and finite acceptance angles on any measurement must be treated carefully (see, for example, Brinkmann and Trajmar (1981), Zetner, Trajmar and Csanak (1990), Bedersen and Kieffer (1971), and Wagenaar and de Heer (1985)). It is discussed in the following section.

The entrance optics to the analyser is usually chosen so that it can be tuned to nearly constant transmission over a large energy range. The gun optics is also usually chosen so that the beam can be focussed at the same position with the same image size over a wide energy range. This is necessary in order to avoid distortion of peak shapes and resonance features and inaccuracies in the cross-section measurements. After transmission through the analyser a simple lens transfers the electrons to the surface of an electron multiplier, usually a channeltron, which is operated in the pulse count mode.

Static gas targets such as those used by Wagenaar and de Heer (1985) are usually unsuitable for differential cross-section measurements. These days scattering experiments are carried out in a crossed-beam arrangement. A large variety of beam sources are used. These range from effusion from simple orifices or capillary arrays to supersonic nozzles, from ovens

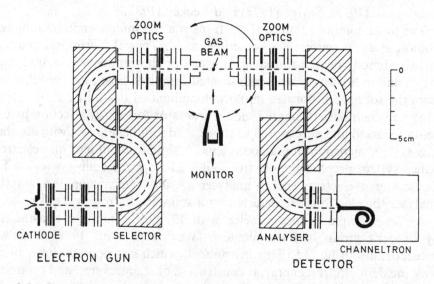

Fig. 2.5.   Double tandem electron spectrometer for low energy electron–atom differential cross-section measurements (Weyhreter *et al.*, 1988).

for targets with low vapour pressure at room temperature to gas discharges for beams of metastable atoms and free radicals. More recently, intense cold atomic beams have been produced by laser cooling and trapping (e.g. Metcalf and Straten, 1994). Beams of excited atoms can be prepared by optical or electron impact excitation or by gas discharges. These, as well as high-energy beam sources, are discussed fully in the literature, (e.g. Pauly (1988)). State-selected beams (see, for example, Bergmann, 1988, and Reuss, 1988), laser-excited aligned and oriented beams (e.g. Düren, 1988) and magnetically spin-polarized beams (e.g. Iannotta, 1988) can all be prepared routinely and reliably with present technology. The discussion of these techniques, as well as the detection of the beams, velocity selection etc., is largely outside the scope of this book. They are discussed fully in the literature, for example Scoles (1988) and references therein. A brief discussion of the preparation of polarised and laser-excited atom sources is given in section 2.6.

The electron-impact spectrometer, such as that shown in fig. 2.5, can be operated in a number of distinct modes. In the energy-loss mode the impact energy E_0 and scattering geometry are kept fixed and the count rate is measured as a function of the energy loss $E_L = E_0 - E_i$ where E_i is the energy of the scattered electrons leaving the target in state i. The features in such energy-loss spectra are related to the energy levels of the target and the scattered intensities are related to the corresponding cross sections. If the energy loss E_L is kept fixed and the incident energy is varied the energy dependence of the cross section for a particular channel can be measured. This can reveal structure, such as resonances and cusps, in the cross section of the chosen process.

Energy-loss spectra can also be obtained by fixing the energy E_i of the emitted electrons and varying the incident energy and energy loss simultaneously. This gives information on each energy-loss feature at the same impact energy above its own threshold. Finally, of course, in the measurement of the differential cross section E_0, E_i and E are all kept fixed and the scattering geometry (θ) is varied.

2.3.3 Scattering geometry effects

In an ideal scattering experiment the collisions are assumed to occur at a fixed point in space. In practice the collision volume is finite and the part viewed by the detecting system generally depends on the scattering angle. Care must therefore be taken in relating the scattered particle intensity to the cross section.

In a typical crossed-beam experiment the electron beam intersects the target atom beam at 90°. This is shown schematically in fig. 2.6, where

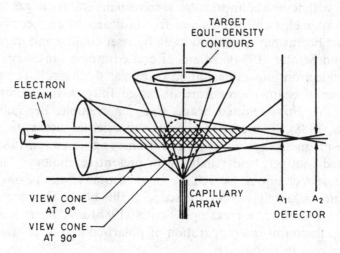

Fig. 2.6. Typical scattering geometry (vertical cut) showing the intersection of the atomic beam, electron beam, and viewing cone of the analyser. A_1 and A_2 are the apertures in the detector defining the viewing cone.

the atom beam is assumed to come from a capillary tube or array. With a capillary array the beam typically has a full width at half maximum of about a half of that with a single capillary tube of the same throughput (Buckman *et al.*, 1993). In either case the target flux is highest on the axis of the beam and falls off uniformly away from the axis. Since the beam diverges from the orifice, the flux also depends on the distance from the orifice. The electron beam on the other hand can usually be considered as essentially parallel in the vicinity of the collision region. Its flux also depends on the distance from the axis of the beam, being maximum along the axis. The intersection of these two beams within the view angle of the detector represents the effective collision volume as seen by the detector (fig. 2.6).

Thus the relation (2.1) between the differential cross section and the number N_i of scattered particles detected per second with nominal energy E_i at the nominal angle $\bar{\Omega}(\bar{\theta}, \bar{\phi})$ becomes

$$N_i(\overline{E}_0, \overline{E}_i, \overline{\Omega}) = \int_{\mathbf{r}} \int_{E_0} \int_{E_i} \int_{\Omega} \rho(\mathbf{r}) f(E_0, \mathbf{r}) \xi(E_i, \mathbf{r})$$

$$\times \frac{d^3\sigma(E_0, E_i, \Omega)}{dE_i d\Omega} d\Omega dE_0 dE_i d^3 r, \qquad (2.15)$$

where $\rho(\mathbf{r})$ is the density distribution of target atoms, $f(E_0, \mathbf{r})$ is the energy and spatial distribution of the incident electron flux, and $\xi(E_i, \mathbf{r})$ is the response function of the detector for electrons of energy E_i scattered by a target atom at position $\mathbf{r}$.

A detailed knowledge of the instrumental function and the target density and electron beam flux distributions is required to extract the cross section from (2.15). In general, however, some simplifying assumptions can be made. For excitation of a discrete final state i it is usually possible to integrate over the energy loss profile assuming constant system response over the natural line width to give

$$\frac{d\sigma_i(E_0, \Omega)}{d\Omega} = N_i \tilde{V}_{\text{eff}}^{-1},$$ (2.16)

where

$$\tilde{V}_{\text{eff}} = \int_r \int_{E_0} \int_{\Omega} \rho(\mathbf{r})f(E_0,\mathbf{r})\xi(\overline{E}_i,\mathbf{r})dE_0 d\Omega d^3r$$ (2.17)

includes all the instrumental and geometrical factors. Similarly if the cross section is slowly varying in energy and angle with respect to the system response (as for example in continuum processes at high energy), the cross section evaluated at the nominal angle and energies may be removed from the integral to give

$$\frac{\overline{d^3\sigma}}{dE_i d\Omega} = N_i(E_0, E_i, \Omega)\tilde{V}_{\text{eff}}^{-1}.$$ (2.18)

In general, relation (2.17) can be simplified by assuming that the energy distribution of electrons is independent of $\mathbf{r}$ and that the detector efficiency depends only on E_i and not on $\mathbf{r}$, i.e.

$$f(E_0, \mathbf{r}) = f(E_0)f(\mathbf{r})$$

and

$$\xi(E_i, \mathbf{r}) = \xi(E_i).$$ (2.19)

In this case the integrals over the energy distributions and the coordinates can be separated to give

$$\tilde{V}_{\text{eff}}(E_i, \theta, \phi) = C(E_i)V_{\text{eff}}(\theta, \phi),$$ (2.20)

with

$$V_{\text{eff}}(\theta, \phi) = \int_r \int_{\Omega(\mathbf{r})} \rho(\mathbf{r})f(\mathbf{r})d\Omega(\mathbf{r})d^3r$$ (2.21)

and

$$C(E_i) = \int_{E_i} \int_{E_0} f(E_0)\xi(E_i)dE_i dE_0.$$ (2.22)

$V_{\text{eff}}(\theta, \phi)$ can be calculated from a knowledge of the density distributions of the intersecting beams and the acceptance angles of the detector. Its angular dependence has been calculated for a number of frequently used scattering geometries and beam sources by Brinkmann and Trajmar (1981). For a static gas target the correction has the familiar $\sin\theta$ form down to low scattering angles, while for a well-collimated supersonic beam V_{eff} is essentially independent of θ and ϕ. Appreciable corrections are usually needed for target beams generated by orifices, tubes, and even capillary arrays. With a knowledge of the angular dependence of V_{eff} the observed scattered intensity distribution can be converted to relative differential cross sections.

Absolute values of the overlap integral V_{eff} are quite difficult to obtain since it is necessary to determine both the atomic beam density and electron beam flux distributions. The use of rectangular beams of constant flux density throughout the interaction volume greatly simplifies the problem. With a carefully-designed Faraday cup it is then possible to determine the electron flux distribution. The direct measurement of the target gas density is often very difficult in beam–beam experiments (see for instance Bedersen and Kieffer (1971) and Scoles (1988)). Even if V_{eff} is known absolutely, one still needs to determine $C(E_i)$ to obtain absolute cross sections. The usual practice is to determine the angular dependence of V_{eff} and to determine the energy dependence of $C(E_i)$ and then to use a normalisation procedure to fix the absolute scale.

2.3.4 Normalisation procedures

For atomic targets a convenient way of determining the average target density and $C(E_i)$ is to carry out a low-energy elastic scattering differential cross-section measurement and to use a phase-shift analysis to determine the absolute cross section.

For potential scattering, the scattering amplitude can be expressed in terms of phase shifts for partial waves (section 4.4.2 and equn. (4.88)). At low energies the first few partial waves (characterised by orbital angular momentum $L \leq L_0$) generally dominate the scattering. However, the small-angle scattering often requires the inclusion of many higher-order partial waves (Williams, 1975). The higher partial waves and small-angle scattering arise from the long-range part of the scattering potential, which is given by the polarisation potential $V(r) \sim -\alpha/2r^4$. For this potential the higher-order phase shifts can be calculated with sufficient accuracy by the Born approximation (see section 7.5.4), and the contribution for $L > L_0$ can be summed analytically to give a contribution $f_B^{L_0}$ to the scattering amplitude, which can be added to the contributions for $L \leq L_0$, to give (see equn.(4.88))

$$f(\theta) = \frac{1}{2ik} \left(\sum_{L=0}^{L_0} (2L+1)[\exp(2i\delta_L) - 1]P_L(\cos\theta) + 2ikf_B^{L_0}(\theta) \right). \quad (2.23)$$

The differential cross section is given by the absolute square of this amplitude (4.48). Trial values of the phase shifts δ_L are then varied to give best fit to the measured relative differential cross section. Once the phase shifts are known, the cross section is determined on an absolute scale by taking the absolute square of (2.23).

It is often better to do the normalisation on an elastic scattering resonance (section 4.6), where the decay to the ground state can occur through only one partial wave. This method was first introduced by Gibson and Dolder (1969). Interference with the nonresonant background generally introduces some complications (see Williams and Willis, 1975). The modifications required in the analysis due to spin are discussed fully by Andrick and Bitsch (1975).

Once the absolute cross section at a particular energy has been determined, a knowledge of the energy dependence of the response function $C(E_i)$ and of the electron beam flux may be used to derive absolute cross sections at other energies and for inelastic processes (Williams and Willis, 1975).

Another commonly-used normalisation procedure is to use the relative flow technique. In this method the elastic differential cross section for a particular species may be obtained by comparing the scattered intensity under the same conditions with that from another target with a known cross section. It is important to ensure, for both the gas under study and the reference gas, that the electron flux density and distribution, the detector efficiency, and the target beam flux distribution are the same for both gases during the measurement.

Under conditions of molecular flow through a capillary or a capillary array, the unknown differential cross section is given by (e.g. Brinkmann and Trajmar, 1981).

$$\frac{d\sigma(\theta)}{d\Omega} = \frac{d\sigma_r(\theta)}{d\Omega} \frac{I(\theta)}{I_r(\theta)} \frac{P_r}{P}, \qquad (2.24)$$

where P is the gas pressure in the reservoir before the capillary, I the scattered intensity, and the subscript r refers to the reference gas. Molecular flow is difficult to achieve with reasonable flow rates, i.e. reasonable data acquisition times. At higher flow rates with back pressures up to a few torr the expression

$$\frac{d\sigma(\theta)}{d\Omega} = \frac{d\sigma_r(\theta)}{d\Omega} \frac{I(\theta)}{I_r(\theta)} \frac{N_r}{N} \left(\frac{M_r}{M}\right)^{1/2}, \qquad (2.25)$$

must be used (Trajmar and Register, 1984). Here N is the total flow rate and the average thermal velocity has been replaced by the target mass under the assumption that the two gases are at the same temperature.

2.4 Ionisation

If the excitation is to the continuum, then a number of multidifferential cross sections are possible. Consider, for instance, the single ionisation of an atom A:

$$e(\mathbf{k}_0) + A \longrightarrow e(\mathbf{k}_i) + e(\mathbf{k}_\mu) + A_\mu^+, \tag{2.26}$$

where $\mathbf{k}_i$, $\mathbf{k}_\mu$, and $\mathbf{k}_0$ are the momenta of the free electrons. Here the subscript i denotes the final continuum state of the target atom and the momentum of the 'scattered' electron. The state i is defined by the momentum $\mathbf{k}_\mu$ of the 'ejected' electron and the state μ of the residual ion A_μ^+ (which may be in the continuum). The indistinguishability of the electrons is accounted for in calculations (see section 10.1). For experiments which do not resolve angular momentum projections or observe polarisations, this collision process is described by the so-called triple differential cross section or the (e,2e) cross section. It is discussed fully in chapters 10 and 11.

2.4.1 Double-differential cross sections

If only one free electron, say the 'scattered' electron i, is detected and its energy and angular dependence are measured, then the double differential cross section is obtained. This is given by

$$\frac{d^3\sigma}{d\Omega_i dE_i} = \sum_\mu \int d\Omega_\mu \frac{d^5\sigma_\mu}{d\Omega_\mu d\Omega_i dE_i}, \tag{2.27}$$

where the final ion states μ have been summed over (or integrated if above the double ionisation threshold) and the angle of emission of the undetected electron has been integrated over in the triple-differential cross section. Such cross sections are measured (e.g. Müller-Fiedler, Jung and Ehrhardt, 1986; Goruganthu and Bonham, 1986; Oda, 1975) with instruments similar to those discussed in section 2.3 (e.g. fig. 2.5). A momentum selected projectile beam crosses a beam of target atoms at 90°, and the emitted electrons are analysed according to their energies and momenta. The double-differential cross sections, which are discussed in chapter 10, show some general features. The fast scattered electrons are emitted into a narrow cone in the forward direction, with the angle of the cone being smaller for larger impact energies. The slow ejected electrons are emitted almost isotropically. The cross sections are normally put on an absolute scale by extrapolating the generalised oscillator strength for the incident electrons to zero momentum transfer (see section 2.2.2).

Accurate absolute measurements of double-differential cross sections are quite difficult to make. Measurements carefully taken by competent investigators often differ significantly. Kim (1983) gave a recommended set of cross sections obtained after using various asymptotic boundary conditions and scaling laws to check the consistency of the available data. More recently Rudd (1991) proposed a different interpolation procedure to determine recommended cross sections for hydrogen and helium.

2.4.2 Single-differential cross sections

The single-differential cross section is obtained by integrating the double-differential cross section over all angles of emission of the electron

$$\frac{d\sigma_i}{dE_i} = \int d\Omega_i \frac{d^3\sigma_i}{d\Omega_i dE_i}. \tag{2.28}$$

The form of this cross section is shown in fig. 2.7. It represents the energy loss spectrum integrated over all angles of the outgoing electrons. Due to the indistinguishability of these electrons, the single differential cross section must be symmetric about $(E_i + E_\mu)/2 = (E_0 - \epsilon_0)/2$. The resonances shown in the continuum part of the spectrum in fig. 2.7 are schematic representations of autoionising transitions. At sufficiently high impact energies E_0 the ion may be left with sufficient energy to emit further electrons, leading to a distortion of the spectrum shown in fig. 2.7.

Single-differential cross sections are difficult to obtain in a direct measurement and they are usually obtained by numerical integration of the double-differential cross sections over all angles (2.28). The single-differential cross section describes the energy distribution of secondary electrons and is therefore important in modelling radiation damage, in studies of stellar and upper atmospheric phenomena, plasma fusion work,

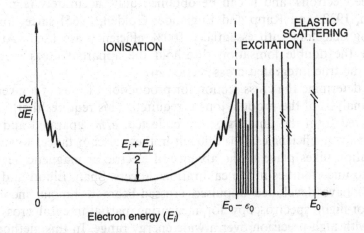

Fig. 2.7. Schematic representation of the single-differential cross section. The threshold energy for ionisation is given by ϵ_0, the ground-state separation energy.

electrical discharges and laser physics. The other single-differential cross section, $d\sigma/d\Omega$, obtained from integrating the double-differential cross sections over all energies, has little physical importance and has not been measured.

2.4.3 Total-ionisation cross section

Integration of the single-differential cross section over energy from zero to the maximum energy $E_i + E_\mu = E_0 - \epsilon_0$ gives the integral or total-ionisation cross section. Since each ionising collision results in two indistinguishable outgoing electrons

$$\sigma_I = \frac{1}{2} \int_0^{E_0 - \epsilon_0} \frac{d\sigma_i}{dE_i} dE_i. \tag{2.29}$$

Total-ionisation cross sections are easiest to obtain by direct measurement (see section 2.2.2). Generally a well-collimated beam of nearly-monoenergetic electrons is passed through a gas or vapour target, and the positive ions formed are essentially all collected. It is necessary to use a thin target to ensure no secondary ionisation is produced. Then the collected positive current is given by (see equns. (2.8) and (2.11)).

$$I^+ = I(n\ell)_{\text{eff}} \sigma_{al}, \tag{2.30}$$

where the apparent-ionisation cross section is the weighted sum of the various possible multi-ionisation cross sections

$$\sigma_{al} = \sigma^+ + 2\sigma^{2+} + 3\sigma^{3+} + ..., \tag{2.31}$$

where σ^{n+} is the partial cross section for producing n electrons.

The cross section σ_{al} represents a useful cross section for the production of free electrons, and it can be obtained quite accurately (e.g. Tate and Smith, 1932, and Rapp and Englander-Golden, 1965) since nondiscriminating detectors with essentially 100% efficiency are used. At energies below the double ionisation threshold the apparent cross sections will yield the true integrated cross section σ_I.

To determine the cross section for production of ions of a given charge, e/m analysis of the product ions is required. This requires that the ions be extracted from the target gas, are collected, e/m separated and detected with known efficiencies. It is difficult in these circumstances to ensure that ionisation takes place in the absence of electric or magnetic fields, which lead to uncertainties in the calibration process. Shah, Elliott and Gilbody (1987) have developed a pulsed crossed-beam technique incorporating time-of-flight spectroscopy for measuring partial-integral cross sections σ^{n+} with high precision over a wide energy range. In this method a short pulse of electrons is passed through a thermal energy beam of atoms in a high-vacuum region. Immediately after the transit of the electron beam

through the target beam, the slow ions of different charges are swept out
of the collision region by a pulsed electric field, and selectively identified by
their characteristic flight time to a particle multiplier detector. Great care
has to be taken to ensure high, equal and constant extraction efficiency.
The cross sections are then normalised to an accurate known cross section
at a suitable point.

2.4.4 Triple-differential or (e,2e) cross sections

The so-called triple-differential cross section is the measure of the probability that in an (e,2e) event an incident electron of momentum k_0 and
energy E_0 produces two electrons of energy and momenta E_i, E_μ and k_i,
k_μ in the solid angles $d\Omega_i(\theta_i, \phi_i)$ and $d\Omega_\mu(\theta_\mu, \phi_\mu)$ respectively. It is the
most kinematically-complete description of ionisation and provides the
most sensitive test of the theory of this process if no polarisations and no
angular momentum projections are observed. Many final-state configurations are observed. If the momenta all lie in a plane the kinematics is said
to be coplanar, otherwise it is referred to as noncoplanar. The kinematics
can also be chosen to be symmetric ($E_i = E_\mu = (E_0 - \epsilon_\mu)/2, \theta_i = \theta_\mu$) or
highly asymmetric ($E_i \gg E_\mu, \theta_i \sim 0$).

The measurement of an (e,2e) cross section requires the coincident
detection of two electrons with well-defined energies and angles of emission. Fig. 2.8 shows schematically the coincidence spectrometer used by
Weigold, Zheng and von Niessen (1991) and Zheng et al. (1990). Not
shown are the Helmholtz coils and magnetic shielding used to null out
the local magnetic fields. Two hemispherical analysers each with a five
element cylindrical zoom and retarding lens system are mounted on two
independently-rotatable horizontal turntables. The analysers can also be
rotated in the vertical direction about the interaction region by two in-
vacuum stepping motors. A third horizontal turntable is used for mounting
an electron gun (not shown). In the configuration shown in fig. 2.8 an
electron gun is mounted vertically below the analysers, the incident beam
being along the axis of rotation of the turntables.

Target gases or condensible vapours are admitted to the interaction
region via suitable nozzles, multichannel arrays or collimating apertures.
The system has also been designed to allow the interaction region to be
illuminated by light from a tuneable laser (Zheng et al., 1990), permitting
the investigation of (e,2e) collisions with oriented and excited atoms. The
laser, atomic and electron beams all intersect at right angles.

Read and co-workers have recently used a coincidence spectrometer with extremely flexible geometry (Murray, Turton and Read, 1992;
Hawley-Jones et al., 1992). This is achieved by a design in which the

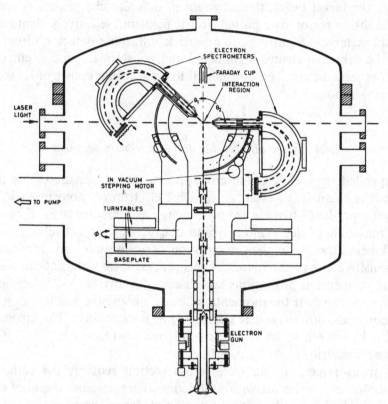

Fig. 2.8.   Schematic diagram of an (e,2e) coincidence spectrometer (Weigold *et al.*, 1991).

two electrostatic analysers rotate independently about a common vertical axis while the electron gun rotates about a horizontal axis that lies in the plane of rotation of the analysers. The Faraday cup and the hypodermic needle used to produce the atomic beam rotate together with the electron gun and associated monochromator. The gun, Faraday cup, atomic beam and analysers are all directed towards the point of intersection of the two rotation axes. A photomultiplier views the interaction region. It and the Faraday cup are used for focussing and alignment purposes. A feature of this spectrometer is the comprehensive computer interface which controls and optimises all aspects of the spectrometer, from tuning the electron beam and analysers and setting of the analyser positions to subsequent data collection (Murray *et al.*, 1992). This spectrometer yields more reliable and consistent results than was previously possible with manual operation.

Coincidences are typically characterised by extremely low count rates. In order to increase the count rates, the earlier single-channel coincidence spectrometers have been replaced by multiparameter detection techniques. These have either been concerned with obtaining data at a number of angles simultaneously (Moore *et al.*, 1978) or by gathering data with good energy resolution over a range of outgoing energies simultaneously (Cook *et al.*, 1984; McCarthy and Weigold, 1988, 1991; Lower and Weigold, 1989). In the latter method electrons are passed through an energy-dispersing element before impinging on a position-sensitive detector which determines the spatial coordinates of the detected electrons. The position-sensitive detector usually consists of a chevron-mounted pair of microchannel plates followed by a suitable position-sensitive anode. The anode is usually a resistive anode which determines position arrival by charge division (see fig. 2.8 and Fraser and Mathieson, 1981), but it could also be treated as an RC delay line which determines position by measuring differences in pulse rise times (Parkes, Evans and Mathieson, 1974) or wedge-and-strip anode whose operation is also based on charge division (Martin *et al.*, 1981). From the measured arrival position the energy of the detected electron can be inferred.

Storer *et al.* (1994) and Weigold (1993) have recently described a coincidence spectrometer (see fig. 2.9) which uses two-dimensional position-sensitive detectors placed behind angle and energy-dispersing analysers. This spectrometer can simultaneously measure with high resolution over a wide range of angles and energies. The position of arrival of electrons traversing the analysers is determined by charge division between the four

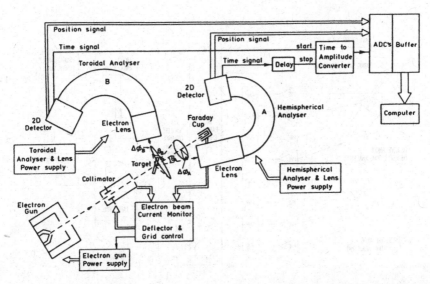

Fig. 2.9. Schematic view of the coincidence spectrometer of Storer *et al.* (1994) showing the hemispherical and toroidal energy and azimuthal angle analysers.

corners of a two-dimensional resistive anode. All the data acquisition is via an online computer, which determines the angle of emission and energy of each detected electron, and corrects the coincidence timing information for the flight time of the electrons through the two analysers. This timing correction reduces the width of the coincidence peak significantly and therefore improves the signal-to-noise ratio correspondingly.

The charges collected from the four corners of each of the two resistive anodes are amplified and fed into a gated nine-parallel-input analogue-to-digital converter (ADC) unit (fig. 2.10). Similarly fast pulses from the backs of the microchannel plates are transmitted with suitable delays to a time to amplitude converter (TAC) and the time difference output pulses are fed to one of the inputs of the analogue to digital converter unit. A gate signal from the time-to-amplitude converter opens the gates whenever two electrons are detected within 100 ns of each other.

The multiparameter data-acquisition system has a tenth parallel data channel which is used to store the setting of the beam energy in the form of the difference between the incident energy and the mean of the summed emitted energies. A dual-channel buffer memory module, one for each analogue to digital converter, allows data conversion and subsequent storage in one 8K 16 bit word buffer memory, while the second 8K duplicate buffer is being read by the online computer. The buffer memory is configured as a queue, with data being written and read in the same order.

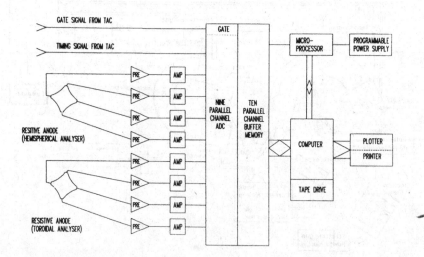

Fig. 2.10. Schematic of the 'slow' pulse logic circuit of Storer *et al.* (1994). Not shown is the fast timing and gating circuitry.

The position of arrival, and hence the energy and angle of emission, is then calculated for each pair of electrons and the corresponding time-of-flight of each electron through the analysers is also calculated and used to correct the coincidence spectrum. The software allows the data to be displayed in a number of ways, and is also used to check the calibre of the data and to provide instrumental checks.

A time window Δt_c is usually set to measure coincidence events and another $\Delta t_b = R\Delta t_c$ is set to measure the uncorrelated or random background events. If the background timing spectrum is flat, the number of true coincidence counts N_t accumulated in time T is given by

$$N_t = N_c - N_b/R, \tag{2.32}$$

where N_c and N_b are the number of counts within the coincidence and background windows respectively. Distortions of TAC background spectra from the ideal flat distribution are treated in detail by Coleman (1979). The statistical accuracy in the true coincidence counts is given by

$$\Delta N_t = [N_c + N_a(1 + 1/R)]^{1/2}, \tag{2.33}$$

where $N_a = N_b/R$ is the number of accidental coincidence counts in the coincidence timing window.

If n is the target gas density and I the incident current, the number of true and accidental coincidences in time T can be written respectively as

$$N_t = C_1 n I T$$
$$N_a = \dot{N}_s \dot{N}_e \Delta t_c T = C_2 \Delta t_c (nI)^2 T, \tag{2.34}$$

where $\dot{N}_s$ and $\dot{N}_e$ are the scattered and ejected count rates respectively and C_1 and C_2 are fully determined by the physics of the process and the overall efficiency of the instrumental arrangement. The signal-to-background ratio in a coincidence experiment is therefore given by

$$r = C_1/C_2 nI \Delta t_c. \tag{2.35}$$

Thus reducing the incident current improves the signal-to-background ratio but it also reduces the true coincidence count rate and therefore increases the statistical uncertainty δ, where (from (2.33) and (2.34))

$$\delta = \frac{\Delta N_t}{N_t} = \frac{1}{C_1}\left[\frac{1}{T}\left(\frac{C_1}{nI} + C_2\left(1 + \frac{1}{R}\right)\Delta t_c\right)\right]^{1/2}. \tag{2.36}$$

The advantage of small timing resolution Δt_c and large window ratio R is immediately obvious, as is the usual dependence on the accumulation time T. Increasing nI also improves the experiment until the second term in (2.36) depending on C_2 becomes dominant. Increasing C_1, by increasing say the angular acceptance, or decreasing C_2 by using constraints such as energy conservation (Lower and Weigold, 1989), will also lead to improved results. This is discussed in some detail by McCarthy and Weigold (1991).

The measurement of absolute multidimensional cross sections, such as (e,2e), is usually difficult to achieve with high accuracy due to the various experimental difficulties associated with low pressure gas targets and electron optics. Relative cross sections to different ion states can be obtained with much greater accuracy than the absolute values. The relationship between the (e,2e) differential cross section and the experimentally observable parameters is given by (see section 2.3.3)

$$\dot{N}_t(\bar{E}_0, \bar{E}_i, \bar{E}_\mu, \bar{\Omega}_i, \Omega) = \int_{r_<} \int_{E_0} \int_{E_i} \int_{E_\mu} \int_{\Omega_i} \int_{\Omega_\mu} \rho(\mathbf{r}) f(E_0, \mathbf{r}) \xi_i(E_i, \mathbf{r})$$

$$\times \xi_\mu(E_\mu, \mathbf{r}) \frac{d^5\sigma_\mu}{d\Omega_\mu d\Omega_i dE_i} \delta(E_0 - \epsilon_\mu - E_i - E_\mu) d\Omega_\mu d\Omega_i dE_0 dE_\mu dE_i d^3r,$$

$$(2.37)$$

where $\dot{N}_t$ is the true coincidence count rate and $\mathbf{r}_<$ indicates that the integral over $\mathbf{r}$ is taken over the overlap between the collision regions $\mathbf{r}_i$ and $\mathbf{r}_\mu$ seen by the detectors i and μ respectively. The other parameters are discussed in section 2.3.3. The absolute scale is inferred from a direct measurement of the quantities appearing in (2.37).

If the cross section is slowly varying in energy and angle with respect to the detector response functions, and the detector efficiency depends only on $E_{i,\mu}$ and not on $\mathbf{r}$, and the energy distribution of the incident electrons is independent of $\mathbf{r}$, then using (2.19) we have

$$\frac{d^5\sigma_\mu}{d\Omega_\mu d\Omega_i dE_i} = \frac{\dot{N}_t}{(NI\Delta\Omega_i\Delta\Omega_\mu)_{\text{eff}} \bar{\xi}(E_i, E_\mu)\Delta E_c}, \qquad (2.38)$$

where

$$(NI\Delta\Omega_i\Delta\Omega_\mu)_{\text{eff}}(\theta_i, \phi_i, \theta_\mu, \phi_\mu) =$$

$$\int_{r_<} \int_{\Omega_i(\mathbf{r})} \int_{\Omega_\mu(\mathbf{r})} \rho(\mathbf{r}) f(\mathbf{r}) d\Omega_i(\mathbf{r}) d\Omega_\mu(\mathbf{r}) d^3r \qquad (2.39)$$

and

$$\bar{\xi}(E_i, E_\mu)\Delta E_c =$$

$$\int_{E_0} \int_{E_i} \int_{E_\mu} f(E_0)\xi_i(E_i)\xi_\mu(E_\mu)$$

$$\times \delta(E_0 - \epsilon_\mu - E_i - E_\mu) dE_0 dE_i dE_\mu. \qquad (2.40)$$

ΔE_c is the effective coincidence energy resolution and $\bar{\xi}$ the effective coincidence detection efficiency.

The difficulty in determining absolute (e,2e) cross sections accurately can be inferred from the difficulty inherent in determining accurately the quantities appearing in (2.37)–(2.40). Beaty *et al.* (1977) and Stefani, Camilloni and Giardini-Guidoni (1978) used crossed-beam techniques which resulted in large uncertainties. Using a gas cell target and fixed scattering angles of 45°, van Wingerden *et al.* (1979,1981) were able to obtain absolute cross sections with errors of only 20%.

The alternative approach has been to normalise the data to theory in regions where the theory is believed to be accurately understood, Lahmam-Bennani *et al.* (1987). If the (e,2e) cross section is measured over a large enough angular range it can be integrated to give the double-differential cross section (2.27), which is in turn put on an absolute scale by, for instance, using the Bethe sum rule (Lahmam-Bennani *et al.*, 1980), which assumes the validity of the Born approximation. Alternatively the double-differential cross section can be normalised to known elastic-scattering cross sections (Avaldi *et al.*, 1987*a,b*) or to the optical oscillator strength at zero momentum transfer (Müller-Fiedler *et al.*, 1986, and Avaldi *et al.*, 1987*a,b*). The (e,2e) cross section can be normalised to the optical dipole oscillator strength in the limit of zero momentum transfer (Lassettre *et al.*, 1969). This is only feasible for high incident energies and small E_μ (Leung and Brion, 1985; Jung *et al.*, 1985).

None of the above methods of normalisation provide accurate cross sections for general kinematics in the low-energy regime, particularly near threshold. Rösel *et al.* (1992) have introduced a new technique to overcome some of these limitations. They arrange the detector geometries and viewing angles to ensure that both detectors are viewing the complete intersection region of the incident electrons and atomic beams. They then determine the effective target number and incident current by measuring the total number of ions produced under identical conditions and normalising to a known total ionisation cross section. The effective solid angles and detector efficiencies (2.38–2.40) are then obtained by normalisation to a known double-differential cross section. The double-differential cross section is usually obtained by methods discussed in the previous paragraph. To do this normalisation the energy resolution $\sqrt{\Delta E_0^2 + \Delta E_i^2}$ for the double-differential cross section measurements must be accurately determined. This is usually done by measuring the width of the elastic peak in an energy-loss spectrum. The coincidence energy resolution given in (2.40) can be simplified if Gaussian profiles are assumed for the incident and detected electrons. As discussed by Lahmam-Bennani *et al.* (1985), the full width at half maximum of the coincidence resolution is given by the incident energy resolution convoluted with the product of the two spectrometer energy resolutions.

$$\Delta E_c^2 = \Delta E_0^2 + \left(\frac{1}{\Delta E_i^2} + \frac{1}{\Delta E_\mu^2} \right)^{-1}. \tag{2.41}$$

On the other hand the energy width of a binding-energy spectrum obtained by varying one of the energies is given by

$$\Delta E_\epsilon = (\Delta E_0^2 + \Delta E_i^2 + \Delta E_\mu^2)^{1/2}. \tag{2.42}$$

2.5 Polarised electrons

An ensemble of electrons is said to be polarised if there is a preferential orientation of the electron spins. If there are $N_\uparrow$ electrons with spins parallel to a particular direction or axis of quantisation and $N_\downarrow$ with spins antiparallel to that direction, then the component of the electron polarisation vector $\mathbf{P} = (P_x, P_y, P_z)$ in that direction is defined by

$$P = \frac{N_\uparrow - N_\downarrow}{N_\uparrow + N_\downarrow}. \tag{2.43}$$

$|\mathbf{P}|$ is called the degree of polarisation.

2.5.1 Polarised electron sources

The performance characteristics of various sources of polarised electrons have been reviewed extensively (e.g. Kessler, 1985). These sources depend either on the spin dependence of the interaction in the process in which the free electrons are produced, or on the pre-polarisation of one element

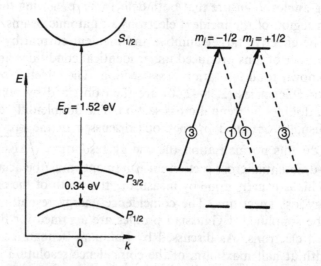

Fig. 2.11.   The one-electron energy (E) *vs* momentum (k) diagram for GaAs near $k = 0$ showing the band gap E_g and the spin–orbit splitting of 0.34 eV in the valence P band. Transitions with σ^+ and σ^- light from the $P_{3/2}$ band are indicated by solid and dashed lines respectively. The circled numbers represent the relative transition probabilities.

of the system involved in the production process. An example of the former is Mott scattering, which will be discussed in the next section. An example of the latter situation is the extraction of polarised electrons from a ferromagnet. The principal methods are based on the spin–orbit interaction, which splits energy levels in atoms or solids and results in photoelectrons that may have significant polarisation (Fano, 1969; Kessler, 1985; Heinzmann, 1987).

A polarised electron source for crossed electron–atom beam scattering experiments should ideally have both a high figure of merit and a high brightness, which is the current density emitted into the unit solid angle. The figure of merit is $\zeta = PI_e^{1/2}$, where I_e is the current in the beam. The statistical accuracy of a spin-dependent measurement is proportional to $1/\zeta$. It is also generally advantageous to have a source with good energy resolution and one which has the capability of frequent polarisation reversal in a manner which does not change the electron optics.

The most widely-used source is the photoemission GaAs source (Garwin, Pierce and Siegmann, 1974; Pierce *et al.*, 1980), the principle of which is shown in fig. 2.11. The spin–orbit interaction splits the P-valence band at the Γ point ($k = 0$ in the plot of one-electron energy E *vs* momentum k) into fourfold degenerate $P_{3/2}$ and twofold degenerate $P_{1/2}$ bands with a separation of 0.34 eV.

The $S_{1/2}$ conduction band is separated from the $P_{3/2}$ band by $E_g = 1.52$ eV at Γ. Thus light with energy just above the gap energy can only promote electrons from the $P_{3/2}$ band to the $S_{1/2}$ band. For circularly-polarised light the selection rules are $\Delta m_j = \pm 1$ for right-hand (σ^+) and left-hand (σ^-) light respectively. These transitions are shown in fig. 2.11 by solid and dashed lines respectively. The relative transition probability from the $m_j = 3/2$ state is, however, three times greater than from the $m_j = 1/2$ state. Thus there are three times as many electrons with $m_j = -1/2$ than with $m_j = +1/2$ with σ^+ light and vice versa for σ^- light. If the GaAs crystal is coated with a caesium and oxygen layer it is possible to reduce the positive electron affinity of the GaAs crystal to negative values, which allows the electrons to escape. The theoretical limit of $P=0.5$ based on the transition probabilities shown in fig 2.12 in practice is never achieved. Spin-flip scattering processes in the crystal lower the polarisation to typically 0.3–0.35. It has recently become possible to grow strained GaAs crystals which have the m_j degeneracy removed (Nakanishi *et al.*, 1991, and Maruyama *et al.*, 1992). With these crystals it is in principle possible to have $P \sim 1$, with measured values of P being as high as 0.9 (Maruyama *et al.*, 1992).

The operation of a GaAs source is shown schematically in fig 2.12. Right-hand (or left-hand) circularly-polarised light is incident on a negative-affinity GaAs crystal. The negative affinity is achieved by coating the crystal with caesium and oxygen in ultra-high vacuum conditions (Pierce *et al.*, 1980). The longitudinally-polarised electrons are electrostatically deflected by 90°, which does not affect the spin direction, thereby becoming transversely polarised. The polarisation of the electron beam can be simply reversed by reversing the polarisation of the light. The currents produced by these sources are generally less than 10 μA. Details of the preparation of a suitable GaAs photocathode as well as the operation and performance of the source can be found in the review by Pierce *et al.* (1980).

2.5.2 Analysis of polarised beams

The device commonly used for polarisation measurement, the Mott detector, is based on the well-known fact that electrons can be polarised and their polarisation analysed by high energy large-angle scattering from a heavy atom (Mott, 1929). An electron moving with velocity **v** with respect to a charged scattering centre sets up an electric field **E** given by

$$\mathbf{E} = -\frac{1}{r}\frac{dV}{dr}\mathbf{r}. \tag{2.44}$$

In the rest frame of the electron there appears a magnetic induction

$$\mathbf{B} = -\frac{1}{c}\mathbf{v} \times \mathbf{E} = \frac{1}{c}\mathbf{E} \times \mathbf{p}, \tag{2.45}$$

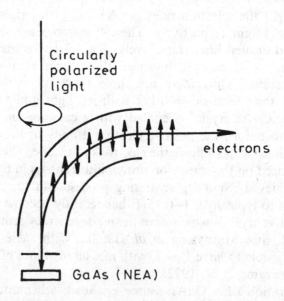

Fig. 2.12. Schematic of a GaAs source of transversely-polarised electrons using photemission from a negative electron affinity GaAs crystal.

where **p** is the electron momentum (atomic units are used). Due to the electron spin **S** a spin–orbit contribution to the energy arises, given by

$$V_{SL} = - \boldsymbol{\mu} \cdot \mathbf{B} = -\frac{1}{c}\mathbf{S} \cdot \mathbf{B} = -\frac{1}{c^2}\mathbf{S} \cdot [\mathbf{E} \times \mathbf{p}], \qquad (2.46)$$

$\boldsymbol{\mu}$ is the magnetic dipole moment of the electron associated with its spin operator **S**. Thus

$$V_{SL} = \frac{1}{2}\frac{1}{c^2}\frac{1}{r}\frac{dV}{dr}(\mathbf{S} \cdot \mathbf{L}), \qquad (2.47)$$

where $\mathbf{L} = \mathbf{r} \times \mathbf{p}$ is the orbital angular moment and the factor of $1/2$ which has been included arises from a relativistic kinematic effect, the Thomas precession, not included in the above non-relativistic derivation. The relativistic derivation is given in section 3.5.2.

The resulting potential for an electron moving in the field of a charged nucleus depends not only on its spin direction but also whether it passes the nucleus on the right (**L** positive) or left (**L** negative) sides. As illustrated in fig. 2.13, for a spin-up electron $\mathbf{S} \cdot \mathbf{L}$ and V_{SL} are positive (negative) depending on whether the electron passes on the right (left) of the nucleus. Thus the electrons with spin up which pass the nucleus on the left are scattered to the right with a stronger force than the ones passing on the right and scattered to the left and vice versa for spin-down electrons.

Hence the intensity of scattering to the right is greater than that to the left for spin up particles, and vice versa. The polarisation of an incident beam of electrons can be determined from the scattering intensities I_ℓ and I_r to the left and right by

$$P = \frac{1}{S}\frac{I_\ell - I_r}{I_\ell + I_r}. \qquad (2.48)$$

The analysing power (Sherman function) S is a complicated function of the electron energy, scattering angle and the atomic number Z. From (2.47) and the form of the Coulomb potential $V(r)$, it can be seen that the

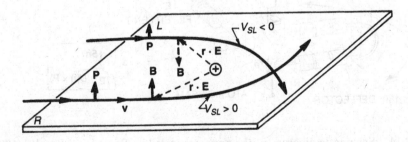

Fig. 2.13. Scattering of two electrons with spin up (out of scattering plane) by a nucleus.

spin—orbit term will be largest for high Z targets and close approaches to the nucleus (high-energy large-angle scattering). The precise measurement of S is difficult, with $\pm 5\%$ being the best uncertainty normally achieved in a careful experiment (e.g. Fletcher, Gay and Lubell, 1986).

The construction and operation of Mott polarimeters (or detectors) has been fully described by Kessler (1985). In their most common configuration the detected electrons are scattered through $\pm 120°$ from a thin gold foil target at 100—120 keV. The analysing power S of the foil is calibrated by measuring the asymmetry for a number of foil thicknesses and extrapolating to zero thickness, where S has been reliably calculated to be 0.39 for 100 keV electrons. Multiple scattering reduces S for thicker foils.

A variation of the Mott analyser, dubbed 'Mini-Mott' because of its relatively small size, has two electrodes in the vacuum chamber. Electrons are accelerated to the inner electrode, scattered from the gold foil, and then decelerated as they travel to the outer electrode, which is nominally at ground potential. This arrangement has the advantages that it is easier to discriminate against inelastically-scattered electrons, thus reducing multiple scattering effects, and that the detectors can be operated at or near ground potential. This type of analyser has been realised in cylindrical

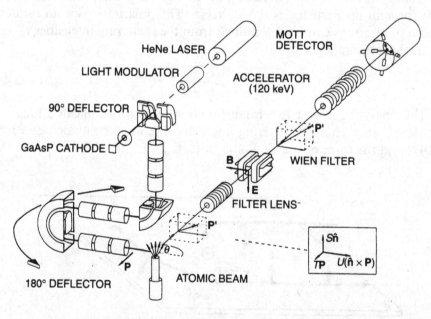

Fig. 2.14. Schematic diagram of the apparatus used by Berger and Kessler (1986) for their complete determination of the scattering amplitude for elastic scattering of electrons by xenon.

(Hodge *et al.*, 1979) and spherical forms (Gray *et al.*, 1984; Campbell *et al.*, 1985). In the spherical geometry it operates effectively in the 20–60 keV range with a typical figure of merit $S^2 I/I_0$ of 2×10^{-5}.

Typical experimental arrangements for polarised electron–atom collision studies are shown in figs. 2.14 and 2.17. Fig. 2.14 shows the arrangement used by Berger and Kessler (1986) in their complete analysis of the scattering amplitudes, with respect to their magnitudes and relative phases, for electron scattering from xenon. Circularly-polarised light from a He–Ne laser produced the polarised photoelectrons from a *p*-doped $GaAs_{0.6}P_{0.4}(100)$ crystal. The electrons were accelerated to 100 eV and passed through a 90° electrostatic deflector, focussing lenses, and a deflection system rotatable about the atomic beam axis. Electrons elastically scattered through an angle of θ passed through a filter lens and a Wien filter, accelerated to 120 keV and entered a Mott detector for polarisation analysis. The GaAsP crystal has a larger band gap (1.82 eV) than GaAs at the Γ point, which allows the use of the simple and reliable He–Ne laser as the light source.

The magnetic coil shown in figs. 2.15 and 2.16 was used to orient the electron polarisation vector $\hat{\mathbf{P}}$ parallel to the axis of the analysing target–Mott detector system. The deflection system is part of the differential pumping stage which is necessary for the maintenance of the required ultra-high vacuum in the source chamber.

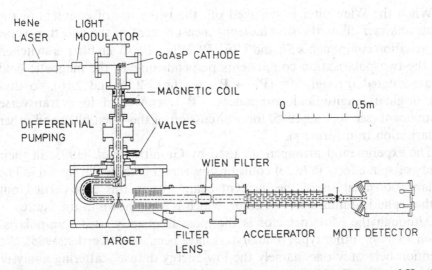

Fig. 2.15. The vertical cross section of the apparatus used by Berger and Kessler (1986). The pumping system is not shown.

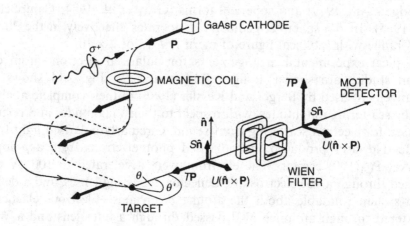

Fig. 2.16. The behaviour of the polarisation vector in the experiment of Berger and Kessler (1986).

When the polarisation **P** of the incident electron beam is resolved into the components $P_\parallel$ and P_n to the scattering plane, the polarisation **P'** of the scattered beam is given by (Kessler, 1985)

$$\mathbf{P'} = \frac{(P_n + S)\hat{\mathbf{n}} + T\mathbf{P} + U(\hat{\mathbf{n}} \times \mathbf{P})}{1 + P_n S}, \qquad (2.49)$$

where $\hat{\mathbf{n}}$ is the unit vector normal to the scattering plane. In the experiment of Berger and Kessler (1986) $P_n = 0$ and $\mathbf{P} = \mathbf{P}_\parallel$, so that

$$\mathbf{P'} = S\hat{\mathbf{n}} + T\mathbf{P} + U(\hat{\mathbf{n}} \times \mathbf{P}). \qquad (2.50)$$

When the Wien filter is switched off, the two pairs of counters in the Mott analyser allow the simultaneous measurement of the two transverse polarisation components $S\hat{\mathbf{n}}$ and $U(\hat{\mathbf{n}} \times \hat{\mathbf{P}})$. When the Wien filter is switched on the two polarisation components perpendicular to the magnetic field **B** are rotated through 90° (**P'** → **P''**, see figs. 2.15 and 2.16), so that the original longitudinal component $T\mathbf{P}$ is converted to a transverse component (see Kessler,1985, for a discussion of the Wien filter and other polarisation transformers).

The experimental arrangement used by Granitza *et al.* (1993) in their study of spin effects in (e,2e) collisions on xenon is shown in fig. 2.17. The polarisation analysis of the incident beam was in this case carried out with a spherical mini-Mott detector, which is also shown in the figure.

Although the Mott detector is the most frequently used spin-polaris-ation analyser, other types of analysers have been used (Kessler, 1985). We mention here only one, namely the low-energy diffuse scattering analyser of Unguris, Pierce and Celotta (1986). This compact and efficient analyser is based on low-energy (150 eV) diffuse scattering from a high-Z target,

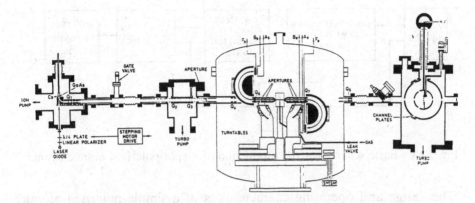

Fig. 2.17. Schematic diagram of the apparatus used by Granitza *et al.* (1993) to investigate spin effects in (e,2e) collisions on xenon. The source, differential pumping stage, collision chamber and mini-Mott detector are shown.

such as an evaporated polycrystalline gold film, opaque to the incident beam. By collecting scattered electrons over a large solid angle, a figure of merit $S^2 I/I_0 = 10^{-4}$ is achieved in this analyser with an analysing power $S = 0.11$. Because of its compact size this analyser is superior for many applications.

2.6 Polarised atom sources

The preparation of spin-polarised atomic beams began with the pioneering work of Gerlach and Stern (1924). It was extensively developed in the 1950s during studies of the hyperfine structure of atoms (Ramsey, 1956) and later by nuclear physicists who were interested in the study of the spin dependence of nuclear forces.

The basic idea in this work was to use inhomogeneous magnetic fields to state select an atomic beam. The inhomogeneous fields used were usually either the dipole field of the Stern—Gerlach type or hexapole fields. More recently with the advent of lasers it has been possible to state select atoms in a beam by means of optical pumping. With optical pumping it is possible to prepare beams in state-selected excited states as well as the ground state. The basic scheme for producing polarised beams of atoms or molecules is sketched in fig. 2.18.

2.6.1 State selection by inhomogeneous magnetic fields

State selection and focussing of atoms by magnetic fields has been extensively reviewed over several decades (e.g. Ramsay, 1956; Reuss, 1988; Ianotta, 1988), and only a brief description of a source of spin-polarised hydrogen atoms will be given here.

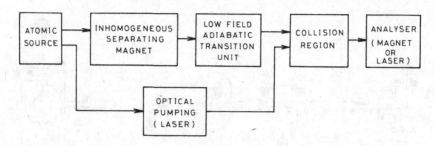

Fig. 2.18   Basic scheme for the production of a spin-polarised atomic beam.

The design and operating characteristics of a simple polarised atomic hydrogen beam particularly suitable for crossed-beam experiments has been given by Chan *et al.* (1988). The hydrogen atoms are produced from molecular hydrogen by a radio-frequency discharge source. Microwave and d.c. discharges and high-temperature tungsten or tantalum ovens can also be used to produce ground-state beams of atomic hydrogen. The beam is then collimated and passed through a permanent hexapole magnet, chopped and passed through another permanent hexapole magnet. The atoms with $m_J = -1/2$ are defocussed and deflected out of the beam entirely by the inhomogeneous magnetic field while those with $m_J = +1/2$ execute elliptical trajectories about the axis and are transmitted.

Two components contribute to the total angular momentum **F** of the atom, the nuclear angular momentum (spin) **I** and the electronic angular momentum **J**, so that

$$\mathbf{F} = \mathbf{I} + \mathbf{J}. \tag{2.51}$$

With **I** and **J** we have associated a nuclear and electronic magnetic moment respectively. In zero or weak external magnetic field the nuclear magnetic moment μ_I interacts with the electronic magnetic field $\mathbf{B}_J$ with an energy $-\mu_I \cdot \mathbf{B}_J$. This coupling between **I** and **J** leads to a splitting of the atomic level W_J into hyperfine levels W_F, one for each allowed value of **J**. This is shown schematically in fig. 2.19. In a strong external magnetic field **B** the coupling between **I** and **J** breaks down and they precess independently about **B**. Since the electronic magnetic moment μ_J is much larger than the nuclear one (due to their respective gyromagnetic ratios) the electronic energy $-\mu_J \cdot \mathbf{B}$ is much larger than the nuclear one $-\mu_I \cdot \mathbf{B}$. The resulting energy level diagram, called a Breit–Rabi diagram, is shown in fig. 2.19.

The basic properties of Stern–Gerlach (dipole), quadrupole and hexapole magnets for deflecting and focussing of atomic beams have been discussed in detail in the literature (Reuss, 1988; Hughes *et al.*, 1972). A hexapole state-selecting magnet has the advantage that for $m_J = +1/2$ atoms the cylindrical beam shape is preserved in addition to acting as a

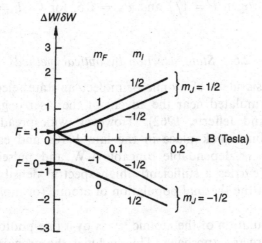

Fig. 2.19. Breit–Rabi hyperfine energy level diagram for ground-state hydrogen, showing the magnetic energy of the atom ΔW normalised to the hyperfine splitting at zero field (δW).

lens. This means that by a suitable choice of parameters the atom density can be optimised at the interaction region. A small solenoid placed axially at the exit of the hexapoles is used to align the spins with the axis of motion of the atoms, so that the atomic spins are perpendicular to the intersecting electron beam. A magnetic spin rotator (Iannotta, 1988) at the exit to the interaction region adiabatically rotates the atomic spins into the transverse orientation, in preparation for polarisation analysis by a Stern–Gerlach dipole magnet (Ramsay, 1956). The Stern–Gerlach polarimeter permits detailed scanning of the beam profile and employs an electromagnet, which allows it to be demagnetised. This is essential for both beam intensity monitoring and centroid determination. The atoms and molecules transmitted by the Stern–Gerlach unit are detected by a quadrupole mass analyser, which determines the atomic and molecular fractions in the beam.

The atomic polarisation state selection is greater than 0.99 in the high field regions. However the magnetic field must generally be kept low in the collision region (≤ 100 mGauss). The ground-state hyperfine interaction reduces the effective polarisation of the atom to one half of its high field state. In the low fields region the electron spin is again coupled to the nuclear spin I. Thus the low-field electronic polarisation of a one-electron atom after perfect high-field state selection is given by

$$P_A = 1/(2I + 1). \tag{2.52}$$

For atomic hydrogen $I = 1/2$ and $P_A = 0.5$, for Cs $I = 7/2$ and $P_A = 0.125$.

2.6.2 State selection by optical methods

Most of the basic ideas of atomic or molecular state selection by optical means were formulated near the advent of the laser (e.g. Kastler, 1950, and Dehmelt and Jefferts, 1962). However, widespread application of the technique only took place in the late 1970s and early 1980s with the availability of dependable tuneable CW and pulsed lasers. Only laser radiation carries a sufficiently high spectral density for significant manipulation of the thermal population of atomic (or molecular) levels in a beam.

Selective population of the atomic levels by a one-photon process is the most straightforward approach. This includes the preparation of excited states of atoms. With polarised radiation the atomic ensemble will be oriented or aligned. Sequential two-photon absorption can be used to excite high-lying states, including Rydberg levels (Feneuille and Jacquinot, 1981) and the ground state (Baum, Caldwell and Schröder, 1980). Since the excited states involved in optical pumping are shortlived (of the order of 10 ns for optically-allowed dipole transitions), rate equations must be used to establish the equilibrium populations. These have been discussed in detail by many authors (e.g. McClelland and Kelley, 1985; Bussert,

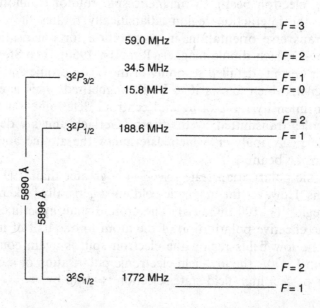

Fig. 2.20.   Section of the hyperfine resolved level diagram for sodium.

1986; and Farrell, MacGillivray and Standage, 1988). Most of the work has been done with alkali-metal atoms. The relevant hyperfine resolved levels for sodium are shown in fig. 2.20. From this it can be seen that the band width of the laser must be narrow in order to resolve the hyperfine transitions. With present technology this can be assumed to be typically between 1 and 10 MHz. To achieve this over the long period required for most scattering experiments, active stabilisation of the laser is necessary. This can conveniently be done by arranging the laser to cross the atomic beam at right angles (Doppler-free excitation) and using the fluorescence from the collision volume as the feedback signal for optimisation. As the laser frequency drifts, the fluorescence region will move from the central region of the atomic beam, where the atoms have zero transverse velocities, to a region where the Doppler shift due to the transverse velocity of the atoms compensates for the drift.

In order to obtain a well-defined m_F (or m_J) distribution it is necessary to use polarised light. The selection rules for dipole radiation are

$$\Delta F = 0, \pm 1, \ \Delta m_F = 0, \pm 1, \tag{2.53}$$

and for linearly and for circularly right- and left-hand polarised light

$$\Delta m_F = 0, \pm 1 \tag{2.54}$$

respectively. Thus pumping with circularly-polarised σ^+ or σ^- light on a two level system, such as $\text{Na}(3^2S_{1/2}, F = 2 \to 3^2P_{3/2}, F = 3)$ enriches the population in the $m_F = +F$ or $m_F = -F$ state respectively. That this is so for both the excited state and the ground state follows from the selection rules (2.53, 2.54). In principle complete ground-state polarisation can be achieved, but in practice the hyperfine splitting of the ground-state level prevents all of the atoms participating in the pump process, and the beam is not totally polarised. In order to achieve $P_A = 1$ a Stern–Gerlach magnet can be used to remove the $F = 1$ ground-state atoms (Dreves *et al.*, 1981). Alternatively an acousto-optical modulator can be used to frequency-shift part of the laser beam, and thus by crossing the atomic beam at right angles with the two superimposed laser beams, complete polarisation of the atoms can be achieved (Baum *et al.*, 1989), all of the atoms accumulating in the $m_F = +F_{\text{max}}$ or $m_F = -F_{\text{max}}$ levels with σ^+ or σ^- light, respectively. A radio-frequency field of the appropriate frequency (1772 MHz for sodium) can also be used to transfer the $F = 1$ to the $F = 2$ population while laser pumping the $F = 2$ ($^2S_{1/2}$) to $F' = 3$ ($^2P_{3/2}$) transition with σ^+ or σ^- light (Dreves *et al.*, 1983).

The polarisation of the beam can be measured by a number of techniques. Probably the most precise method is fluorescent monitoring. Information on the polarisation of the ground state can be obtained from the polarisation of the fluorescent light (Fischer and Hertel, 1982). Similarly the optical pumping can be monitored by observing the changes in the properties of transmitted light. Often a separate weak probing beam is employed. The anisotropy of γ-ray emission by polarised radioactive atoms in the beam has also been used to determine the polarisation of the beam (e.g. Bonn *et al.*, 1975). The detection of γ-rays is often cleaner and easier than optical photons. Standard nuclear magnetic resonance techniques are also sometimes used (Kuize, Wu and Happler, 1988).

The excited state in an optically-pumped two-level transition is obviously also polarised if σ^+ or σ^- light is used. However, due to the short lifetime (~ 10 ns for dipole transitions) the atomic beam will contain fewer excited atoms than ground-state atoms. It is important to ensure that the pumping laser has high power to ensure a sufficiently high and well-defined stationary population of the excited state. This makes the use of continuous-beam lasers preferable over pulsed lasers in most experimental arrangements for collision studies. The population of the various levels in optical pumping needs to be calculated by the use of rate equations. A detailed analysis for the case of the pumping of Na has been given by McLelland and Kelley (1985) and Farrell *et al.* (1988).

The alkali metals, in particular sodium and potassium, have played a dominant role in providing state-selected collision partners through optical pumping. The subject has been summarised in a number of excellent articles such as Hertel and Stoll (1978), Fischer and Hertel (1982) and Düren (1988). The reason why the sodium and potassium transitions $^2S_{1/2}$, $F = 2 \rightarrow {}^2P_{3/2}$, $F = 3$ are particularly suitable examples of the optical pumping technique is due to a number of factors. Firstly the efficiency of excitation is high, with 62% of the beam particles being involved in the excitation. Secondly the selection rule for spontaneous decay ($\Delta F = 0, \pm 1$) prevents loss to the $F = 1$ ground state. Thirdly the separation between the $^2P_{3/2}$, $F = 3$ state and its neighbour, the $^2P_{3/2}$, $F = 2$ state, is larger than the line width of the laser. Some small loss to the $F = 1$ ground state does occur due to the long tail of the Lorentzian line shape for the $^2P_{3/2}$, $F = 2$ excited state.

Because of the short lifetime of the excited state, it is obvious that the collision region must coincide largely with the optically-pumped region of the atomic beam. Generally the arrangement is chosen so that the laser, atomic, and incident electron beams all intersect at right angles (Hertel and Stoll, 1974, 1978; Zheng *et al.*, 1990). Multifrequency techniques, which can be implemented by the use of acousto-optical modulators, can lead to the excitation of an excited state from both ground states (Ertmer *et al.*, 1985).

Metastable states have also been studied, in particular those of rare-gas atoms. The beam emerging from a gas discharge or electron impact source unfortunately generally contains atoms in more than one metastable state.

For example, in a helium discharge metastable atoms in both the 2^1S_0 and 2^3S_1 states are formed. The 2^1S_0 state can be depopulated by exposing the beam to light from a helium discharge lamp, which emits light of sufficient intensity in the $2^1P_1 \to 2^1S_0$ transition to excite all atoms in the 2^1S_0 state to the 2^1P_1 level, which in turn decays to the 1^1S_0 ground state (Fry and Williams, 1969). By circularly polarising the $2^3P - 2^3S$ radiation in a discharge lamp, atoms can be accumulated in the $m_F = 1$ state of the 2^3S level (Riddle *et al.*, 1981). This provides a polarised beam of metastable states.

2.7 Electron–photon correlation experiments

2.7.1 Electron–photon coincidences –
angular and polarisation correlations

Electron–photon coincidence experiments have given a new insight into the excitation of atoms (see chapter 8). In a typical (e,e'γ) experiment, electrons scattered with a fixed energy loss into a given narrow cone about a certain angle are detected in delayed coincidence with photons emitted into a narrow cone in some specified direction. Angular-correlation measurements are then made by mapping the photon coincidence count rate as a function of emission direction (Eminyan *et al.*, 1974). Alternatively the polarisation of the emitted photon can be measured (polarisation correlation), which yields similar information to the angular correlation measurement.

The coincidence technique has been discussed in detail in section 2.4.4 and much of that discussion is valid for electron–photon coincidence measurements. The coincidence technique offers the important advantage of eliminating photon contributions from excited atoms produced by cascading rather than by direct excitation. This depends on the band pass of either the photon detector or electron detector being sufficiently narrow to isolate the excited state being studied.

One hazard in (e,e'γ) measurements which does not arise in (e,2e) measurements is the potential for radiation trapping. If the photon transition from the excited state is to the ground state, other ground-state atoms in the atomic beam can resonantly re-absorb the emitted photon within the interaction region or within the viewing angle of the photon detector. The re-emitted photon on subsequent decay will have characteristics (angular or polarisation dependence) which differ from that due to the photon emitted in the electron impact ionisation process. This can seriously degrade the data and great care must be taken to eliminate the possibility of resonance trapping (McConkey, van der Burgt and Corr, 1992).

The first electron–photon coincidence experiment with polarised electrons was carried out by Wolcke *et al.* in 1984. The Münster group recently extended this work on the excitation of the 6^3P_1 sublevels in Hg from the 6^1S_0 ground state (Goeke, Hanne and Kessler, 1989, and Sohn and Hanne, 1992). As discussed in chapter 8, in a conventional (e,e'γ) coincidence experiment it is possible to get information on the population of magnetic sublevels of the excited atoms. With polarised incident electrons the reflection symmetry in the scattering plane is, in general, broken. As a result the charge-cloud distribution of the excited state may be tilted (by an angle different from 0° or 90°) with respect to the scattering plane. Such a charge-cloud distribution will in general emit elliptically-polarised photons. It is, however, possible to determine the parameters that characterise the anisotropic charge cloud distribution of the radiating atoms from measurements of the linear polarisation of the decay photons.

Fig. 2.21 shows schematically the apparatus used by the Münster group (Sohn and Hanne, 1992). Circular-polarised light from an He–Ne laser is used to produce longitudinally-polarised electrons from a negative affinity GaAsP cathode, the source being in an ultra-high-vacuum chamber. The electrons are made transversely polarised (in the z-direction) by electrostatic deflection through 90° after which they pass through a magnetic

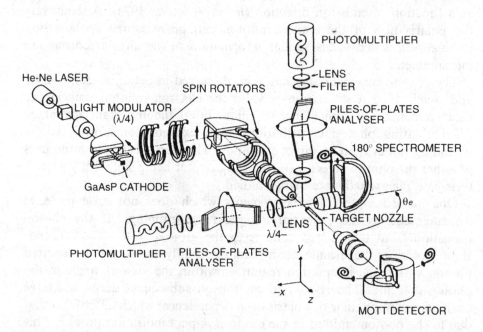

Fig. 2.21. Schematic diagram of the apparatus used by Sohn and Hanne (1992) for (e,e'γ) coincidence measurements on Hg ($6^1S_0 \rightarrow 6^3P_1$) using polarised electrons.

coil system. This coil system can be used to rotate the polarisation vector by 90° into the y direction, which is perpendicular to the scattering plane. The electrons then enter a differential pumping stage, are deflected electrostatically through another 90° and pass through a second magnetic lens which does not rotate the polarisation vector. The polarised electron beam is then decelerated from its transport energy (200 eV) to the desired collision energy (8 and 15 eV) and focussed on the mercury beam. Electrons that have excited the 6^3P_1 state of mercury are analysed and detected by a channeltron in coincidence with the ultra-violet 4.9 eV photons from the $6^3P_1 - 6^1S_0$ transition. The photon detector consists of a pile-of-quartz-plates linear polarisation analyser, a wavelength filter and a photomultiplier. The light is observed either in the y direction (perpendicular to the scattering plane) or in the $-x$ direction (in the scattering plane). Quarter-wave plates, when inserted in front of the linear-polarisation filters, are used for determining the helicity $\sigma^\pm$ (or circular polarisation) of the emitted photons. The observed coincidence rates are functions of the electron polarisation $\mathbf{P}_e(= P_x \text{ or } P_y)$ of the incident electron beam and the helicity and the linear polarisation are aligned along angles $\alpha = 0°$, 45°, 90° and 135° with respect to the z (incident electron) axis. The shape and size of the charge-cloud distribution are found to be sensitive functions of the electron polarisation vector.

2.7.2 Time evolution of excited states

The coincidence measurements discussed in the previous section were concerned with the total coincidence signal, i.e. the signal obtained when the decay of a particular ensemble of states is integrated over. These states are produced in a very short time ($\sim 10^{-15}$s) in electron impact excitation, and can sometimes evolve in a complicated way. In the absence of internal fields (e.g. the n^1P states of helium) each of the $|\ell m\rangle$ states decays with the same exponential time dependence $\exp(-\gamma t)$, and the coincidence technique can be used to yield the decay constant γ of the excited state (see Imhof and Read, 1977, and references therein). However, if the excited state is perturbed by an internal (or external) field before decay, then the exponential decay is modulated sinusoidally giving rise to the phenomenon of quantum beats (Blum, 1981).

Quantum beats have been observed in a variety of experiments, particularly in beam–foil measurements. Teubner *et al.* (1981) were the first to observe quantum beats in electron–photon coincidence measurements, using sodium as a target. The 'zero-field' quantum beats observed by them are due to the hyperfine structure associated with the $3^2P_{3/2}$ excited state (see fig. 2.20). The coincidence decay curve showed a beat pattern superimposed on the exponential decay given by

$$I(t) = I_0 + \sum_i B_i(\cos(E_F - E_{F'})t + \phi_{FF'}). \qquad (2.55)$$

The B_i contain information on the collision dynamics, in particular on the alignment and orientation of the excited states.

Heck and Williams (1987) observed quantum beats in the decay of the $n = 2$ states of atomic hydrogen. In the presence of an external electric field the $2s$ and $2p$ states can be mixed and their correlation measured. With an applied field of 250 Vcm^{-1} the modulation periods should be 0.1 ns and 0.6 ns. They were able to observe the second beat period corresponding to interference between the states which reduce to $2s_{1/2}$ and $2p_{1/2}$ in the field-free limit. Williams and Heck (1988) were able to use the technique to determine many of the $n = 2$ state multipoles (section 8.2.4).

2.7.3 Superelastic scattering from laser-excited targets

An interesting and alternative technique to electron–photon coincidence measurement of coherence effects in excitation processes is superelastic scattering from laser-excited targets. This technique, first developed by Hertel and Stoll (1974, 1978), can be thought of as the time inverse of the $(e,e'\gamma)$ coincidence experiment,

$$e_0 + A_0 \longrightarrow A^* + e_i$$
$$ \big\lfloor\!\!\longrightarrow A_0 + \gamma, \qquad (2.56)$$

in which laser light excites target atoms, a beam of monoenergetic incident electrons (e_i) de-excites these atoms and is superelastically scattered in the process. One advantage of this technique in those cases in which it can be applied is that the count rates are typically several orders of magnitude higher than in the equivalent coincidence experiments. The polarisation of the laser may also be varied, allowing the preparation of excited atoms in various mixtures of angular momentum states.

McClelland, Kelley and Celotta (1986) were the first to measure super-elastic scattering in the configuration where the spins of both the incoming electron and target atom were polarised, ensuring that the transitions studied are transitions between well-characterised pure quantum states. In particular they studied the superelastic scattering of spin polarised electrons from the $m_F = 3$ and $m_F = -3$ states of Na $3^2P_{3/2}$ atoms (or the $m_\ell = +1$ and $m_\ell = -1$ states on making the conventional assumption

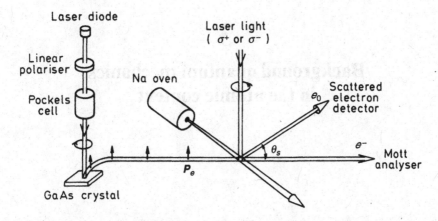

Fig. 2.22. Schematic of the polarised electron—polarised atom scattering apparatus of McClelland *et al.* (1986, 1989).

that the nuclear spin plays no significant dynamic role in the collision process).

Their experimental arrangement is shown schematically in fig. 2.22. Transversely-polarised electrons from a GaAs source are superelastically scattered from excited sodium atoms in a beam of sodium atoms pumped by σ^+ and σ^- circularly-polarised light (see section 2.6). The polarisation of the incident electrons is measured by a Mott analyser. Spin asymmetries were measured as a function of the incident electron spin (up or down) for both the $m_\ell = +1$ and $m_\ell = -1$ states. From these measurements it is possible to extract both triplet and singlet contributions to $L_\perp$, the angular momentum transferred in the collision perpendicular to the scattering plane, and the ratio of triplet to singlet cross sections (see chapters 8 and 9). Hertel, Kelley and McClelland (1987) developed a general framework using density matrices for the analysis of collisional alignment and orientation, studied by the scattering of spin-polarised electrons from laser-excited atoms.

3

Background quantum mechanics in the atomic context

The basic structure of quantum mechanics and its relationship to physical measurements have been beautifully developed in the textbook of Dirac (1958), which is one of the classics of physics literature. Here we summarise the quantum mechanical ideas necessary for the background to our development of the theory of electron—atom collisions. We introduce notation relevant to our subject.

3.1 Basic mathematical constructions

3.1.1 States

A physical system, which may be part of a larger system, is associated with a linear vector space whose elements are ket vectors

$$|A\rangle.$$

For each space there is a dual space of bra vectors

$$\langle B|$$

whose complex scalar products with the ket vectors are denoted by a bracket

$$\langle B|A\rangle = \langle A|B\rangle^*. \tag{3.1}$$

The letters A and B denote quantities characterising the vectors. They refer to dynamical aspects of the state of the system. The vectors are called state vectors, a term which we abbreviate to states. The length of a state vector, considered apart from related state vectors, has no physical significance. We are free to choose it, a process known as normalisation.

The states are of two kinds, which require rather different treatments. The specifying quantities A may be members of a discrete set counted by an integer i, in which case the state is denoted

$$|i\rangle,$$

50

or they may be part of a continuum, that is they may take all values in a continuous range of a variable x, in which case the state is denoted

$$|x\rangle.$$

3.1.2 Observables

The spaces of interest in physics are spanned by the eigenstates $|\alpha'\rangle$ of real (that is self-adjoint) linear operators α, whose eigenvalues α' are real. Such operators are called observables. The eigenvalue equation is

$$\alpha|\alpha'\rangle = \alpha'|\alpha'\rangle. \tag{3.2}$$

An important property of the eigenstates of real linear operators belonging to different eigenvalues is that they are orthogonal.

$$\langle\alpha'|\alpha''\rangle = 0, \ \alpha' \neq \alpha''. \tag{3.3}$$

States can be simultaneously eigenstates of more than one observable. In this case the observables commute. If the simultaneous eigenstates of a set of commuting observables form a complete set they span a new space which is the direct product space of the spaces spanned by the eigenstates of each of the observables in the set. The dimension of the new space is the product of the dimensions of the spaces spanned by the eigenstates of the individual observables.

The set of commuting observables can be considered as a new observable. In this way we can extend the space associated with a particular dynamical property of a system to a more general dynamical property of that system or to dynamical properties of a larger system that includes it.

Examples of interest in atomic physics are the x, y, z coordinates of an electron. They can be considered as individual observables or the set of all three may be considered as an observable $\mathbf{r}$, the position of the electron. The simultaneous eigenstates are

$$|x\rangle|y\rangle|z\rangle \equiv |\mathbf{r}\rangle. \tag{3.4}$$

Another relevant example is the eigenstate of the momenta of each of two electrons.

$$|\mathbf{p}_1\rangle|\mathbf{p}_2\rangle \equiv |\mathbf{p}_1\mathbf{p}_2\rangle. \tag{3.5}$$

Canonically-conjugate observables do not commute. Corresponding to a generalised position coordinate q there is a generalised momentum p. The commutation law is

$$qp - pq = i\hbar. \tag{3.6}$$

3.1.3 Representations

A set of numbers representing the states or observables is called a representation. In the geometrical space of three-dimensional vectors **r** a set of numbers representing **r** is the set of three coordinates x', y', z' in a system of orthogonal axes. We may consider the system of unit vectors $\hat{\mathbf{x}}', \hat{\mathbf{y}}', \hat{\mathbf{z}}'$ in the directions of the axes as a basis for the representation of **r**. A coordinate is the scalar product of **r** with one of the unit vectors. A different basis is provided by a rotated set of axes. A vector is changed into a new vector by operating with a 3×3 matrix. This concept is easily extended to the spaces of quantum mechanics.

We first consider the space spanned by the eigenbras $\langle i|$ of an observable that has discrete eigenvalues. A state $|A\rangle$ is represented by the set of numbers (coordinates) which are the scalar product of $|A\rangle$ with each of the $\langle i|$. The representative of $|A\rangle$ is the set

$$\langle i|A\rangle.$$

It is convenient to normalise the basis states so that they obey the orthonormality relation

$$\langle i'|i\rangle = \delta_{i'i}. \tag{3.7}$$

The set $|i\rangle$ is an orthonormal basis.

The representative of an observable α is the scalar product of the bra $\langle i'|$ with the ket $\alpha|i\rangle$ formed by operating with α on $|i\rangle$. It is sometimes called a matrix element

$$\langle i'|\alpha|i\rangle.$$

Consider two basis states $|i\rangle$ and $|i'\rangle$. In view of the orthonormality relation (3.7) we can expand $|i'\rangle$ in the form

$$|i'\rangle = \Sigma_i |i\rangle \delta_{ii'} = \Sigma_i |i\rangle \langle i|i'\rangle. \tag{3.8}$$

We may consider $\Sigma_i |i\rangle \langle i|$ as an operator that operates on $|i'\rangle$. Equn. (3.8) shows that it is the unit operator

$$\Sigma_i |i\rangle \langle i| = 1. \tag{3.9}$$

Any partial sum of (3.9) projects the corresponding states from the complete set. It is a projection operator.

Equn. (3.9) is a special case of the *representation theorem* or *closure theorem*, which is one of two important theorems that are used frequently in formal quantum mechanics.

We now extend the theorem to a space spanned by the eigenstates $|x\rangle$ of an observable whose eigenvalues form a continuum. An example is the momentum **p** of a free electron. The representation theorem becomes

$$|x'\rangle = \int dx|x\rangle\langle x|x'\rangle, \tag{3.10}$$

where the integration is over the range of x. The orthonormality relation is

$$\langle x|x'\rangle = \delta(x - x'). \tag{3.11}$$

This is called the 'delta function', although it is never used as a function, since it is infinite if $x = x'$. It is better to consider it as a notation for an integral.

$$|x'\rangle = \int dx|x\rangle\delta(x - x'). \tag{3.12}$$

In the language of functions the definition of the delta function is

$$f(y) = \int dxf(x)\delta(x - y). \tag{3.13}$$

We will need the representation of the delta function

$$\lim_{\epsilon \to 0+} \frac{2\epsilon}{(x-y)^2 + \epsilon^2} = 2\pi\delta(x - y), \tag{3.14}$$

which is a symbolic statement of the limit of the corresponding integral

$$\lim_{\epsilon \to 0+} \frac{1}{2\pi} \int_{\infty}^{\infty} dx \frac{2\epsilon}{x^2 + \epsilon^2} f(x) = f(0). \tag{3.15}$$

A simple extension is the three-dimensional delta function $\delta(\mathbf{x} - \mathbf{y})$, whose definition is

$$f(\mathbf{y}) = \int d^3x f(\mathbf{x})\delta(\mathbf{x} - \mathbf{y}), \tag{3.16}$$

where the integration is over the whole three-dimensional geometrical space of the vector **x**.

Examples of representations in common use in atomic reaction theory are the coordinate and momentum representations where, if the system under study is a single electron, the basis states are the eigenstates $\langle \mathbf{r}|$ and $\langle \mathbf{p}|$ of the position and momentum of the electron respectively. Examples of discrete representations are also important. They will be left until later.

For some important observables the complete set of eigenstates contains some discrete states $|i\rangle$ and some continuum states $|x\rangle$. A general statement of the representation theorem is

$$\Sigma_i|i\rangle\langle i| + \int dx|x\rangle\langle x| = 1. \tag{3.17}$$

For ease of notation we often use the formal device of representing the
sum and integral of the representation theorem (3.17) by a sum over a set
i, which is understood in part as a discrete notation for a continuum

$$\Sigma_i |i\rangle\langle i| = 1. \tag{3.18}$$

Whenever it becomes necessary to treat the continuum explicitly we return
to the continuum notation which replaces part of the sum by an integral.

The second theorem of vital importance in formal quantum mechanics
concerns a function $f(\alpha)$ of an observable α. We call it the *function theorem*.
The theorem is stated as follows.

If $|\alpha'\rangle$ is an eigenstate of an observable α belonging to the eigenvalue
α', that is if

$$\alpha|\alpha'\rangle = \alpha'|\alpha'\rangle,$$

then a function $f(\alpha)$ of α obeys the eigenvalue equation

$$f(\alpha)|\alpha'\rangle = f(\alpha')|\alpha'\rangle. \tag{3.19}$$

The proof is simple for a function that can be expanded as a power series.
It is obviously true for linear combinations. For powers of α it is proved
by successive application of the following process

$$\alpha^2|\alpha'\rangle = \alpha(\alpha|\alpha'\rangle) = \alpha(\alpha'|\alpha'\rangle) = \alpha'\alpha|\alpha'\rangle = \alpha'^2|\alpha'\rangle.$$

For functions that cannot be expanded everywhere as a power series the
theorem becomes a definition that is used to complete the function at the
necessary points.

The use of the function theorem can be seen in conjunction with the
representation theorem. We choose the spectral representation of the
observable α, that is the representation in which the basis states are the
eigenstates (corresponding to the eigenvalue spectrum) of α.

$$f(\alpha) = \Sigma_{\alpha'} f(\alpha)|\alpha'\rangle\langle\alpha'| = \Sigma_{\alpha'} f(\alpha')|\alpha'\rangle\langle\alpha'|. \tag{3.20}$$

The theorem replaces the operator $f(\alpha)$, which cannot be described in a
computational algorithm, by numbers $f(\alpha')$, which can. This is done at
the expense of introducing a large set of terms in the sum over α'.

3.1.4 The Schrödinger equation

The observable corresponding to the total energy of a dynamical system
is the Hamiltonian H.

$$H = K + V, \tag{3.21}$$

where K is the total kinetic energy and V is the total potential energy. It
is of central importance in physics. Its eigenstates $|\Psi\rangle$ are the quantum

states of the system. The eigenvalue equation for H is the Schrödinger equation.

$$(E - H)|\Psi\rangle = 0. \tag{3.22}$$

$|\Psi\rangle$ is not fully specified by (3.22). Its complete specification includes the appropriate boundary conditions.

One-body systems of particular interest are the free spinless particle and the spinless particle bound in a Coulomb potential U_C (the nonrelativistic hydrogen atom). Spin is introduced in section 3.3.2. The state of a free spinless particle is an eigenstate of momentum. It is completely specified by the momentum $\mathbf{p}$. The corresponding Schrödinger equation is

$$(E - K)|\mathbf{p}\rangle = 0. \tag{3.23}$$

The state of a spinless hydrogen atom is completely specified by the principal quantum number n, the orbital angular momentum quantum number ℓ and the magnetic (projection) quantum number m. The Schrödinger equation is

$$(\epsilon_{n\ell m} - K - U_C)|n\ell m\rangle = 0. \tag{3.24}$$

3.1.5 Representations of particular interest

Here we introduce the notation to be used for the representations of some state vectors and observables that are basic to our development of the theory of electron–atom systems. We express all quantities in atomic units

$$\hbar = m = e = 1,$$

where m and e are respectively the mass and charge of the electron.

The coordinate representation of the state vector of a system is its (coordinate) wave function.

$$\text{Free spinless particle (plane wave)} : \langle \mathbf{r}|\mathbf{p}\rangle = (2\pi)^{-3/2} e^{i\mathbf{p}\cdot\mathbf{r}}. \tag{3.25}$$

$$\text{Hydrogen atom} \qquad\qquad : \langle \mathbf{r}|n\ell m\rangle = \psi_{n\ell m}(\mathbf{r}). \tag{3.26}$$

The momentum representation of the state vector of a system is its momentum wave function.

$$\text{Free spinless particle} : \langle \mathbf{q}|\mathbf{p}\rangle = \delta(\mathbf{q} - \mathbf{p}). \tag{3.27}$$

$$\text{Hydrogen atom} \quad : \langle \mathbf{q}|n\ell m\rangle = \phi_{n\ell m}(\mathbf{q}). \tag{3.28}$$

It is often useful to transform from one representation to another. Transformations are effected by the representation theorem. The transformation between the momentum and coordinate representations is the

Fourier transformation. For the hydrogen atom this is

$$\phi_{n\ell m}(\mathbf{q}) = \langle \mathbf{q}|n\ell m\rangle = \int d^3r \langle \mathbf{q}|\mathbf{r}\rangle\langle \mathbf{r}|n\ell m\rangle$$

$$= (2\pi)^{-3/2} \int d^3r e^{-i\mathbf{q}\cdot\mathbf{r}} \psi_{n\ell m}(\mathbf{r}). \qquad (3.29)$$

A useful representation of the delta function is given by using the representation theorem together with (3.25).

$$\delta(\mathbf{r}' - \mathbf{r}) = \langle \mathbf{r}'|\mathbf{r}\rangle = \int d^3p \langle \mathbf{r}'|\mathbf{p}\rangle\langle \mathbf{p}|\mathbf{r}\rangle = (2\pi)^{-3} \int d^3p e^{i\mathbf{p}\cdot(\mathbf{r}'-\mathbf{r})}. \qquad (3.30)$$

To formulate and solve the Schrödinger equation for a bound atom it is usual to express the relevant observables in the coordinate representation. Observables for the motion of an electron are represented by differential operators.

$$x-\text{momentum} : \langle x'|p_x|x\rangle = \langle x'| - i\frac{\partial}{\partial x}|x\rangle = -i\delta(x' - x)\frac{\partial}{\partial x}. \qquad (3.31)$$

$$\text{Momentum} \quad : \langle \mathbf{r}'|\mathbf{p}|\mathbf{r}\rangle = -i\delta(\mathbf{r}' - \mathbf{r})\,\nabla. \qquad (3.32)$$

$$\text{Kinetic energy} : \langle \mathbf{r}'|\tfrac{1}{2}p^2|\mathbf{r}\rangle = -\tfrac{1}{2}\delta(\mathbf{r}' - \mathbf{r})\nabla^2. \qquad (3.33)$$

Potential energy observables are represented by functions of geometrical vectors such as $\mathbf{r}$.

$$\text{Coulomb potential } U_C : \langle \mathbf{r}'|U_C|\mathbf{r}\rangle = \delta(\mathbf{r}' - \mathbf{r})\frac{1}{r}. \qquad (3.34)$$

The Coulomb potential is a special case of a local, central potential.

$$\text{Local central potential } U : \langle \mathbf{r}'|U|\mathbf{r}\rangle = \delta(\mathbf{r}' - \mathbf{r})U(r). \qquad (3.35)$$

In problems involving the interaction of an electron with a system containing electrons an operator that frequently appears is a nonlocal potential.

$$\text{Nonlocal potential } V : \langle \mathbf{r}'|V|\mathbf{r}\rangle = V(\mathbf{r}', \mathbf{r}). \qquad (3.36)$$

In scattering problems the initial and final detected particles are free if the target is uncharged. The momentum representation of their motion observables is particularly simple.

$$\text{Momentum} : \langle \mathbf{p}'|\mathbf{p}|\mathbf{p}\rangle = \delta(\mathbf{p}' - \mathbf{p})\mathbf{p}. \qquad (3.37)$$

$$\text{Kinetic energy} : \langle \mathbf{p}'|\tfrac{1}{2}p^2|\mathbf{p}\rangle = \delta(\mathbf{p}' - \mathbf{p})\tfrac{1}{2}p^2. \qquad (3.38)$$

The momentum representation of a local, central potential is of interest. Its calculation illustrates the use of the representation theorem and the delta function and also some calculus manipulations that are common.

$$\langle \mathbf{p}' | U | \mathbf{p} \rangle = \int d^3r \int d^3r' \langle \mathbf{p}' | \mathbf{r}' \rangle \langle \mathbf{r}' | U | \mathbf{r} \rangle \langle \mathbf{r} | \mathbf{p} \rangle$$

$$= (2\pi)^{-3} \int d^3r \int d^3r' e^{i(\mathbf{p}\cdot\mathbf{r}-\mathbf{p}'\cdot\mathbf{r}')} \delta(\mathbf{r}' - \mathbf{r}) U(r)$$

$$= (2\pi)^{-3} \int d^3r \, e^{i\mathbf{K}\cdot\mathbf{r}} U(r), \qquad (3.39)$$

where

$$\mathbf{K} = \mathbf{p} - \mathbf{p}'.$$

By defining

$$x = \hat{\mathbf{K}} \cdot \hat{\mathbf{r}}$$

we can simplify this further.

$$\langle \mathbf{p}' | U | \mathbf{p} \rangle = (2\pi)^{-3} \int_0^{2\pi} d\phi \int_{-1}^1 dx \int_0^\infty dr \, r^2 \, e^{iKrx} U(r)$$

$$= (2\pi^2)^{-1} K^{-1} \int_0^\infty dr \, r \, \sin Kr \, U(r). \qquad (3.40)$$

Note that the momentum representation of a local, central potential depends only on $K = |\mathbf{p} - \mathbf{p}'|$.

In the special case of a Coulomb potential we perform the integration in (3.40) by introducing a convergence factor $e^{-\mu r}$ and taking the limit $\mu \to 0$ after the integration. We write (3.40) in the form

$$\langle \mathbf{p}' | U_C | \mathbf{p} \rangle = (2\pi^2 K)^{-1} \lim_{\mu \to 0} \text{Im} \int_0^\infty dr \, r \, e^{-(\mu-iK)r} \frac{1}{r}$$

$$= (2\pi^2 K)^{-1} \lim_{\mu \to 0} \text{Im} \frac{1}{\mu - iK}$$

$$= (2\pi^2 K^2)^{-1}. \qquad (3.41)$$

3.1.6 Time development

The Hamiltonian H of a system may be represented in terms of the differential operator $\partial/\partial t$. This is the analogue in special relativity of the representation of the momentum in terms of ∇. Operating with the observable H on an arbitrary time-dependent state $|\Psi(t)\rangle$ we have

$$i\frac{\partial}{\partial t}|\Psi(t)\rangle = H|\Psi(t)\rangle. \qquad (3.42)$$

This is the Schrödinger equation of motion.

We may consider $|\Psi(t)\rangle$ as the result of operating on $|\Psi(t_0)\rangle$ with the time-development operator T.

$$|\Psi(t)\rangle = T|\Psi(t_0)\rangle. \tag{3.43}$$

Since $|\Psi(t)\rangle$ is an arbitrary state for all t we may write an operator equation for the result of substituting (3.43) in (3.42),

$$i\frac{\partial}{\partial t}T = HT, \tag{3.44}$$

which can be integrated using the initial condition $T=1$ for $t = t_0$.

$$T = e^{-iH(t-t_0)}. \tag{3.45}$$

For a dynamical system the total energy E is constant in time. Choose

$$|\Psi(t_0)\rangle = |\Psi\rangle, \tag{3.46}$$

where $|\Psi\rangle$ is the state of the system, given by the Schrödinger equation (3.22). $|\Psi\rangle$ is an eigenstate of H with eigenvalue E. This choice results in

$$|\Psi(t)\rangle = e^{-iE(t-t_0)}|\Psi\rangle. \tag{3.47}$$

Since $|\Psi(t)\rangle$ differs from $|\Psi\rangle$ by only a phase factor it always remains an eigenstate of H. An eigenstate of the Hamiltonian of a system is called a stationary state.

3.2 Physical interpretation

Suppose we have a dynamical system whose state $|\Psi\rangle$ can be calculated by solving the Schrödinger equation with the appropriate boundary conditions. An experiment is set up to observe a particular dynamical property of the system corresponding to an observable whose eigenstates are $|\Phi\rangle$.

The experiment observes an ensemble of events, each with the same initial conditions (as nearly as can be physically achieved). The number n of observations of the system when it is in a particular state $|\Phi\rangle$ is recorded. The number n, suitably normalised, is taken as an estimate of the probability of finding the system in the state $|\Phi\rangle$. The standard error of the estimate is $n^{1/2}$.

The physical interpretation of quantum mechanics is as follows. The probability amplitude f of finding the system in the state $|\Phi\rangle$ is

$$f = \langle\Phi|\Psi\rangle. \tag{3.48}$$

The corresponding probability P is

$$P = N^{-1}|f|^2. \tag{3.49}$$

The normalisation N is

$$N = \langle\Psi|\Psi\rangle. \tag{3.50}$$

We normally abbreviate 'probability amplitude' to 'amplitude'.

3.2.1 *Examples of probability amplitudes*

Bound states are normalisable. If we represent all the coordinates of the system by x then the coordinate representation of N is

$$N = \int dx \langle \Psi | x \rangle \langle x | \Psi \rangle. \tag{3.51}$$

The integrand is the probability of finding the system with coordinates x. If the integrand tends rapidly enough to zero when all parts of the system are remote from the centre of mass then the integral is convergent. $|\Psi\rangle$ is defined by choosing $N=1$. The total probability is 1.

Unbound systems, such as an electron scattered by a hydrogen atom, are not normalisable, since there is a finite probability of finding the electron anywhere in space. The normalisation of the states of an unbound system will be discussed in chapter 6 on formal scattering theory.

For a bound one-electron system the amplitude for finding the electron at the point $\mathbf{r}$ is

$$\langle \mathbf{r} | \Psi \rangle.$$

The observation of an electron at a sufficiently well-defined point is impossible experimentally, since we cannot resolve the necessary distances of order 10^{-9}cm. It is possible to observe the probability of finding an electron with momentum $\mathbf{p}$ using a good, but necessarily imperfect, probe (McCarthy and Weigold, 1983). The corresponding amplitude is

$$\langle \mathbf{p} | \Psi \rangle.$$

Kinetic energies corresponding to atomic values of p are of order 10 eV and it is easy to observe momenta with accuracy better than 10%. The momentum $\mathbf{p}$ of a bound electron is observed by knocking it out of its bound state with an electron of high momentum $\mathbf{p}_0$ and observing the final momenta $\mathbf{p}_A$ and $\mathbf{p}_B$. We then have $\mathbf{p} = \mathbf{p}_A + \mathbf{p}_B - \mathbf{p}_0$ by conservation of momentum, assuming that the collision is the only way of transferring momentum. The function $|\langle \mathbf{p} | \Psi \rangle|^2$ observed in this way for the hydrogen atom is shown in fig. 3.1.

The most important type of amplitude is a transition amplitude for an observable V that changes the state $|\Psi\rangle$ of a system to $V|\Psi\rangle$. The amplitude for finding the transformed system in the state $|\Phi\rangle$ is

$$\langle \Phi | V | \Psi \rangle.$$

Such an amplitude is often called a matrix element, even if the states are eigenstates of observables with continuous eigenvalues.

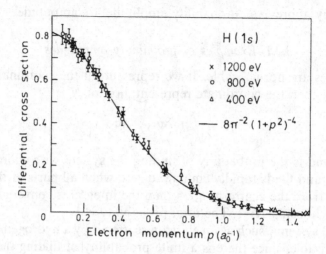

Fig. 3.1. The probability of finding an electron with absolute momentum p in a hydrogen atom, observed by measuring the complete kinematics of ionisation events at the total energies shown (Lohmann and Weigold, 1981). The curve shows the square of the momentum-space wave function.

3.2.2 A physical beam of electrons

Consider a system consisting of a beam of electrons and a target, whose individual state vectors are $|\psi\rangle$ and $|\Phi\rangle$ respectively. The state vector of the whole system is $|\Psi\rangle$. According to the interpretation (3.48) the amplitude for observing the system in the state $|\Phi\psi\rangle$ is

$$\langle\psi\Phi|\Psi\rangle.$$

We now consider the state vector $|\psi\rangle$ of the beam. The momentum of each electron is defined as nearly as possible by the experiment to be $\mathbf{p}$ in the $\hat{\mathbf{z}}$ direction. The coordinate representation of its state vector might be considered to be an eigenstate of momentum.

$$\langle\mathbf{r}|\psi\rangle = \langle\mathbf{r}|\mathbf{p}\rangle = (2\pi)^{-3/2}e^{i\mathbf{p}\cdot\mathbf{r}}. \tag{3.52}$$

The probability of finding the electron would then be the same anywhere in space and in particular at any point on the z axis.

At first sight it would appear to be impossible to represent one electron by such a wave function because an electron is localised in space, say at a point on the z axis $\mathbf{z}_m$, even if we do not observe $\mathbf{z}_m$. This is true also of a clump of electrons in a beam, which may have a maximum density at $\mathbf{z}_m$. We may consider $\mathbf{z}_m$ as being given for example by the time of flight of the clump from a source position $\mathbf{z}_0$. A localised beam clump may be represented by a linear combination of terms (3.52) called a wave packet.

$$\langle \mathbf{r}|\psi\rangle = (2\pi)^{-3/2} \int d^3p F(\mathbf{p}, \mathbf{p}_m, W) e^{i\mathbf{p}\cdot(\mathbf{r}-\mathbf{z}_m)}, \tag{3.53}$$

where $\mathbf{p}_m$ is the central momentum of the wave packet and W is its width.

The probability of observing the system is

$$N|\langle\psi\Phi|\Psi\rangle|^2 = N \int d^3r' \int d^3r \int d\tau' \int d\tau \langle\psi|\mathbf{r}'\rangle\langle\Phi|\tau'\rangle$$
$$\times \langle\mathbf{r}'\tau'|\Psi\rangle\langle\Psi|\tau\mathbf{r}\rangle\langle\tau|\Phi\rangle\langle\mathbf{r}|\psi\rangle, \tag{3.54}$$

where the normalisation N is discussed in section 6.4.

The wave-packet structure of the beam affects only the factor

$$\langle\psi|\mathbf{r}'\rangle\langle\mathbf{r}|\psi\rangle = (2\pi)^{-3} \int d^3p' \int d^3p F^*(\mathbf{p}', \mathbf{p}_m, W) F(\mathbf{p}, \mathbf{p}_m, W)$$
$$\times e^{i(\mathbf{p}\cdot\mathbf{r}-\mathbf{p}'\cdot\mathbf{r}')} e^{i(\mathbf{p}-\mathbf{p}')\cdot\mathbf{z}_m}. \tag{3.55}$$

In fact in a normal experiment we have no knowledge of $\mathbf{z}_m$ at all on a scale of positions that are comparable to the position characteristics of an electron–atom system, which are of the order of 10^{-7}cm. All we know is that the electrons are in the apparatus, whose scale is of the order 10 cm. We must therefore integrate (3.55) over $\mathbf{z}_m$, obtaining the factor

$$(2\pi)^{-3} \int d^3z_m e^{i(\mathbf{p}-\mathbf{p}')\cdot\mathbf{z}_m} = \delta(\mathbf{p}-\mathbf{p}'). \tag{3.56}$$

We now have

$$\langle\psi|\mathbf{r}'\rangle\langle\mathbf{r}|\psi\rangle = \int d^3p |F(\mathbf{p}, \mathbf{p}_m, W)|^2 e^{-i\mathbf{p}\cdot\mathbf{r}'} e^{i\mathbf{p}\cdot\mathbf{r}} \tag{3.57}$$

and the probability of observing the system is

$$N|\langle\psi\Phi|\Psi\rangle|^2 = N \int d^3p |F(\mathbf{p}, \mathbf{p}_m, W)|^2 |\langle\mathbf{p}\Phi|\Psi\rangle|^2. \tag{3.58}$$

We may thus consider a beam experiment as a collection of beam experiments, each having an eigenstate of momentum as its initial state and a weight $|F(\mathbf{p}, \mathbf{p}_m, W)|^2$ in the collection. The weight is taken into account in estimating the experimental error.

3.3 Angular momentum

Since atoms are strongly affected by the central potential of the nucleus, an important part in electron–atom collision theory is played by states that are invariant under rotations. From the general dynamical principle that invariance under change of a dynamical variable implies a conservation law for the canonically-conjugate variable we expect rotational invariance to imply conservation of angular momentum. Hence angular momentum is very important in atomic theory. We summarise important results in the theory of angular momentum. The reader is referred to Merzbacher (1970) for an elementary treatment and to Brink and Satchler (1971) for more advanced details. Useful relationships are given by Rotenberg *et al.* (1959).

3.3.1 Orbital angular momentum

The orbital angular momentum observable **L** is defined by

$$\mathbf{L} = \mathbf{r} \times \mathbf{p}. \tag{3.59}$$

By applying the commutation rule (3.6) we obtain the commutation rules for the cartesian components of **L**. We continue to use atomic units in which $\hbar = 1$. The rules are

$$L_x L_y - L_y L_x = iL_z \tag{3.60}$$

and two other equations obtained by cyclic permutations of x, y, z. We may also obtain the operator identities

$$L^2 = (\mathbf{r} \times \mathbf{p}) \cdot (\mathbf{r} \times \mathbf{p}) = r^2 p^2 - \mathbf{r}(\mathbf{r} \cdot \mathbf{p}) \cdot \mathbf{p} + 2i\mathbf{r} \cdot \mathbf{p} \tag{3.61}$$

and

$$L^2 L_z - L_z L^2 = 0. \tag{3.62}$$

The representation of the kinetic energy of an electron in spherical polar coordinates is obtained from (3.61). It is

$$K = \tfrac{1}{2}p^2 = -\tfrac{1}{2}\nabla_r^2 + \tfrac{1}{2}L^2/r^2, \tag{3.63}$$

$$\nabla_r^2 = \frac{1}{r^2}\frac{\partial}{\partial r}\left(r^2\frac{\partial}{\partial r}\right). \tag{3.64}$$

Since L^2 and L_z are commuting observables they have simultaneous eigenstates $|\ell m\rangle$, which obey the eigenvalue equations

$$L^2|\ell m\rangle = \ell(\ell + 1)|\ell m\rangle,$$
$$L_z|\ell m\rangle = m|\ell m\rangle, \tag{3.65}$$

where ℓ is a positive integer or zero and

$$-\ell \leq m \leq \ell. \tag{3.66}$$

3.3.2 Spherical harmonics

The coordinate representation of $|\ell m\rangle$ is

$$\langle \hat{\mathbf{r}} | \ell m \rangle = Y_{\ell m}(\hat{\mathbf{r}}). \tag{3.67}$$

The functions $Y_{\ell m}(\hat{\mathbf{r}})$ are the spherical harmonics, which are given in terms of the associated Legendre polynomials $P_{\ell m}(x)$. For $m \geq 0$

$$Y_{\ell m}(\theta, \phi) = \left[\frac{2\ell + 1}{4\pi} \frac{(\ell - m)!}{(\ell + m)!}\right]^{1/2} (-1)^m P_{\ell m}(\cos\theta) e^{im\phi}. \tag{3.68}$$

Some important properties of the spherical harmonics are as follows

$$Y_{\ell m}^*(\theta, \phi) = (-1)^m Y_{\ell -m}(\theta, \phi), \tag{3.69}$$

$$Y_{\ell 0}(\theta, 0) = \left[\frac{2\ell + 1}{4\pi}\right]^{1/2} P_\ell(\cos\theta), \tag{3.70}$$

where $P_\ell(x)$ is a Legendre polynomial.

The spherical harmonics have been defined so that they are orthonormal.

$$\int d\hat{\mathbf{r}} \, Y_{\ell'm'}^*(\hat{\mathbf{r}}) Y_{\ell m}(\hat{\mathbf{r}}) = \delta_{\ell\ell'}\delta_{mm'}. \tag{3.71}$$

The addition theorem relates spherical harmonics with different arguments.

$$\Sigma_m Y_{\ell m}^*(\hat{\mathbf{r}}') Y_{\ell m}(\hat{\mathbf{r}}) = \frac{2\ell + 1}{4\pi} P_\ell(\hat{\mathbf{r}}' \cdot \hat{\mathbf{r}}). \tag{3.72}$$

The parity of $Y_{\ell m}(\hat{\mathbf{r}})$ is given by

$$Y_{\ell m}(-\hat{\mathbf{r}}) = (-1)^\ell Y_{\ell m}(\hat{\mathbf{r}}). \tag{3.73}$$

If ℓ is even/odd $Y_{\ell m}(\hat{\mathbf{r}})$ has even/odd parity, that is it is even/odd under space reflection.

Because of the frequent occurrence of the factor $[(2\ell + 1)/4\pi]^{1/2}$ in equations involving spherical harmonics, simpler equations are obtained for the renormalised spherical harmonics

$$C_{\ell m}(\hat{\mathbf{r}}) = \left[\frac{4\pi}{2\ell + 1}\right]^{1/2} Y_{\ell m}(\hat{\mathbf{r}}). \tag{3.74}$$

3.3.3 Total angular momentum and spin

A more-general type of angular momentum operator $\mathbf{J}$ obeys the same commutation rules (3.60) as $\mathbf{L}$. These rules follow from rotational invariance. They are

$$J_x J_y - J_y J_x = iJ_z \tag{3.75}$$

and two other equations obtained by cyclic permutations of x, y, z. The simultaneous eigenstates of J^2 and J_z satisfy the equations

$$J^2|jm\rangle = j(j+1)|jm\rangle,$$

$$J_z|jm\rangle = m|jm\rangle, \tag{3.76}$$

where j is a half integer or zero and

$$-j \le m \le j. \tag{3.77}$$

We sometimes abbreviate the description of $|jm\rangle$ by calling it an 'eigenstate of $\mathbf{J}$ with eigenvalue j'. Atomic states are characterised by the total angular momentum quantum number j.

A special case of the observable $\mathbf{J}$ is the spin $\mathbf{s}$ of an electron, which has the eigenvalue $j = s = 1/2$. It is conveniently expressed in terms of the observable $\boldsymbol{\sigma}$.

$$\mathbf{s} = \tfrac{1}{2}\boldsymbol{\sigma}. \tag{3.78}$$

The spin has no coordinate representation. It constitutes three degrees of freedom of an electron that are independent of the position or momentum degrees of freedom. The corresponding space is spin space. It has dimension 2. The spin representation of the eigenstates may be written either as spin wave functions or as vectors in spin space called spinors.

$$\langle \boldsymbol{\sigma}|\tfrac{1}{2}\tfrac{1}{2}\rangle = \chi^{1/2}_{1/2}(\boldsymbol{\sigma}) = \begin{pmatrix} 1 \\ 0 \end{pmatrix},$$

$$\langle \boldsymbol{\sigma}|\tfrac{1}{2}-\tfrac{1}{2}\rangle = \chi^{1/2}_{-1/2}(\boldsymbol{\sigma}) = \begin{pmatrix} 0 \\ 1 \end{pmatrix}. \tag{3.79}$$

The spin observable has a matrix representation in terms of the Pauli matrices, which operate on states in spin space.

$$\sigma_x = \begin{pmatrix} 0 & 1 \\ 1 & 0 \end{pmatrix}, \quad \sigma_y = \begin{pmatrix} 0 & -i \\ i & 0 \end{pmatrix}, \quad \sigma_z = \begin{pmatrix} 1 & 0 \\ 0 & -1 \end{pmatrix}. \tag{3.80}$$

3.3.4 Vector addition of angular momenta

In many cases we encounter systems made up of two or more parts, each with angular momentum. For example they may be two electrons or the orbital and spin angular momentum of a single electron. Suppose the total angular momentum $\mathbf{J}$ is the vector sum of two angular momenta $\mathbf{J}_1$ and $\mathbf{J}_2$.

$$\mathbf{J} = \mathbf{J}_1 + \mathbf{J}_2. \tag{3.81}$$

The $(2j_1 + 1)$-dimensional space of $\mathbf{J}_1$ is spanned by the eigenstates $|j_1 m_1\rangle$ of J_1^2 and J_{1z}; similarly for $\mathbf{J}_2$.

The eigenvectors $|jm\rangle$ of J^2 and J_z are elements of the product space of simultaneous eigenvectors of the commuting observables $J_1^2, J_{1z}, J_2^2, J_{2z}$. The $(2j_1 + 1)(2j_2 + 1)$-dimensional product space is partitioned into subspaces of dimension $2j + 1$, where

$$|j_1 - j_2| \le j \le j_1 + j_2. \tag{3.82}$$

Each subspace is labelled by its value of j. The dimensions of all the j-subspaces sum to $(2j_1 + 1)(2j_2 + 1)$. The j-subspace is spanned by $2j + 1$ vectors $|jm\rangle$ with vector index m, where

$$-j \le m \le j. \tag{3.83}$$

Vectors $|jm\rangle$ are linear combinations of the simultaneous eigenvectors

$$|j_1 j_2 m_1 m_2\rangle = |j_1 m_1\rangle |j_2 m_2\rangle \tag{3.84}$$

that also span the j-subspace. They may be expanded using the representation theorem.

$$|jm\rangle = \Sigma_{m_1 m_2} |j_1 j_2 m_1 m_2\rangle \langle j_1 j_2 m_1 m_2|jm\rangle. \tag{3.85}$$

The expansion coefficients are the Clebsch–Gordan coefficients. We often abbreviate the meaning of (3.85) by saying that it expresses the coupling of the angular momenta j_1 and j_2 to j.
Equation (3.81) implies that

$$J_z = J_{1z} + J_{2z}. \tag{3.86}$$

Hence the eigenvalues of these observables are additive.

$$m = m_1 + m_2. \tag{3.87}$$

This relationship means that one of the sums in (3.85) is redundant. We include it, however, for symmetry and consider that the Clebsch–Gordan coefficient contains a factor $\delta_{m(m_1 + m_2)}$. It also means that $j_1 + j_2 + j$ is an integer.

Some important properties of the Clebsch–Gordan coefficients follow.

They are real.

$$\langle j_1 j_2 m_1 m_2|jm\rangle = \langle jm|j_1 j_2 m_1 m_2\rangle. \tag{3.88}$$

Two orthonormality relations are given by the representation theorem.

$$\Sigma_{m_1 m_2} \langle j'm'|j_1 j_2 m_1 m_2\rangle \langle j_1 j_2 m_1 m_2|jm\rangle = \langle j'm'|jm\rangle$$
$$= \delta_{j'j}\delta_{m'm}, \tag{3.89}$$

$$\Sigma_{jm} \langle j_1 j_2 m_1' m_2'|jm\rangle \langle jm|j_1 j_2 m_1 m_2\rangle = \langle j_1 j_2 m_1' m_2'|j_1 j_2 m_1 m_2\rangle$$
$$= \delta_{m_1' m_1}\delta_{m_2' m_2}. \tag{3.90}$$

The properties of the Clebsch–Gordan coefficients under rearrangement of indices are best described in terms of the 3-j symbols in section 3.3.5.

An important example of vector addition (or coupling) occurs in the construction of the state of an electron including the spin. This is

$$|\ell jm\rangle = \Sigma_{\mu\nu}|\ell\mu\rangle|\tfrac{1}{2}\nu\rangle\langle\ell\tfrac{1}{2}\mu\nu|jm\rangle. \tag{3.91}$$

We keep the quantum number ℓ in the notation for the state because it is necessary to keep track of the parity, which is a property of coordinate or momentum space (represented here by the orbital eigenstate) and has nothing to do with spin space. The coordinate-spin representation of (3.91) may be called a jj coupling function because of its use in states for systems of several electrons where the total angular momentum is obtained by vector addition of the angular momenta **J** of each electron.

$$\langle\hat{\mathbf{r}}\sigma|\ell jm\rangle = \Sigma_{\mu\nu}\langle\ell\tfrac{1}{2}\mu\nu|jm\rangle Y_{\ell\mu}(\hat{\mathbf{r}})\chi_{\nu}^{1/2}(\sigma). \tag{3.92}$$

3.3.5 *The 3-j, 6-j and 9-j symbols*

Because of the large number of angular momentum observables involved in the state vectors of systems of more than one electron, it helps greatly in keeping track of the quantum numbers to express the coupling of angular momenta in terms of the symbols of Wigner, which have useful symmetries.

In order to understand formal relationships the basic coupling of two angular momenta is expressed in terms of the Clebsch–Gordan coefficients. In setting up the algebra for computation it is more convenient to use the 3-j symbols. Calculations of the Wigner symbols and the Clebsch–Gordan coefficients are found in subroutine libraries. See, for example, Soper (1989).

The 3-j symbol is expressed in terms of the Clebsch–Gordan coefficient by

$$\langle j_1 j_2 m_1 m_2|jm\rangle = (-1)^{j_1-j_2+m}\,\hat{j}\begin{pmatrix} j_1 & j_2 & j \\ m_1 & m_2 & -m \end{pmatrix}, \tag{3.93}$$

where we use the notation

$$\hat{j} \equiv (2j+1)^{1/2}. \tag{3.94}$$

The j indices obey the triangle inequality (3.82) and (3.87) means that the m indices sum to zero.

The symmetry properties of the 3-j symbols are

no change made under an even permutation of columns,

multiply by $(-1)^{j_1+j_2+j}$ for an odd permutation of columns,

multiply by $(-1)^{j_1+j_2+j}$ for a change of sign of all the m indices.

Transition amplitudes involving states of several coupled angular momenta involve products of Clebsch–Gordan coefficients with a sum over the m indices. Such amplitudes cannot depend on the m indices, which are changed by changing the arbitrary directions of the coordinate axes. The m-independence of such amplitudes is expressed by m-independent coefficients, the 6-j symbols.

$$\begin{Bmatrix} j_1 & j_2 & j_3 \\ \ell_1 & \ell_2 & \ell_3 \end{Bmatrix} = \sum_{\text{all } m,n} (-1)^S \begin{pmatrix} j_1 & j_2 & j_3 \\ m_1 & m_2 & m_3 \end{pmatrix} \begin{pmatrix} j_1 & \ell_2 & \ell_3 \\ m_1 & n_2 & -n_3 \end{pmatrix}$$

$$\times \begin{pmatrix} \ell_1 & j_2 & \ell_3 \\ -n_1 & m_2 & n_3 \end{pmatrix} \begin{pmatrix} \ell_1 & \ell_2 & j_3 \\ n_1 & -n_2 & m_3 \end{pmatrix}, \tag{3.95}$$

$$S = \ell_1 + \ell_2 + \ell_3 + n_1 + n_2 + n_3.$$

The symmetry properties of the 6-j symbols are

no change made under interchange of columns or of any two numbers in the bottom row with the corresponding two numbers in the top row.

Sometimes the potential observables giving transitions between two-electron states do not depend on the spins of the individual electrons. Potentials such as the two-electron Coulomb potential act only on the coordinate or momentum-space degrees of freedom and therefore couple orbital angular momentum states $|\ell_1\mu_1\rangle$, $|\ell_2\mu_2\rangle$. The state of the total angular momentum J involves coupling the orbital angular momenta to L and the spins to S and then coupling the resultant states to J. This is called the LS coupling representation.

Magnetic potentials act through the interaction of magnetic moments and magnetic fields. The one relevant to atomic collisions is the spin–orbit potential, which couples the orbital and spin angular momenta of one electron to j_1 and those of the other electron to j_2. The resultant states are coupled to the total angular momentum quantum number J. Such states belong to the jj coupling representation. The angular momentum eigenstate of the first electron for example is the state $|\ell_1 j_1 m_1\rangle$ of (3.91).

The coefficients for the transformation from the LS representation to the jj representation are proportional to the 9-j symbols. The transformation is

$$|(\ell_1\ell_2)L(s_1 s_2)S : J\rangle = \sum_{j_1 j_2} \widehat{L}\widehat{S}\widehat{j_1}\widehat{j_2} \begin{Bmatrix} \ell_1 & \ell_2 & L \\ s_1 & s_2 & S \\ j_1 & j_2 & J \end{Bmatrix} |(\ell_1 s_1)j_1(\ell_2 s_2)j_2 : J\rangle. \tag{3.96}$$

The m indices are not explicitly shown in (3.96) because transition amplitudes do not depend on them. The elimination of the m indices is discussed further in section 3.3.7 in the context of the Wigner–Eckart theorem.

3.3.6 Spherical tensors

Any quantity that transforms under rotations in the same way as the total angular momentum eigenstate $|jm\rangle$ is called a spherical tensor T_m^j. A tensor operator of rank k is written

$$\mathbf{T}^k.$$

It is an operator with $2k + 1$ components

$$T_q^k, \quad -k \le q \le k.$$

Tensor notation is very useful in considering the computation of amplitudes involving angular momentum coupling.

Examples of spherical tensors are

$$Y_M^L(\hat{\mathbf{r}}) = Y_{LM}(\hat{\mathbf{r}}) = \langle \hat{\mathbf{r}}|LM \rangle,$$
$$C_M^L(\hat{\mathbf{r}}) = (4\pi)^{1/2}\hat{L}^{-1}Y_M^L(\hat{\mathbf{r}}),$$
$$\chi_v^{1/2}(\boldsymbol{\sigma}),$$
$$\mathcal{Y}_m^{\ell j}(\hat{\mathbf{r}}, \boldsymbol{\sigma}) = \Sigma_{\mu v}\langle \ell \tfrac{1}{2}\mu v|jm\rangle Y_\mu^\ell(\hat{\mathbf{r}})\chi_v^{1/2}(\boldsymbol{\sigma}). \tag{3.97}$$

Adjoints of tensor operators require some discussion. The adjoint of a linear operator is the transposed complex conjugate.

$$T_q^{k\dagger} = (-1)^{p+k-q}T_{-q}^k. \tag{3.98}$$

Here p is an arbitrary integer, which does not affect the transformation properties under rotations. It is impossible to make a consistent universal choice of p and still keep some of the notation conventions necessary for easy comparison of the present notation with published work. If k is a half odd integer, p is defined to be zero. If k is an integer, p is defined to be $-k$.

The tensor product is the tensor generalisation of the basic angular momentum coupling definition (3.84,3.85). We combine two tensors $R_{q'}^{k'}$ and S_q^k to form a tensor T_Q^K in the following way.

$$[R_{q'}^{k'} \times S_q^k]_Q^K = T_Q^K = \Sigma_{qq'}R_{q'}^{k'}S_q^k\langle k'kq'q|KQ\rangle. \tag{3.99}$$

The scalar product is a special case of the tensor product with $k' = k$, $K = 0$, k integer. It uses a slightly different convention. We put

$$S_q^k = U_q^{k\dagger} = (-1)^q U_{-q}^k. \tag{3.100}$$

The scalar product is written as

$$\mathbf{R}^k \cdot \mathbf{U}^k = \Sigma_q(-1)^q R_q^k U_{-q}^k. \tag{3.101}$$

An example of the scalar product that is used in electron–atom collision theory is the multipole expansion of the two-electron Coulomb potential.

$$v(\mathbf{r}',\mathbf{r}) = 1/|\mathbf{r}' - \mathbf{r}|$$

$$= \Sigma_\lambda v_\lambda(r',r)\mathbf{C}^\lambda(\hat{\mathbf{r}}') \cdot \mathbf{C}^\lambda(\hat{\mathbf{r}}), \tag{3.102}$$

$$v_\lambda(r',r) = r_<^\lambda/r_>^{\lambda+1}, \tag{3.103}$$

where $r_<$ and $r_>$ mean the lesser and greater of r',r respectively.

3.3.7 Matrix elements and the Wigner–Eckart theorem

We have remarked that the invariance of transition amplitudes under rotations of the (arbitrary) coordinate axes requires that they are independent of the m indices.

The dependence on the m indices of the amplitude for the transition between states $|JM\rangle$ and $|J'M'\rangle$ of total angular momentum due to a tensor operator T_Q^K has a remarkably simple form in which the indices M',M and Q all appear in a single 3-j symbol. It is given by the Wigner–Eckart theorem.

$$\langle J'M'|T_Q^K|JM\rangle = (-1)^{J'-M'} \begin{pmatrix} J' & K & J \\ -M' & Q & M \end{pmatrix} \langle J' \parallel \mathbf{T}^K \parallel J\rangle. \tag{3.104}$$

The m-independent amplitude $\langle J' \parallel \mathbf{T}^K \parallel J\rangle$ is called the reduced matrix element.

The simplest example of the Wigner–Eckart theorem is given by the Gaunt integral over three spherical harmonics, which is the matrix element for the transition between eigenstates $|\ell m\rangle$ and $|\ell' m'\rangle$ of a single orbital angular momentum observable due to a tensor operator Y_M^L. We prefer to use the renormalised tensor operator C_M^L, which simplifies the expression.

$$\langle \ell'm'|C_M^L|\ell m\rangle = (4\pi)^{1/2}\hat{L}^{-1} \int d\hat{\mathbf{r}} Y_{\ell'm'}^*(\hat{\mathbf{r}})Y_{LM}(\hat{\mathbf{r}})Y_{\ell m}(\hat{\mathbf{r}})$$

$$= \widetilde{\ell\ell'}^{-1}\langle L\ell Mm|\ell'm'\rangle\langle \ell'0|L\ell 00\rangle. \tag{3.105}$$

Using the rules of section 3.3.5 this reduces to

$$\langle \ell'm'|C_M^L|\ell m\rangle = (-1)^{\ell'-m'} \begin{pmatrix} \ell' & L & \ell \\ -m' & M & m \end{pmatrix} \langle \ell' \parallel \mathbf{C}^L \parallel \ell\rangle,$$

$$\langle \ell' \parallel \mathbf{C}^L \parallel \ell\rangle = (-1)^{\ell'}\hat{\ell}'\hat{\ell} \begin{pmatrix} \ell' & L & \ell \\ 0 & 0 & 0 \end{pmatrix}. \tag{3.106}$$

Another useful reduced matrix element is for the interaction of electrons whose states involve spin–orbit coupling.

$$\langle \ell'j' \parallel \mathbf{C}^L \parallel \ell j\rangle = (-1)^{j'+\frac{1}{2}}\hat{j}'\hat{j} \begin{pmatrix} j' & L & j \\ \frac{1}{2} & 0 & -\frac{1}{2} \end{pmatrix} \tfrac{1}{2}[1+(-1)^{\ell'+L+\ell}]. \tag{3.107}$$

The last factor in (3.107) expresses the fact that the transition represented by the amplitude must preserve parity. This is automatically ensured in (3.106) by the property

$$\begin{pmatrix} \ell' & L & \ell \\ 0 & 0 & 0 \end{pmatrix} = 0 \quad \text{if} \quad \ell' + L + \ell \quad \text{is odd.} \tag{3.108}$$

The two-electron spin-angular matrix element of the Coulomb potential (3.102) is given by

$$\langle j_1' j_2' j' m' | \mathbf{C}^\lambda \cdot \mathbf{C}^\lambda | j_1 j_2 j m \rangle = \delta_{j'j} \delta_{m'm} (-1)^{j_1 + j_2' + j} \begin{Bmatrix} j & j_2' & j_1' \\ \lambda & j_1 & j_2 \end{Bmatrix}$$
$$\times \langle j_1' \| \mathbf{C}^\lambda \| j_1 \rangle \langle j_2' \| \mathbf{C}^\lambda \| j_2 \rangle. \tag{3.109}$$

Here we have omitted the orbital (parity) indices ℓ for brevity of notation.

3.3.8 Time reversal

The operation of time reversal interchanges the initial and final states of a colliding system. The invariance of collision amplitudes under time reversal is the principle of detailed balance. It is observed to hold for electron–atom collisions. We are interested in finding the form of the time-reversal operator θ and its effect on electron states.

If H is real the Schrödinger equation of motion (3.42) is

$$H | \Psi^*(t) \rangle = H^* | \Psi^*(t) \rangle = -i \frac{\partial}{\partial t} | \Psi^*(t) \rangle = i \frac{\partial}{\partial (-t)} | \Psi^*(t) \rangle. \tag{3.110}$$

Therefore $| \Psi^*(t) \rangle$ obeys the time-reversed Schrödinger equation of motion. The time-reversal operator θ is the complex conjugation operator u.

$$\theta = u. \tag{3.111}$$

Examples of the operation of time reversal on spin-independent states are given in the coordinate representation.

The free particle is represented by

$$\langle \mathbf{r} | \theta | \mathbf{p} \rangle = (2\pi)^{-3/2} e^{-i\mathbf{p}\cdot\mathbf{r}} = \langle \mathbf{r} | -\mathbf{p} \rangle. \tag{3.112}$$

Time reversal changes the direction of the momentum.

The eigenstate of orbital angular momentum is

$$\langle \hat{\mathbf{r}} | \theta | \ell m \rangle = Y_{\ell m}^*(\hat{\mathbf{r}}) = (-1)^m \langle \hat{\mathbf{r}} | \ell \, -m \rangle. \tag{3.113}$$

Time reversal changes the sign of the m index and multiplies by $(-1)^m$.

A magnetic potential may not be real. An example is the spin–orbit potential. This is due to the coupling of the magnetic moment of the electron to the solenoid magnetic field caused by the orbital motion of its charge about a centre. It is $V_s(r)\boldsymbol{\sigma} \cdot \mathbf{L}$. The coordinate representation of $\boldsymbol{\sigma} \cdot \mathbf{L}$ is

$$-i\boldsymbol{\sigma} \cdot (\mathbf{r} \times \nabla).$$

We look for an operator U that has the property that $U|\Psi^*(t)\rangle$ obeys (3.110) when H is $\boldsymbol{\sigma} \cdot \mathbf{L}$. That is we require

$$U(\boldsymbol{\sigma} \cdot \mathbf{L})^* U^{-1} U|\Psi^*(t)\rangle = i\frac{\partial}{\partial(-t)} U|\Psi^*(t)\rangle. \tag{3.114}$$

Putting $\mathbf{J} = \frac{1}{2}\boldsymbol{\sigma}$ in the angular momentum commutation rules (3.75) we can verify that

$$U = \sigma_y, \tag{3.115}$$

where the matrix representation of σ_y is given by (3.80). The time-reversal operator for spin states is

$$\theta = \sigma_y u. \tag{3.116}$$

We find the effect of θ on the spin eigenstates (3.79) by using their representation as vectors in spin space.

$$\theta \begin{pmatrix} 1 \\ 0 \end{pmatrix} = \begin{pmatrix} 0 & -i \\ i & 0 \end{pmatrix} \begin{pmatrix} 1 \\ 0 \end{pmatrix} = i \begin{pmatrix} 0 \\ 1 \end{pmatrix},$$

$$\theta \begin{pmatrix} 0 \\ 1 \end{pmatrix} = \begin{pmatrix} 0 & -i \\ i & 0 \end{pmatrix} \begin{pmatrix} 0 \\ 1 \end{pmatrix} = -i \begin{pmatrix} 1 \\ 0 \end{pmatrix}. \tag{3.117}$$

This is summarised by

$$\theta \chi_\nu^{1/2}(\boldsymbol{\sigma}) = (-1)^\nu \chi_{-\nu}^{1/2}(\boldsymbol{\sigma}). \tag{3.118}$$

The effect of θ on the angular momentum state of an electron with spin is seen by using the relations (3.91,3.93) and the symmetry of the 3-j symbol under change of sign of all the m indices. As we did in (3.98) we introduce an arbitrary phase factor independent of j and m, obtaining

$$\theta|\ell jm\rangle = (-1)^{j-m}|\ell j \ -m\rangle. \tag{3.119}$$

In general, time reversal changes the sign of the m index and multiplies by a phase factor.

3.4 The Pauli exclusion principle

In the problem of N identical electrons interacting with each other in the Coulomb potential of a nucleus, the Pauli exclusion principle plays a crucial role. If we consider the possible states of one electron interacting with the rest of the system the exclusion principle means that a state cannot be occupied by more than one electron.

In this section we set up an algebraic structure to ensure that the many-electron states we will consider obey the exclusion principle.

3.4.1 Independent-particle configurations

We consider the subspace of N-electron states with the same total angular momentum and parity, denoted respectively by j and ℓ. Since these states are orthogonal to states with different ℓj, the subspaces are independent and the problems of the interaction of states are separate.

Each ℓj subspace is spanned by a complete set of independent-particle configurations. A configuration is a state in the product space of N one-electron states. The one-electron states $|\mu\rangle$ are solutions of one-electron problems. It is obviously sensible to choose a one-electron problem to produce states that are closely related to the states of an electron in the interacting system. For example, bound states should occupy roughly the same volume of coordinate space as the atom. How to choose the bound states will be left until chapter 5. For the purpose of the present discussion the word 'orbital' will refer to a one-electron state.

Orbitals $|\mu\rangle$ are characterised by their quantum numbers $n_\mu, \ell_\mu, j_\mu, m_\mu$. Orbitals in the positive-energy continuum are characterised by momentum p_μ instead of the principal quantum number n_μ. In each case we characterise them by the integer μ. In the present context no confusion is caused by using a discrete notation for the continuum.

$$
\begin{aligned}
|\mu\rangle &= |p_\mu \ell_\mu j_\mu m_\mu\rangle \quad : \text{continuum}, \\
&= |n_\mu \ell_\mu j_\mu m_\mu\rangle \quad : \text{bound}.
\end{aligned} \tag{3.120}
$$

The ith electron is characterised by its position—spin coordinates. The corresponding basis state in the coordinate—spin representation is abbreviated thus.

$$
\langle i| = \langle \mathbf{r}_i \sigma_i|. \tag{3.121}
$$

To describe the N-electron system we select N orbitals from the set $|\mu\rangle$, characterised by integers $\xi = 1, .., N$. An element of the product space is given in the coordinate—spin representation by

$$
P_\eta \prod_{i,\xi=1}^{N} \langle i|\xi\rangle, \tag{3.122}
$$

where P_η is a number that depends on the permutation η of the electrons among the orbitals.

Since the electrons are identical no one permutation is physically distinguishable from another. Consider the element $|\rho\rangle$ of the N-electron product space consisting of the antisymmetric sum of all possible permutations. This is the normalised determinant

$$|\rho\rangle = (N!)^{-1/2}\det \langle i|\xi\rangle. \tag{3.123}$$

Note that continuum orbitals are normalised in the sense discussed in section 6.4. If two orbitals $|\xi\rangle$ are identical the determinant vanishes because it has two equal columns. No two electrons can be in the same orbital. The N-electron state $|\rho\rangle$ therefore obeys the Pauli exclusion principle. We call it a configuration.

The eigenstate $|f_{\ell j}\rangle$ of the N-electron Schrödinger equation for the ℓj subspace may be expanded in configurations

$$|f_{\ell j}\rangle = \Sigma_\rho |\rho\rangle\langle\rho|f_{\ell j}\rangle. \tag{3.124}$$

This is the configuration-interaction representation. It is antisymmetric in all the coordinates and spins. Antisymmetry is a requirement for all N-electron states.

3.4.2 Creation and annihilation operators

We now develop a formalism that enables us to preserve antisymmetry in expressing different N-electron configurations $|\rho\rangle$, obtained by different selections of N orbitals from the set $|\mu\rangle$, in terms of a chosen base configuration $|0\rangle$. We arrange the orbitals $|\mu\rangle$ in increasing energy order, orbitals with the same energy being ordered by increasing quantum numbers. A sensible choice for the configuration $|0\rangle$ is the one in which the first N orbitals are occupied.

The configurations are expressed in terms of the occupation-number representation, where they are characterised by the occupation numbers n_μ of the orbitals $|\mu\rangle$. n_μ is either 1 or 0. The base configuration is

$$|0\rangle = |\underbrace{111..11}_{N \text{ states}}000...\rangle. \tag{3.125}$$

An example of a different configuration, obtained by annihilating an electron in the orbital $|3\rangle$ and creating one in the orbital $|N+2\rangle$ is

$$|\rho\rangle = |\underbrace{110..11}_{N \text{ states}}010..\rangle. \tag{3.126}$$

In general N-electron configurations $|\rho\rangle$ are obtained by annihilating electrons in M orbitals with $n_\mu=1$ in $|0\rangle$ and creating electrons in M different orbitals with $n_\mu=0$ in $|0\rangle$. In other words we change the occupation numbers from 1 to 0 in M orbitals and from 0 to 1 in M different orbitals.

We create or annihilate an electron in a particular orbital $|\alpha\rangle$ by operating on a configuration $|\rho\rangle$ with a creation operator $a_\alpha^\dagger$ or an annihilation

operator a_α. In defining the creation and annihilation operators we char-
acterise the configuration $|\rho\rangle$ simply by the occupation number n_α of the
orbital whose occupation number is to be changed.

$$|\rho\rangle \equiv |n_\alpha\rangle. \tag{3.127}$$

The creation operator $a_\alpha^\dagger$ is defined by

$$a_\alpha^\dagger|n_\alpha\rangle = (-1)^{P_\alpha}(1 - n_\alpha)|n_\alpha + 1\rangle. \tag{3.128}$$

If $n_\alpha=1$, $a_\alpha^\dagger$ gives zero. If $n_\alpha=0$, $a_\alpha^\dagger$ gives an $(N + 1)$-electron configuration
in which $n_\alpha=1$. P_α is the number of occupied orbitals with $\mu < \alpha$. We
understand this by considering the evaluation of the $(N + 1) \times (N + 1)$
determinant $a_\alpha^\dagger|\rho\rangle$. We add terms obtained by multiplying each element in
turn of the column α by its cofactor $N \times N$ determinant. The terms have
alternating signs, the sign of the first term being $(-1)^{P_\alpha}$.

The annihilation operator a_α is defined by

$$a_\alpha|n_\alpha\rangle = (-1)^{P_\alpha}n_\alpha|n_\alpha - 1\rangle. \tag{3.129}$$

If $n_\alpha = 0$, a_α gives zero. If $n_\alpha = 1$, a_α gives an $(N-1)$-electron configuration
in which $n_\alpha = 0$.

In many-body problems we are normally concerned with systems that
have a fixed number of electrons. In order to preserve the electron
number N we must operate on the configuration $|\rho\rangle$ with equal numbers
of creation and annihilation operators. We now develop the algebra of
these operators. First we consider products of creation and annihilation
operators for one orbital $|\alpha\rangle$.

$$\begin{aligned}a_\alpha a_\alpha^\dagger|n_\alpha\rangle &= |n_\alpha\rangle \quad \text{if} \quad n_\alpha = 0, \\ &= 0 \qquad \text{if} \quad n_\alpha = 1. \end{aligned} \tag{3.130}$$

$$\begin{aligned}a_\alpha^\dagger a_\alpha|n_\alpha\rangle &= 0 \qquad \text{if} \quad n_\alpha = 0, \\ &= |n_\alpha\rangle \quad \text{if} \quad n_\alpha = 1. \end{aligned} \tag{3.131}$$

We therefore have the commutation rule

$$a_\alpha a_\alpha^\dagger + a_\alpha^\dagger a_\alpha = 1. \tag{3.132}$$

In considering changes of occupation number in two different orbitals
$|\alpha\rangle$ and $|\beta\rangle$ we characterise the configuration $|\rho\rangle$ by two occupation numbers n_α and n_β.

$$|\rho\rangle \equiv |n_\alpha n_\beta\rangle. \tag{3.133}$$

The operation of a_α and $a_\beta^\dagger$ gives a non-zero result only if $n_\alpha = 1$, $n_\beta = 0$.

$$a_\alpha a_\beta^\dagger|n_\alpha n_\beta\rangle = (-1)^{P_\alpha}(-1)^{P_\beta}|(n_\alpha - 1)(n_\beta + 1)\rangle. \tag{3.134}$$

When these operators act in the reverse order we have

$$a_\beta^\dagger a_\alpha |n_\alpha n_\beta\rangle = a_\beta^\dagger (-1)^{P_\alpha} |(n_\alpha - 1)n_\beta\rangle. \tag{3.135}$$

The number of occupied orbitals with $\mu < \beta$ for the $(N-1)$-electron configuration on the right-hand side of (3.135) is one fewer than for the original configuration $|\rho\rangle$. We apply the definition (3.128) of the creation operator $a_\beta^\dagger$ to (3.135) to obtain

$$a_\beta^\dagger a_\alpha |n_\alpha n_\beta\rangle = (-1)^{P_\beta - 1}(-1)^{P_\alpha} |(n_\alpha - 1)(n_\beta + 1)\rangle = -a_\alpha a_\beta^\dagger |n_\alpha n_\beta\rangle. \tag{3.136}$$

The commutation rules (3.132,3.136) are combined into a set of rules that constitute the algebra for ensuring antisymmetry in a many-fermion problem.

$$a_\alpha a_\beta^\dagger + a_\beta^\dagger a_\alpha = \delta_{\alpha\beta}. \tag{3.137}$$

$$a_\alpha a_\beta + a_\beta a_\alpha = 0, \tag{3.138}$$

$$a_\alpha^\dagger a_\beta^\dagger + a_\beta^\dagger a_\alpha^\dagger = 0. \tag{3.139}$$

Note that equation (3.131) means that

$$N_\alpha = a_\alpha^\dagger a_\alpha \tag{3.140}$$

is a real linear operator whose eigenvalue is n_α. It is the number operator. This means that $a_\alpha^\dagger$ and a_α are adjoints of each other and justifies the adjoint notation.

3.4.3 Symmetric operators

Electron–atom collisions involve $(N + 1)$-electron configurations with N electrons initially in the ground state of the target atom and at least one electron in an orbital of the scattering continuum, for example an eigenstate of momentum or plane-wave state. These configurations are transformed by one- or two-electron operators, which are additive for electrons and symmetric in the electron coordinates. One-electron operators include the kinetic energy K and one-electron potentials V such as the electron–nucleus potential. The electron–electron potential is a two-electron operator. We find a form for the operators, the second-quantised form, which gives a method for the calculation of antisymmetrised matrix elements.

We first consider the symmetric one-electron operator T, which is the sum of operators $t_i, i = 0, N$, for each electron. A useful example of t_i is the bare nucleus Hamiltonian $K_i + V_i$, where V_i is the electron–nucleus potential. The second-quantised form for T is found by considering matrix elements for $(N + 1)$-electron determinants $|\rho'\rangle$, $|\rho\rangle$ of orbitals selected from a set $|\mu\rangle$, $|\nu\rangle$. We introduce the position–spin coordinate x_i, which stands for the set $\mathbf{r}_i, \sigma_i$, and the notation $\bar{x}_i$, which stands for all the position–spin coordinates except x_i. The total coordinate set is x. We expand the matrix element $\langle\rho'|T|\rho\rangle$ in these coordinates and introduce the expansion of each determinant in the product of the elements $\langle x_i|\nu\rangle$ of the *i*th row and their cofactors $\langle\bar{x}_i|a_\nu|\rho\rangle$, remembering that normalisation requires a factor $(N+1)^{-1/2}$.

$$\langle\rho'|T|\rho\rangle$$

$$= \sum_{i=0}^{N} \int dx_i' \int dx_i \int d\bar{x}_i' \int d\bar{x}_i \langle\rho'|\bar{x}_i' x_i'\rangle\langle x_i'\bar{x}_i'|t_i|\bar{x}_i x_i\rangle\langle x_i\bar{x}_i|\rho\rangle$$

$$= \sum_{i=0}^{N}(N+1)^{-1} \sum_{\mu\nu} \int dx_i' \int dx_i \int d\bar{x}_i \langle\rho'|a_\mu^\dagger|\bar{x}_i\rangle\langle\mu|x_i'\rangle\langle x_i'|t_i|x_i\rangle$$

$$\times \langle x_i|\nu\rangle\langle\bar{x}_i|a_\nu|\rho\rangle$$

$$= \Sigma_{\mu\nu}\langle\rho'|a_\mu^\dagger a_\nu|\rho\rangle\langle\mu|t|\nu\rangle. \tag{3.141}$$

The second-quantised form of the one-electron operator T is therefore

$$T = \Sigma_{\mu\nu}a_\mu^\dagger a_\nu\langle\mu|t|\nu\rangle. \tag{3.142}$$

We now consider the special case where the orbitals $|\mu\rangle$, $|\nu\rangle$ come from the complete orthonormal set of eigenstates $|\alpha\rangle$ of t.

$$t|\alpha\rangle = t_\alpha'|\alpha\rangle. \tag{3.143}$$

The form (3.142) of T now becomes the operator that measures the total value of T, in view of (3.140).

$$T = \Sigma_\alpha a_\alpha^\dagger a_\alpha t_\alpha' = \Sigma_\alpha N_\alpha t_\alpha'. \tag{3.144}$$

We now revert to the more-general set of orbitals $|\mu\rangle$, $|\nu\rangle$ and expand them in the set $|\alpha\rangle$, obtaining

$$T = \Sigma_\alpha\Sigma_{\mu\nu}a_\mu^\dagger a_\nu\langle\mu|\alpha\rangle t_\alpha'\langle\alpha|\nu\rangle. \tag{3.145}$$

Comparing (3.144, 3.145) we find the following operator transformation

$$a_\alpha^\dagger = \Sigma_\mu a_\mu^\dagger\langle\mu|\alpha\rangle$$

$$a_\alpha = \Sigma_\nu a_\nu\langle\alpha|\nu\rangle. \tag{3.146}$$

The symmetric two-electron operator V is given in terms of the electron–electron potentials $v_{\alpha\beta}$ by

$$V = \sum_{\alpha<\beta} v_{\alpha\beta} = \tfrac{1}{2}\Sigma_{\alpha\beta}(N_\alpha N_\beta - N_\alpha \delta_{\alpha\beta})v_{\alpha\beta}. \qquad (3.147)$$

From the definition (3.140) of N_α and the commutation rules (3.137,3.138) we find

$$N_\alpha N_\beta - N_\alpha \delta_{\alpha\beta} = a_\alpha^\dagger a_\beta^\dagger a_\beta a_\alpha. \qquad (3.148)$$

On applying the transformation (3.146) we obtain the required form for V.

$$V = \tfrac{1}{2}\sum_{\mu\nu\rho\sigma} a_\mu^\dagger a_\nu^\dagger a_\rho a_\sigma \langle\mu\nu|v|\rho\sigma\rangle. \qquad (3.149)$$

The convention for writing two-electron matrix elements is that μ and σ belong to one electron, ν and ρ to the other.

Note that the commutation rules (3.137–3.139) and the symmetric operators (3.142,3.149) have been derived from properties of determinants. We have not assumed that the orbitals $|\mu\rangle$, $|\nu\rangle$ are orthogonal. In evaluating matrix elements care must be taken to keep track of the scalar products of orbitals that are not orthogonal, such as bound orbitals and plane waves. The N-electron target configurations are conveniently normalised by (3.123). The normalisation of the continuum orbitals is discussed in chapter 6.

3.5 The Dirac equation

The relativistically-covariant description of the motion of an electron in quantum mechanics was first given by Dirac (1928). We consider the relativistic motion of an electron in the potential of an atom.

3.5.1 The free electron

The Schrödinger equation of motion (3.42) for an electron is

$$i\frac{\partial}{\partial t}|\psi(t)\rangle = H|\psi(t)\rangle. \qquad (3.150)$$

For a free electron the Hamiltonian H is the total energy, given in atomic units by

$$H = K + c^2 = c(p^2 + c^2)^{1/2}. \qquad (3.151)$$

Note that we are departing from the more-usual relativistic convention by featuring the kinetic energy K rather than the total relativistic energy. This enables us to maintain a consistent notation where we use E for the eigenvalue of K. We use the value of the fine-structure constant $e^2/\hbar c$ to find c in atomic units.

$$c = 137.035\ 99. \tag{3.152}$$

If we consider the coordinate representation of p^2 (3.33) we see that (3.150) is not covariant because the time derivative appears linearly while the space derivatives in ∇^2 do not. In a covariant equation $\mathbf{p}$ must occur linearly in H. A sufficiently-general form is

$$H = c\boldsymbol{\alpha} \cdot \mathbf{p} + \beta c^2. \tag{3.153}$$

The eigenvalue equation for H is the Dirac equation:

$$[E - c\boldsymbol{\alpha} \cdot \mathbf{p} + c^2(1 - \beta)]|\psi\rangle = 0. \tag{3.154}$$

We find relations satisfied by $\boldsymbol{\alpha}$ and β by requiring $|\psi\rangle$ to be a solution of the relativistic Schrödinger equation obtained from (3.151).

$$[(E + c^2)^2 - c^2 p^2 - c^4]|\psi\rangle = 0. \tag{3.155}$$

Multiplying (3.154) on the left by $E + c\boldsymbol{\alpha} \cdot \mathbf{p} + c^2(1 + \beta)$ we obtain

$$[E + c\boldsymbol{\alpha} \cdot \mathbf{p} + c^2(1 + \beta)][E - c\boldsymbol{\alpha} \cdot \mathbf{p} + c^2(1 - \beta)]|\psi\rangle = 0. \tag{3.156}$$

Expressing the operators in (3.156) in cartesian components and comparing (3.155, 3.156) we find the following relations.

$$\alpha_x^2 = \alpha_y^2 = \alpha_z^2 = \beta^2 = 1,$$
$$\alpha_x\alpha_y + \alpha_y\alpha_x = \alpha_y\alpha_z + \alpha_z\alpha_y = \alpha_z\alpha_x + \alpha_x\alpha_z = 0,$$
$$\alpha_x\beta + \beta\alpha_x = \alpha_y\beta + \beta\alpha_y = \alpha_z\beta + \beta\alpha_z = 0. \tag{3.157}$$

The Dirac matrices are α_x, α_y, α_z, β. The simplest representation has four dimensions. We choose β to be diagonal.

$$\boldsymbol{\alpha} = \begin{pmatrix} 0 & \sigma \\ \sigma & 0 \end{pmatrix}, \quad \beta = \begin{pmatrix} 1 & 0 \\ 0 & -1 \end{pmatrix}, \tag{3.158}$$

where the Pauli matrices σ are given by (3.80). $|\psi\rangle$ is a four-component spinor.

3.5.2 Electron in a central potential

We choose a reference frame in which the nucleus is stationary and the vector potential $\mathbf{A}$ is zero, and consider a scalar potential $V(r)$. The Dirac equation becomes

$$[E - V - c\boldsymbol{\alpha} \cdot \mathbf{p} + c^2(1 - \beta)]|\psi\rangle = 0. \tag{3.159}$$

It is interesting to consider the angular momentum. The total angular momentum $\mathbf{J}$ is conserved, since the potential is rotationally invariant. In order to consider the time conservation of operators it is useful to use the Heisenberg equation of motion for time-dependent operators

$$v_t = T^{-1}vT, \tag{3.160}$$

which follows from (3.42–3.44). Here T is the time-development operator. The equation is

$$i\frac{\partial v_t}{\partial t} = v_t H_t - H_t v_t. \tag{3.161}$$

Conserved operators commute with the Hamiltonian. We want to know if the orbital angular momentum $\mathbf{L}$ (3.59) is conserved. Consider L_x, using the commutation relations (3.6) and equn. (3.161).

$$\begin{aligned}
i\frac{\partial L_x}{\partial t} &= L_x H - H L_x \\
&= c\boldsymbol{\alpha} \cdot [(y p_z - z p_y)\mathbf{p} - \mathbf{p}(y p_z - z p_y)] \\
&= ic(\alpha_y p_z - \alpha_z p_y).
\end{aligned} \tag{3.162}$$

$\mathbf{L}$ therefore does not commute with H although, of course, it commutes with $V(r)$. We now consider the four-dimensional quantity

$$\Sigma = \begin{pmatrix} \sigma & 0 \\ 0 & \sigma \end{pmatrix}. \tag{3.163}$$

Its rate of change is given by

$$i\frac{\partial \Sigma_x}{\partial t} = -2ic(\alpha_y p_z - \alpha_z p_y), \tag{3.164}$$

which follows from (3.153,3.158,3.163). The quantity

$$\mathbf{J} = \mathbf{L} + \mathbf{S}, \tag{3.165}$$

where

$$\mathbf{S} = \tfrac{1}{2}\Sigma, \tag{3.166}$$

is therefore conserved. $\mathbf{S}$ represents an intrinsic angular momentum of the electron, which is called spin. The spin thus arises naturally from the Dirac equation.

It is convenient to write the four-component spinor $|\psi\rangle$ as two two-component spinors

$$|\psi\rangle = \begin{pmatrix} |\psi_1\rangle \\ |\psi_2\rangle \end{pmatrix}. \tag{3.167}$$

Using (3.158) the Dirac equation (3.159) becomes

$$[E - V - c\begin{pmatrix} 0 & \sigma \cdot \mathbf{p} \\ \sigma \cdot \mathbf{p} & 0 \end{pmatrix} + c^2 \begin{pmatrix} 0 & 0 \\ 0 & 2 \end{pmatrix}]\begin{pmatrix} |\psi_1\rangle \\ |\psi_2\rangle \end{pmatrix} = 0. \tag{3.168}$$

This may be written as two coupled two-component equations.

$$(E - V)|\psi_1\rangle - c\boldsymbol{\sigma} \cdot \mathbf{p}|\psi_2\rangle = 0,$$
$$(E - V + 2c^2)|\psi_2\rangle - c\boldsymbol{\sigma} \cdot \mathbf{p}|\psi_1\rangle = 0. \tag{3.169}$$

Note that the small component $|\psi_2\rangle$ is of order v/c times $|\psi_1\rangle$.

In order to understand the boundary conditions for solving these equations we consider the plane-wave ($V = 0$) solutions. The free-electron state is given in the coordinate–spin representation by

$$\langle \mathbf{r}\boldsymbol{\sigma}|\nu\mathbf{p}\rangle = \begin{pmatrix} 1 \\ \dfrac{c\boldsymbol{\sigma} \cdot \mathbf{p}}{E + 2c^2} \end{pmatrix} \langle \boldsymbol{\sigma}|\tfrac{1}{2}\nu\rangle\langle \mathbf{r}|\mathbf{p}\rangle, \tag{3.170}$$

where $\langle \boldsymbol{\sigma}|\tfrac{1}{2}\nu\rangle$ is the two-component spinor (3.79).

We now consider the coupled equations (3.169) in more detail. Eliminating $|\psi_2\rangle$ we have

$$(E - V)|\psi_1\rangle = \frac{1}{2}(\boldsymbol{\sigma} \cdot \mathbf{p})(1 + \frac{E - V}{2c^2})^{-1}(\boldsymbol{\sigma} \cdot \mathbf{p})|\psi_1\rangle. \tag{3.171}$$

This equation may be written to first order in $(E - V)/2c^2$ as a two-component Schrödinger equation for $|\psi_1\rangle$.

$$(E - K - U_R)|\psi_1\rangle = 0, \tag{3.172}$$

where K is the non-relativistic kinetic energy

$$K = \tfrac{1}{2}p^2 \tag{3.173}$$

and U_R is the relativistic potential

$$U_R = V - \frac{p^4}{8c^2} - \frac{1}{4c^2}\frac{dV}{dr}\frac{\partial}{\partial r} + \frac{1}{2c^2}\frac{1}{r}\frac{dV}{dr}\mathbf{S} \cdot \mathbf{L}. \tag{3.174}$$

For $E=1000$ eV the use of (3.172) involves neglecting terms of order 10^{-6}. Apart from V the last term in (3.174) is the most significant. It is the spin–orbit coupling term, which can change the sign of the spin projection of a scattering electron. The second term is the relativistic correction to the kinetic energy. The third term has no classical analogue.

4

One-electron problems

Methods for calculating collisions of an electron with an atom consist in expressing the many-electron amplitudes in terms of the states of a single electron in a fixed potential. In this chapter we summarise the solutions of the problem of an electron in different local, central potentials. We are interested in bound states and in unbound or scattering states. The one-electron scattering problem will serve as a model for formal scattering theory and for some of the methods used in many-body scattering problems.

4.1 Particle in a cubic box

This problem is important in formal scattering theory. A cubic box of side L is represented by a potential $V(\mathbf{r})$, which has a constant value (say zero) inside the box and is infinite at the box boundary. This means that the particle cannot be found outside the box. The coordinate representation of the Schrödinger equation is

$$[E + \tfrac{1}{2}\nabla^2 - V(\mathbf{r})]\psi(\mathbf{r}) = 0. \tag{4.1}$$

The equation can be separated in the x, y, z coordinates, each equation being similar.

$$\psi(\mathbf{r}) = X(x)Y(y)Z(z). \tag{4.2}$$

We consider the x equation for illustration. The origin is at one corner of the box.

$$\left[p_x^2 + \frac{\partial^2}{\partial x^2} \right] X(x) = 2V_x(x)X(x). \tag{4.3}$$

For $0 \leq x \leq L$ the solutions are

$$X(x) = \sin p_x x \quad \text{or} \quad \cos p_x x. \tag{4.4}$$

81

After imposing the boundary condition $X(x) = 0$ at $x = 0$ and L, and requiring continuity if a similar box is placed at L, we obtain a countable set of solutions

$$X(x) = \sin \frac{2\pi n_x}{L} x, \quad n_x = 0, 1, 2, ... \tag{4.5}$$

The solution of the three-dimensional problem is

$$\psi(\mathbf{r}) = \sin \frac{2\pi n_x}{L} x \sin \frac{2\pi n_y}{L} y \sin \frac{2\pi n_z}{L} z. \tag{4.6}$$

The significance of this solution is that placing a box round an experiment involving a free particle changes the continuum of eigenstates of $\mathbf{p}$ into a complete countable set given by

$$p_x = 2\pi n_x/L, \quad \text{etc.} \tag{4.7}$$

A complete countable set of eigenstates spans a Hilbert space, for which the algebra is a simple extension of the linear algebra of a finite space. We have no algebra for the continuum.

4.2 The Schrödinger equation for a local, central potential

The kinetic-energy operator separates in spherical polar coordinates into radial and angular observables given by (3.63). The Schrödinger equation for a local, central potential is therefore

$$[E + \tfrac{1}{2}\nabla_r^2 - \tfrac{1}{2}L^2/r^2 - V(r)]R(r)X(\hat{\mathbf{r}}) = 0. \tag{4.8}$$

We can rewrite the radial equation in terms of a simpler differential operator than ∇_r^2 (3.64) by solving for the function

$$u(r) = rR(r). \tag{4.9}$$

We choose a particular eigenstate (3.65) $Y_{\ell m}(\hat{\mathbf{r}})$ of L^2 and replace L^2 by its eigenvalue, obtaining the radial equation

$$\left[\frac{d^2}{dr^2} - \frac{\ell(\ell+1)}{r^2} + 2(E - V(r)) \right] u_\ell(r) = 0. \tag{4.10}$$

If E is negative, (4.10) is an eigenvalue problem with solutions $u_{n\ell}(r)$ and eigenvalues $\epsilon_{n\ell}$. If E is positive the solution with the correct boundary conditions is a linear combination of angular-momentum eigenstates.

4.3 Bound states in a local, central potential

In atomic physics we encounter one-electron bound states with different types of boundary condition. For the first type the electron is completely confined to a spherical box, near the boundary of which the potential is negligible. The second type involves a potential which falls to zero with increasing r faster than the Coulomb potential Zr^{-1}. The Coulomb potential is a special case, which must be treated separately. Coulomb boundary conditions occur when the potential has an arbitrary form for $r \le r_0$ but is given by Zr^{-1} for $r > r_0$. A Coulomb potential modifies free-particle states everywhere.

4.3.1 The spherical box

We consider a spherical potential which is zero up to $r = a$, where it is infinite. The radial boundary condition results in the radial equation (4.10) being an eigenvalue problem with eigenvalues $\epsilon_{n\ell}$ which are positive with respect to the zero of energy. We define a wave number k by

$$k_{n\ell}^2 = 2\epsilon_{n\ell}. \tag{4.11}$$

The radial equation is

$$\left[\frac{d^2}{dr^2} - \frac{\ell(\ell+1)}{r^2} + k_{n\ell}^2 \right] u_{n\ell}(r) = 0. \tag{4.12}$$

Since (4.12) is a second-order equation $u_{n\ell}$ has two independent forms, which are most easily understood when $\ell = 0$ (s-state). In this case

$$u_{n0}(r) = \sin k_{n0}r \quad \text{or} \quad \cos k_{n0}r. \tag{4.13}$$

For the sine solution the radial wave function

$$R_{n0}(r) = r^{-1}u_{n0}(r) \tag{4.14}$$

is regular at the origin. The cosine solution is called the irregular solution.

The generalisation of $u_{n0}(r)$ to arbitrary positive integers ℓ is given by the regular and irregular Ricatti–Bessel functions $U_\ell(\rho)$ and $V_\ell(\rho)$ respectively. They satisfy (4.12) with

$$\rho = k_{n\ell}r \tag{4.15}$$

and are expressed in terms of the spherical Bessel and Neumann functions $j_\ell(\rho)$ and $n_\ell(\rho)$ by

$$U_\ell(\rho) = \rho j_\ell(\rho),$$
$$V_\ell(\rho) = -\rho n_\ell(\rho). \tag{4.16}$$

The important properties of these functions are given for example by Antosiewicz (1973) and calculations are in subroutine libraries, for example Barnett *et al.* (1974).

The boundary conditions on $u_{n\ell}(r)$ are that it is zero at $r = 0$ and a. This gives a finite set of values $k_{n\ell}$ in analogy with the one-dimensional solution (4.5).

$$u_{n\ell}(r) = U_\ell(k_{n\ell}r). \tag{4.17}$$

The integer n is defined in atomic physics to be the principal quantum number. The number of nodes in the radial function, including those at $r = 0$ and a, is $n-\ell+1$. The functions are labelled by the old spectroscopic notation nx, where x is

$$s, p, d, f, g, ...$$

for ℓ values

$$0,1,2,3,4,... .$$

4.3.2 Short-range potentials

For a short-range potential there is a point r_0 beyond which the potential can be considered to be zero. A bound particle has a negative energy eigenvalue $\epsilon_{n\ell}$ and we define the absolute value of the imaginary wave number $i\beta_{n\ell}$ by

$$\beta_{n\ell}^2 = -\epsilon_{n\ell}. \tag{4.18}$$

The radial equation is

$$\left[\frac{d^2}{dr^2} - \frac{\ell(\ell+1)}{r^2} + 2(\epsilon_{n\ell} - V(r))\right] u_{n\ell}(r) = 0. \tag{4.19}$$

For $r > r_0$ the solution is given by the appropriate linear combination of the regular and irregular Ricatti–Bessel functions of imaginary argument. Putting

$$\rho = i\beta_{n\ell}r \tag{4.20}$$

we see that in the case $\ell = 0$ the solution

$$e^{i\rho} = \cos\rho + i\sin\rho \tag{4.21}$$

decays to zero at long range as $e^{-\beta_{n0}r}$. The generalisation of this to integer ℓ is

$$H_\ell^{(+)}(i\beta_{n\ell}r) = V_\ell(i\beta_{n\ell}r) + iU_\ell(i\beta_{n\ell}r), \tag{4.22}$$

where $H_\ell^{(+)}(\rho)$ is the Ricatti–Hankel function that decays exponentially for large imaginary ρ. Its long-range form is $e^{-\beta_{n\ell}r}$.

The solutions of (4.19) are found by numerical integration of the differential equation up to r_0 with a chosen value of $\beta_{n\ell}$. It is necessary to repeat the solution, hunting for an eigenvalue $\epsilon_{n\ell}$ for which the function $u_{n\ell}(r)$ and its derivative are equal respectively to $H_\ell^{(+)}(i\beta_{n\ell}r_0)$ and its derivative. Alternatively the boundary condition may be given by matching internal and external solutions at two external points r_0 and r_1. Which of these methods is used depends on the algorithm used for solving the differential equation.

For a short-range potential added to a Coulomb potential, such as we might have in a one-electron model of an atom, the external Ricatti–Hankel function of (4.22) is replaced by a solution of the bound Coulomb problem, discussed in the next section.

4.3.3 The Coulomb potential

For the purpose of normal atomic collision problems the state of an electron in the hydrogen atom is given by the solution of the problem of an electron bound in the Coulomb potential. We choose the nucleus to be infinitely massive for simplicity, although there is no difficulty in eliminating the consequent errors of order 10^{-3} by transforming to the system in which the centre of mass is the origin and the particle has the reduced mass $m_e m_H/(m_e + m_H)$.

The radial equation for nuclear charge Z is

$$\left[\frac{d^2}{dr^2} - \frac{\ell(\ell+1)}{r^2} + 2(\epsilon_{n\ell} + Zr^{-1})\right] u_{n\ell}(r) = 0. \tag{4.23}$$

It cannot be solved by the methods of section 4.3.2 because there is no point r_0 beyond which the potential is negligible. The solutions are special functions, which we call hydrogenic orbitals. They are described below.

The normalised hydrogenic orbitals are

$$u_{n\ell}(r) = -\left[\frac{2Z}{n}\frac{(n-\ell-1)!}{2n\{(n+\ell)!\}^3}\right]^{1/2} \rho^{\ell+1} L_{n+\ell}^{2\ell+1}(\rho)e^{-\rho/2}, \tag{4.24}$$

where

$$\rho = \frac{2Z}{n}r \tag{4.25}$$

and $L_{n+\ell}^{2\ell+1}(\rho)$ is an associated Laguerre polynomial.

$$L_{n+\ell}^{2\ell+1}(\rho) = \sum_{k=0}^{n-\ell-1}(-1)^{k+1}\frac{\{(n+\ell)!\}^2}{(n-\ell-1-k)!(2\ell+1+k)!k!}\rho^k. \tag{4.26}$$

The forms of $u_{n\ell}(r)$ for $n=1$ and 2 (see also fig. 4.1) are

$$1s : Z^{3/2}2re^{-Zr},$$
$$2s : Z^{3/2}2^{-3/2}r(2-Zr)e^{-Zr/2},$$
$$2p : Z^{5/2}24^{-1/2}r^2e^{-Zr/2}. \tag{4.27}$$

The energy eigenvalues $\epsilon_{n\ell}$ are given by the Rydberg formula

$$\epsilon_{n\ell} = -\tfrac{1}{2}Z^2/n^2. \tag{4.28}$$

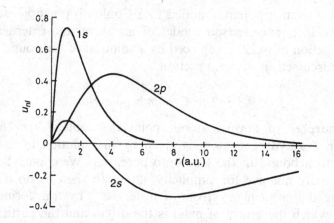

Fig. 4.1.   The three lowest-energy orbitals $u_{n\ell}(r)$ of the hydrogen atom.

4.3.4 Matrix solution for a bound state

The problem of a single electron in an atom may be approximated by
that of an electron in a local, central potential with the Coulomb form at
large distances. The bound radial eigenstates $u_{n\ell}(r)$ of an electron in such
a potential may be expanded in a basis set $f_{i\ell}(r)$ of radial functions, each
of which is square integrable and is normalised.

$$\int_0^\infty dr[f_{i\ell}(r)]^2 = 1. \tag{4.29}$$

We write the radial Schrödinger equation (4.10) in the form

$$(\epsilon_{n\ell} - H_\ell)u_{n\ell}(r) = 0, \tag{4.30}$$

where

$$H_\ell = -\frac{1}{2}\frac{d^2}{dr^2} + \frac{1}{2}\frac{\ell(\ell+1)}{r^2} + V(r). \tag{4.31}$$

We consider the subspace of radial states belonging to the angular-
momentum index ℓ, which we drop from the notation for the states.
$u_{n\ell}(r)$ is the coordinate representation of a state $|n\rangle$.

$$u_{n\ell}(r) = \langle r|n\rangle. \tag{4.32}$$

The basis set of radial states is given by

$$f_{j\ell}(r) = \langle r|j\rangle. \tag{4.33}$$

The expansion of the eigenstate n is written

$$|n\rangle = \Sigma_j|j\rangle\langle j|n\rangle. \tag{4.34}$$

The set of normalised orbitals may not span the ℓ subspace of eigenstates of an electron in the potential. If not, the basis $|j\rangle$ includes the positive-energy continuum, which will be discussed in section 4.4.

The radial Schrödinger equation (4.30) is written using the notation of (4.30,4.32) as

$$(\epsilon_{n\ell} - H_\ell)|n\rangle = 0. \tag{4.35}$$

We use the representation theorem for the basis states $|j\rangle$ and form the matrix element with the bra vector $\langle i|$.

$$\Sigma_j \langle i|\epsilon_{n\ell} - H_\ell|j\rangle \langle j|n\rangle = 0. \tag{4.36}$$

This is formally a matrix eigenvalue problem with eigenvalues $\epsilon_{n\ell}$. The j component of the corresponding eigenvector is $\langle j|n\rangle$.

We consider it as the problem of diagonalising a finite matrix, the computation of which is available in subroutine libraries, for example Anderson *et al.* (1992), by truncating the basis to a finite set of states $|j\rangle$. The eigenvector components $\langle j|n\rangle$ are fully defined by requiring that the eigenvectors $|n\rangle$ are normalised.

Consider the eigenvector components $\langle j|n\rangle$ for lower-energy eigenvalues and for basis states $|j\rangle$ that have large components in eigenvectors for higher-energy eigenvalues. For a well-chosen basis such components are small so that the truncation is justified for low-lying states.

The elements of the Hamiltonian matrix are written as integrals by returning to the coordinate representation.

$$\langle i|\epsilon_{n\ell} - H_\ell|j\rangle = \int_0^\infty dr f_{i\ell}(r)[\epsilon_{n\ell} - H_\ell]f_{j\ell}(r), \tag{4.37}$$

where H_ℓ is given by (4.31).

The hydrogenic orbitals (4.24) are linear combinations of Slater-type orbitals $f_{i\ell}(r)$ whose normalised form is parametrised in terms of positive integers $n_{i\ell}$ and exponents $\zeta_{i\ell}$. It is

$$f_{i\ell}(r) = [(2n_{i\ell})!]^{-1/2}(2\zeta_{i\ell})^{n_{i\ell}+1/2} r^{n_{i\ell}} e^{-\zeta_{i\ell}r}. \tag{4.38}$$

These orbitals are often used as the basis for a more-general problem, since only a relatively-small number of them are needed to represent the low-lying states well and the integrals (4.37) are analytic, making the computation of the Hamiltonian matrix elements fast. The hydrogenic orbitals themselves constitute a better basis since they are orthonormal.

4.4 Potential scattering

The scattering of a spinless electron from a local, central potential is a prototype for scattering problems involving complex targets. The scattering is of course elastic since the potential has no degrees of freedom that can be excited. The problem defines precisely the one-electron continuum functions for positive-energy electrons in the potential.

4.4.1 Differential cross section

In a scattering experiment a beam of electrons of momentum **k** hits a target. We consider the target to be represented by a potential $V(r)$. Electrons are observed by a detector placed at polar and azimuthal angles θ, ϕ measured from the direction of the incident beam, which is the $\hat{z}$ direction in a system of spherical polar coordinates (fig. 4.2). For a central potential the problem is axially symmetric. Relevant quantities do not depend on ϕ. The detector subtends a solid angle

$$d\Omega = \sin\theta d\theta d\phi. \tag{4.39}$$

The observed quantity is the number of electrons detected in a particular energy range $(E, E + dE)$ in unit time. For elastic scattering

$$E = \tfrac{1}{2}k^2. \tag{4.40}$$

The number of detected electrons is proportional to the incident beam flux, which is an accident of the experiment, not a property of the scattering process. The quantity that describes the scattering process is the differential cross section.

$$\frac{d\sigma}{d\Omega} = \frac{\text{number of electrons detected per unit time}}{\text{number of incident electrons per unit area per unit time}}. \tag{4.41}$$

It has the dimensions of area.

The electron in a scattering problem may be found anywhere in space. Its wave function therefore cannot be normalised in the sense that the

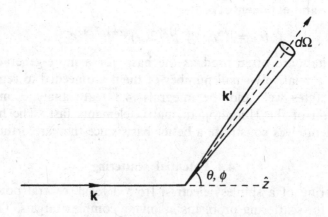

Fig. 4.2. Schematic diagram of an elastic scattering experiment.

probability of finding it somewhere in a finite volume is unity. Instead it is normalised to plane-wave incident flux. If there is no target, the state of the electron is $|\mathbf{k}\rangle$. Its coordinate representation is

$$\langle \mathbf{r}|\mathbf{k}\rangle = (2\pi)^{-3/2} e^{i\mathbf{k}\cdot\mathbf{r}}. \tag{4.42}$$

The probability flux for a wave function $\psi(\mathbf{r})$ is

$$\mathbf{j}(\mathbf{r}) = (2i)^{-1}[\psi^*(\mathbf{r}) \nabla \psi(\mathbf{r}) - \psi(\mathbf{r}) \nabla \psi^*(\mathbf{r})]. \tag{4.43}$$

For $\psi(\mathbf{r}) = e^{i\mathbf{k}\cdot\mathbf{r}}$

$$\mathbf{j}(\mathbf{r}) = \mathbf{k}. \tag{4.44}$$

The wave function of the electron is distorted (fig. 4.3) from its plane-wave form by the potential $V(r)$. It is called a distorted wave $\chi^{(+)}(\mathbf{k}, \mathbf{r})$. The coordinate representation of the scattering state is

$$\langle \mathbf{r}|\chi^{(+)}(\mathbf{k})\rangle = (2\pi)^{-3/2} \chi^{(+)}(\mathbf{k}, \mathbf{r}). \tag{4.45}$$

The distance of the detector from the target is very large on an atomic scale. The target therefore looks like a point and the scattered wave at large distances is a spherical outgoing wave. The superscript $(+)$ in the notation for the state indicates that the large-distance (asymptotic) form of the distorted wave is

$$\chi^{(+)}(\mathbf{k}, \mathbf{r}) \longrightarrow e^{i\mathbf{k}\cdot\mathbf{r}} + f(\theta)\frac{e^{ikr}}{r}, \quad r \longrightarrow \infty. \tag{4.46}$$

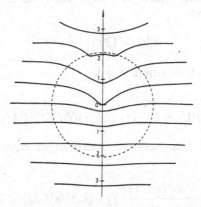

Fig. 4.3. Surfaces of equal phase (wave fronts) at intervals of π in the wave function for the elastic scattering of 200 eV electrons by the static potential of argon, plotted on a plane through the scattering axis. The radial scale is marked in a.u.

For later purposes we are sometimes interested in the time-reversed distorted wave, which has ingoing spherical-wave boundary conditions

$$\chi^{(-)}(\mathbf{k},\mathbf{r}) \longrightarrow e^{i\mathbf{k}\cdot\mathbf{r}} + f^*(\theta)\frac{e^{-ikr}}{r}, \quad r \longrightarrow \infty. \tag{4.47}$$

The amplitude $f(\theta)$ of the scattered wave relative to the incident wave is the scattering amplitude. It is defined as follows. The number of scattered electrons in the volume between r and $r+dr$ in the solid angle $d\Omega$ is

$$|f(\theta)\frac{e^{ikr}}{r}|^2 r^2 dr d\Omega = |f(\theta)|^2 dr d\Omega.$$

The number scattered into $d\Omega$ per unit time is

$$|f(\theta)|^2 k d\Omega,$$

where in atomic units the velocity dr/dt is k. Using (4.41,4.44) we find

$$\frac{d\sigma}{d\Omega} = |f(\theta)|^2. \tag{4.48}$$

The total cross section is

$$\sigma_T = \int d\Omega |f(\theta)|^2. \tag{4.49}$$

4.4.2 Partial-wave expansion

For a central potential it is natural to expand the scattering wave function $\chi^{(+)}(\mathbf{k},\mathbf{r})$ in orbital-angular-momentum eigenstates. We first consider the case of zero potential where the wave function is a plane wave (4.42), which takes the form

$$\langle \mathbf{r}|\mathbf{k}\rangle = (2\pi)^{-3/2} e^{ikr\cos\theta} \tag{4.50}$$

in the coordinate system where $\hat{\mathbf{z}} = \hat{\mathbf{k}}$.

The expansion

$$\langle \mathbf{r}|\mathbf{k}\rangle = \Sigma_{LM} f_{LM}(k,r)\langle\hat{\mathbf{r}}|LM\rangle\langle LM|\hat{\mathbf{k}}\rangle \tag{4.51}$$

is a solution of the free-electron Schrödinger equation which becomes, using (3.63),

$$\Sigma_{LM}\left[\nabla_r^2 + k^2 - \frac{L(L+1)}{r^2}\right] f_{LM}(k,r)\langle\hat{\mathbf{r}}|LM\rangle\langle LM|\hat{\mathbf{k}}\rangle = 0. \tag{4.52}$$

In fact we know from (4.50) that the radial solution is a function of kr and from (4.52) that it is independent of M. Using the transformation (4.9) to simplify the differential operator we find that the radial equation is

$$\left[\frac{d^2}{dr^2} - \frac{L(L+1)}{r^2} + k^2\right] U_L(kr) = 0, \tag{4.53}$$

where $U_L(kr)$ is the regular Ricatti–Bessel function of (4.16).

We now use the addition theorem (3.72) for spherical harmonics to obtain

$$\langle \mathbf{r}|\mathbf{k}\rangle = (2\pi)^{-3/2}4\pi(kr)^{-1}\Sigma_L(2L+1)A_L U_L(kr)P_L(\cos\theta). \tag{4.54}$$

We find the constant A_L by comparing (4.16,4.50,4.54) with the identity (Merzbacher, 1970)

$$e^{ikr\cos\theta} = \Sigma_L(2L+1)i^L j_L(kr)P_L(\cos\theta). \tag{4.55}$$

The expansion (4.51) is the partial-wave expansion. Our final form is

$$\langle \mathbf{r}|\mathbf{k}\rangle = (2/\pi)^{1/2}(kr)^{-1}\Sigma_{LM}i^L U_L(kr)\langle \hat{\mathbf{r}}|LM\rangle\langle LM|\hat{\mathbf{k}}\rangle. \tag{4.56}$$

When the scattering potential $V(r)$ is not zero, the only change is the radial equation, which becomes

$$\left[\frac{d^2}{dr^2} - \frac{L(L+1)}{r^2} - V(r) + k^2\right] u_L(k,r) = 0. \tag{4.57}$$

The distorted wave is

$$\langle \mathbf{r}|\chi^{(+)}(\mathbf{k})\rangle = (2/\pi)^{1/2}(kr)^{-1}\Sigma_{LM}i^L u_L(k,r)\langle \hat{\mathbf{r}}|LM\rangle\langle LM|\hat{\mathbf{k}}\rangle. \tag{4.58}$$

4.4.3 Solution of the radial equation

To solve the radial equation (4.57) we first choose a point r_0 such that the potential has its long-range form for $r > r_0$. The space is divided into internal and external regions by r_0. We rewrite $V(r)$ as

$$V(r) = U(r) - z/r, \tag{4.59}$$

where $U(r) = 0$ for $r > r_0$ and z is the net charge of the target. The Coulomb potential affects the solution at all distances r. The radial equation is integrated numerically from $r = 0$ and matched to its external form, either by equating internal and external functions and their derivatives at r_0 or by equating the functions at two points r_0 and $r_1 > r_0$, depending on the algorithm used for the internal solution.

The external form of the radial wave function is a solution of (4.57) with the potential $-z/r$. The uncharged target (such as an atom) is a special case $z = 0$. It is convenient to rewrite the radial equation in terms of the variable

$$\rho = kr \tag{4.60}$$

using the Coulomb parameter

$$\eta = -z/k. \tag{4.61}$$

The transformed radial equation for the external region is

$$\left[\frac{d^2}{d\rho^2} - \frac{L(L+1)}{\rho^2} - \frac{\eta}{\rho} + 1\right] u_L(\rho) = 0. \tag{4.62}$$

It has two independent, regular and irregular, solutions which are the Ricatti–Bessel functions $U_L(\rho)$ and $V_L(\rho)$ of (4.16) in the case $z = 0$. The solutions are the Coulomb functions (see also fig. 4.4)

$$\text{regular Coulomb function}: F_L(\eta,\rho),$$

$$\text{irregular Coulomb function}: G_L(\eta,\rho).$$

It is not necessary for the physicist to know how to compute the Coulomb functions. They are found in subroutine libraries, for example Barnett *et al.* (1974). A sufficient idea of their form is obtained by putting $\eta = L = 0$ in (4.62), when they are seen to be $\sin\rho$ and $\cos\rho$ respectively. The potential terms dilate or compress the sine and cosine waves, resulting in an overall phase shift at long range.

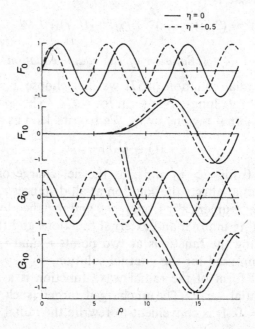

Fig. 4.4. Regular and irregular Coulomb functions for $L = 0$ and 10, $\eta = 0$ and -0.5.

The forms of the Coulomb functions for very small and very large r are important

$$F_L(\eta, \rho) \longrightarrow \frac{\rho^{L+1}}{(2L+1)!!},$$

$$G_L(\eta, \rho) \longrightarrow \frac{(2L-1)!!}{\rho^L}, \quad \rho \longrightarrow 0. \tag{4.63}$$

$$F_L(\eta, \rho) \longrightarrow \sin(\rho - L\pi/2 - \eta \ln 2\rho + \sigma_L),$$

$$G_L(\eta, \rho) \longrightarrow \cos(\rho - L\pi/2 - \eta \ln 2\rho + \sigma_L), \quad \rho \longrightarrow \infty. \tag{4.64}$$

Notice that the term $\eta \ln 2\rho$ in the large-r form is a phase that depends on r at all distances. The large-r form is not valid for any range relevant to computation. The quantity σ_L is the Coulomb phase shift, defined by

$$\sigma_L = \arg \Gamma(L+1+i\eta). \tag{4.65}$$

For a charged target the boundary condition (4.46) is generalised. The incident wave is a Coulomb wave (the solution of the Schrödinger equation for a Coulomb potential) and the scattered wave is outgoing for large r. The form of the outgoing and ingoing solutions is seen from (4.64). We denote these solutions using superscripts ($\pm$) with the same meaning as in (4.46, 4.47).

$$H_L^{(+)}(\eta, \rho) = G_L(\eta, \rho) + i F_L(\eta, \rho),$$

$$H_L^{(-)}(\eta, \rho) = G_L(\eta, \rho) - i F_L(\eta, \rho). \tag{4.66}$$

The external solution for the partial wave L is

$$u_L(k, r) = F_L(\eta, kr) + C_L H_L^{(+)}(\eta, kr), \quad r \geq r_0, \tag{4.67}$$

where C_L is the partial scattering amplitude. The radial equation is solved by matching this to the internal solution, thus determining the complex number C_L.

In computing the internal solution the starting value of $u_L(k, r)$ at very small r is arbitrary. The normalisation is given by (4.67). However, in order to avoid computational problems associated with large and small numbers it is convenient to use the first of equns. (4.63) as the starting condition.

4.4.4 The asymptotic region

Scattered particles are observed at distances of the order 10^7 times the average radius of an atom. At such distances the long-range forms of the wave functions are valid. This is the asymptotic region. It is characterised by different parameters for different purposes. The parameters are determined by the matching of internal and external solutions for each partial wave.

We first consider the case of a real, short-range potential. The radial function is the solution of a real differential equation. However, we know that its external form (4.67) is complex. It must therefore have the form

$$u_L(\rho) = e^{i\phi_L} v_L(\rho), \tag{4.68}$$

where ϕ_L is a real constant and $v_L(\rho)$ is real. We consider only the asymptotic forms of all the wave functions. Using (4.64) in our particular case $\eta = 0$ we write

$$v_L(\rho) \longrightarrow A_L \sin(\rho - L\pi/2) + B_L \cos(\rho - L\pi/2). \tag{4.69}$$

Putting

$$A_L = \cos\delta_L,$$
$$B_L = \sin\delta_L, \tag{4.70}$$

we have

$$v_L(\rho) \longrightarrow \sin(\rho - L\pi/2 + \delta_L). \tag{4.71}$$

The real number δ_L is the phase shift. The real number

$$K_L = \tan\delta_L = B_L/A_L \tag{4.72}$$

is the K-matrix element.

The phase shift is non-zero for only a finite number of partial waves. For large enough L the repulsive term $L(L+1)/r^2$ in the radial equation (4.57) (the centrifugal barrier) is so large that the potential is insignificant. The wave function $u_L(k,r)$ is essentially zero at r_0. The external function has no irregular part so that $B_L = 0$ in (4.69) and $\delta_L = 0$.

Another useful way of understanding the partial wave for a real potential is that it consists of an ingoing wave $H_L^{(-)}(\rho)$ and an outgoing wave whose magnitude is the same because no flux is lost, but whose phase may be given by a phase factor S_L with respect to it. S_L is the S-matrix element. The S matrix is unitary. We write the partial wave $u_L(\rho)$ using the asymptotic form of (4.66).

$$u_L(\rho) \longrightarrow \tfrac{1}{2} i [e^{-i(\rho - L\pi/2)} - S_L e^{i(\rho - L\pi/2)}]. \tag{4.73}$$

Comparing (4.67, 4.73) we have

$$S_L = 1 + 2iC_L, \tag{4.74}$$

$$C_L = \frac{1}{2i}(S_L - 1). \tag{4.75}$$

We now express S_L in terms of the phase shift δ_L. We note that (4.64,4.66,4.67) give the asymptotic form for $u_L(\rho)$

$$e^{i\phi_L}[A_L\sin(\rho - L\pi/2) + B_L\cos(\rho - L\pi/2)] = [(1 + iC_L)\sin(\rho - L\pi/2) + C_L\cos(\rho - L\pi/2)]. \quad (4.76)$$

We therefore have, by comparing coefficients of the sine and cosine solutions,

$$K_L = \frac{C_L}{1 + iC_L}, \quad (4.77)$$

$$C_L = \frac{K_L}{1 - iK_L}. \quad (4.78)$$

Comparing (4.72,4.74,4.78) we find that

$$S_L = \frac{1 - K_L^2}{1 + K_L^2} + \frac{2iK_L}{1 + K_L^2} = e^{2i\delta_L}. \quad (4.79)$$

The phase ϕ_L of the partial wave (4.68) is found by expressing (4.78) in real and imaginary parts and comparing the result with the expression

$$C_L = e^{i\phi_L}B_L, \quad (4.80)$$

obtained from (4.76). We have

$$C_L = B_L\cos\phi_L + iB_L\sin\phi_L = \frac{K_L}{1 + K_L^2} + i\frac{K_L^2}{1 + K_L^2}, \quad (4.81)$$

from which we find (with the appropriate choice of zero phase)

$$\tan\phi_L = K_L,$$
$$\phi_L = \delta_L. \quad (4.82)$$

The phase of the scattering wave function for a real potential is constant and equal to the phase shift.

The asymptotic form of the distorted wave $\chi^{(+)}(\mathbf{k}, \mathbf{r})$ is written using (4.45,4.58) and the addition theorem (3.72).

$$\chi^{(+)}(\mathbf{k}, \mathbf{r}) \longrightarrow (kr)^{-1}\Sigma_L(2L + 1)i^L e^{i\delta_L}\sin(kr - L\pi/2 + \delta_L)P_L(\cos\theta). \quad (4.83)$$

If the potential $V(r)$ is a pure Coulomb potential the asymptotic partial wave is given by the regular Coulomb function (4.64), apart from a constant phase factor. We strictly have no incident plane wave since the Coulomb potential modifies the wave function everywhere. We make the normalisation of the Coulomb distorted wave $\psi_\eta(\mathbf{k}, \mathbf{r})$ analogous to that of (4.83) by choosing the phase factor to be the Coulomb phase shift σ_L.

$$\psi_\eta(\mathbf{k}, \mathbf{r}) \longrightarrow (kr)^{-1}\Sigma_L(2L + 1)i^L e^{i\sigma_L}$$
$$\times \sin(kr - L\pi/2 - \eta\ln 2kr + \sigma_L)P_L(\cos\theta). \quad (4.84)$$

We denote the Coulomb scattering function for all r by a notation analogous to (4.42).

$$\langle \mathbf{r}|\mathbf{k}_\eta^{(+)}\rangle = (2/\pi)^{1/2}(kr)^{-1}\Sigma_{LM}i^L e^{i\sigma_L}F_L(\eta, kr)\langle \hat{\mathbf{r}}|LM\rangle\langle LM|\hat{\mathbf{k}}\rangle. \qquad (4.85)$$

The partial wave form of the scattering amplitude for a short-range potential is found by comparing the asymptotic partial wave form (4.45, 4.58, 4.76) of the distorted wave $\chi^{(+)}(\mathbf{k},\mathbf{r})$ with the asymptotic expression (4.46).

$$e^{i\rho\cos\theta} + f(\theta)k\rho^{-1}e^{i\rho}$$
$$= \rho^{-1}\Sigma_L(2L+1)i^L[(1+iC_L)\sin(\rho-L\pi/2)+C_L\cos(\rho-L\pi/2)]P_L(\cos\theta). \qquad (4.86)$$

Subtracting the asymptotic expression for the plane wave

$$e^{i\rho\cos\theta} \longrightarrow \rho^{-1}\Sigma_L(2L+1)i^L\sin(\rho-L\pi/2)P_L(\cos\theta), \qquad (4.87)$$

and using (4.75,4.79) we find

$$f(\theta) = (2ik)^{-1}\Sigma_L(2L+1)(S_L-1)P_L(\cos\theta) \qquad (4.88a)$$

$$= k^{-1}\Sigma_L(2L+1)e^{i\delta_L}\sin\delta_L P_L(\cos\theta) \qquad (4.88b)$$

$$= k^{-1}\Sigma_L(2L+1)C_L P_L(\cos\theta). \qquad (4.88c)$$

Since $\delta_L = 0$ for $L > L_0$ the scattering amplitude is a finite sum of partial wave terms for $L \leq L_0$.

The total elastic cross section σ_E is obtained from (4.88) by using (3.70) and the orthogonality of the spherical harmonics.

$$\sigma_E = \int d\Omega |f(\theta)|^2 = (\pi/k^2)\Sigma_K(2L+1)|S_L-1|^2 \qquad (4.89a)$$

$$= (4\pi/k^2)\Sigma_L(2L+1)\sin^2\delta_L \qquad (4.89b)$$

$$= (4\pi/k^2)\Sigma_L(2L+1)|C_L|^2. \qquad (4.89c)$$

We now generalise to the case where the potential $V(r)$ has the Coulomb form at long range. We must add the Coulomb scattering amplitude (Schiff, 1955)

$$f_C(\theta) = -\frac{\eta}{2k\sin^2\theta/2}e^{i\eta\ln(\sin^2\theta/2+2i\sigma_0)} \qquad (4.90)$$

to the scattering amplitude for the short-range potential $U(r)$ with Coulomb boundary conditions. This will be fully explained in chapter 6 on formal scattering theory, but it is obvious for $U(r) = 0$.

The asymptotic form of the partial wave is obtained by using the Coulomb forms (4.64) in (4.69) and making arguments analogous to those leading to (4.83). We obtain

$$\chi^{(+)}(\mathbf{k}, \mathbf{r}) \longrightarrow (kr)^{-1}\Sigma_L(2L+1)i^L e^{i(\sigma_L+\delta_L)}\sin(kr - L\pi/2 - \eta\ln 2kr$$
$$+ \sigma_L + \delta_L)P_L(\cos\theta). \tag{4.91}$$

The distorted wave for the full potential $V(r)$ of (4.59) is

$$\langle \mathbf{r}|\chi^{(+)}(\mathbf{k})\rangle = (2/\pi)^{1/2}(kr)^{-1}\Sigma_{LM}i^L e^{i\sigma_L}u_L(k, r)\langle\hat{\mathbf{r}}|LM\rangle\langle LM|\hat{\mathbf{k}}\rangle. \tag{4.92}$$

The scattering amplitude is

$$f(\theta) = f_C(\theta) + (2ik)^{-1}\Sigma_L e^{2i\sigma_L}(2L+1)(S_L-1)P_L(\cos\theta), \tag{4.93}$$

or forms in terms of different scattering parameters analogous to (4.88). The S-matrix element is still defined in terms of the phase shift by

$$S_L = e^{2i\delta_L}. \tag{4.94}$$

The total elastic cross section and the forward ($\theta = 0$) differential cross section are infinite.

We now make a further generalisation to the case where $U(r)$ is a complex potential. A negative imaginary potential absorbs probability flux, so that the S matrix is no longer unitary (4.73,4.79) and the phase shifts are no longer real. In electron–atom scattering a loss of flux from the elastic state occurs when inelastic scattering or ionisation are possible. These are generically termed reactions and the total cross section for exciting them is the total reaction cross section σ_R.

$$\sigma_R = (\pi/k^2)\Sigma_L(2L+1)(1-|S_L|^2) \tag{4.95a}$$

$$= (4\pi/k^2)\Sigma_L(2L+1)(\mathrm{Im}C_L - |C_L|^2). \tag{4.95b}$$

By setting

$$\delta_L = \mu_L + iv_L \tag{4.96}$$

and using (4.75,4.79) we find that

$$\mathrm{Im}C_L - |C_L|^2 = \tfrac{1}{4}(1 - e^{-4v_L}). \tag{4.97}$$

The total reaction cross section (4.95b) is zero if δ_L is real.

The total cross section for a short-range potential is given by (4.89c, 4.95b).

$$\sigma_T = \sigma_E + \sigma_R = (4\pi/k^2)\Sigma_L(2L+1)\mathrm{Im}C_L. \tag{4.98}$$

The imaginary part of the forward scattering amplitude is given by (4.88c).

$$\mathrm{Im}f(0) = k^{-1}\Sigma_L(2L+1)\mathrm{Im}C_L = (k/4\pi)\sigma_T. \tag{4.99}$$

This important relationship is the optical theorem. The imaginary part of
the forward scattering amplitude is proportional to the total cross section.

4.5 Integral equations for scattering

In the previous sections the potential scattering problem has been defined
in terms of a Schrödinger differential equation with outgoing spherical-
wave boundary conditions. The description and computational methods
are analogous to those used for one-electron bound-state problems. In this
section we see that the whole problem in the coordinate representation
can be written in terms of a single integral equation, which in many ways
is easier to understand physically than the differential equation.

A large breakthrough in physical transparency and ease of computation is achieved by expressing the problem as an integral equation in
momentum space. The reason for this is that scattering experiments measure momenta, not positions, so that the momentum-space description
parallels the experiment.

The potential scattering Schrödinger equation is

$$[E^{(+)} - K]|\chi^{(+)}(\mathbf{k})\rangle = V|\chi^{(+)}(\mathbf{k})\rangle. \tag{4.100}$$

The superscript (+) indicates outgoing spherical-wave boundary conditions. We will show that this corresponds to adding to E a small, positive
imaginary part, which will tend to zero. We multiply on the left by the
inverse of the differential operator to obtain the formal solution

$$|\chi^{(+)}(\mathbf{k})\rangle = |\mathbf{k}\rangle + \frac{1}{E^{(+)} - K}V|\chi^{(+)}(\mathbf{k})\rangle. \tag{4.101}$$

The plane wave $|\mathbf{k}\rangle$ has been added to give the correct boundary condition.
When $V = 0$, $|\chi^{(+)}(\mathbf{k})\rangle = |\mathbf{k}\rangle$. The inverse differential operator is the
resolvent or free-particle Green's function operator and is denoted by G_0.

$$|\chi^{(+)}(\mathbf{k})\rangle = |\mathbf{k}\rangle + G_0(E^{(+)})V|\chi^{(+)}(\mathbf{k})\rangle. \tag{4.102}$$

We first replace the resolvent by a number by introducing its spectral
representation and using the function theorem. At the same time we
introduce the coordinate representation.

$$\langle \mathbf{r}|\chi^{(+)}(\mathbf{k})\rangle = \langle \mathbf{r}|\mathbf{k}\rangle + \int d^3r'' \int d^3r' \int d^3k' \langle \mathbf{r}|\mathbf{k}'\rangle \frac{2}{k^{(+)2} - k'^2} \langle \mathbf{k}'|\mathbf{r}'\rangle$$
$$\times \langle \mathbf{r}'|V|\mathbf{r}''\rangle\langle \mathbf{r}''|\chi^{(+)}(\mathbf{k})\rangle. \tag{4.103}$$

The superscript (+) on k denotes the addition of a small positive imaginary
part to k. The coordinate representations of a plane wave, a distorted
wave, and a local, central potential are given by (3.25), (4.25) and (3.35)

respectively. The coordinate-space integral equation is

$$\chi^{(+)}(\mathbf{k},\mathbf{r}) = e^{i\mathbf{k}\cdot\mathbf{r}} + \int d^3r' G_0(E^{(+)};\mathbf{r},\mathbf{r}')V(r')\chi^{(+)}(\mathbf{k},\mathbf{r}'). \qquad (4.104)$$

The Green's function $G_0(E^{(+)};\mathbf{r},\mathbf{r}')$ is defined by

$$G_0(E^{(+)};\mathbf{r},\mathbf{r}') = (2\pi)^{-3}\int d^3k' \frac{2}{k^{(+)2} - k'^2}e^{i\mathbf{k}'\cdot(\mathbf{r}-\mathbf{r}')}. \qquad (4.105)$$

In evaluating the Green's function we give a mathematical meaning to the (+) superscript. It is convenient to make the transformations

$$\boldsymbol{\rho} = \mathbf{r} - \mathbf{r}',$$
$$\boldsymbol{\kappa} = k'\rho,$$
$$\sigma = k\rho. \qquad (4.106)$$

We perform the angular integrations in (4.105) and replace $\sigma^{(+)2}$ by $\sigma^2 + i\epsilon$, where ϵ is a small positive quantity that will tend to zero.

$$G_0(E^{(+)};\mathbf{r},\mathbf{r}') = (2\pi^2\rho)^{-1}\lim_{\epsilon\to 0+}\int_0^\infty d\kappa\frac{\kappa\sin\kappa}{\sigma^2 - \kappa^2 + i\epsilon}. \qquad (4.107)$$

The integrand of (4.107) has poles at $\sigma + i\epsilon/2$ and $-\sigma - i\epsilon/2$. We evaluate the integral by the method of residues. The result is translated into our original notation by inverting the transformation (4.106). It is

$$G_0(E^{(+)};\mathbf{r},\mathbf{r}') = -(2\pi^2\rho)^{-1}\pi e^{i\sigma} = -(2\pi)^{-1}\frac{e^{ik|\mathbf{r}-\mathbf{r}'|}}{|\mathbf{r}-\mathbf{r}'|}, \qquad (4.108)$$

which is an outgoing spherical wave propagating from $\mathbf{r}'$ to $\mathbf{r}$.

If we had replaced ϵ by $-\epsilon$ we would have had an ingoing spherical wave

$$G_0(E^{(-)};\mathbf{r},\mathbf{r}') = -(2\pi)^{-1}\frac{e^{-ik|\mathbf{r}-\mathbf{r}'|}}{|\mathbf{r}-\mathbf{r}'|}. \qquad (4.109)$$

This is the time-reversed Green's function, which gives the equation corresponding to (4.104) for the time-reversed distorted wave $\chi^{(-)}(\mathbf{k},\mathbf{r})$.

The coordinate-space integral equation (4.104) gives great insight into quantum mechanics and its description of the potential-scattering process. We consider the elastic-scattering wave function $\chi^{(+)}(\mathbf{k},\mathbf{r})$ as the probability amplitude for finding the electron at the point $\mathbf{r}$. It is expressed by (4.104) as the sum of probability amplitudes for the electron reaching $\mathbf{r}$ by alternative paths. First it may reach $\mathbf{r}$ directly. This is the plane-wave inhomogeneous term. It may also reach $\mathbf{r}'$ (with amplitude $\chi^{(+)}(\mathbf{k},\mathbf{r}')$) and propagate from $\mathbf{r}'$ to $\mathbf{r}$ as a spherical wave, the amplitude for the propagation occurring being $V(r')$. The integration adds the spherical-wave contributions from all the points $\mathbf{r}'$. The equation (4.104) is equivalent to Huygens's description of wave propagation.

The asymptotic form of (4.104) gives the scattering amplitude (4.46). If $r \gg r'$ (4.108) becomes

$$G_0(E^{(+)};\mathbf{r},\mathbf{r}') \longrightarrow -(2\pi)^{-1}\frac{e^{ik(r-r'\cos\theta)}}{r}, \quad r \gg r', \tag{4.110}$$

where θ is the angle between $\mathbf{r}'$ and $\mathbf{r}$. If $\mathbf{r}$ is the position of the detector then θ is also the angle between $\mathbf{r}'$ and $\mathbf{k}'$, which is the outgoing momentum whose absolute value is k.

The asymptotic form of the scattering wave function (4.104) is

$$\chi^{(+)}(\mathbf{k},\mathbf{r}) \longrightarrow e^{i\mathbf{k}\cdot\mathbf{r}} - (2\pi)^{-1}[\int d^3r' e^{-i\mathbf{k}'\cdot\mathbf{r}'}V(r')\chi^{(+)}(\mathbf{k},\mathbf{r}')]\frac{e^{ikr}}{r}, \tag{4.111}$$

which is the same form as (4.46) with the scattering amplitude given by

$$f(\theta) = -(2\pi)^{-1}\int d^3r' e^{-i\mathbf{k}'\cdot\mathbf{r}'}V(r')\chi^{(+)}(\mathbf{k},\mathbf{r}') \tag{4.112a}$$

$$\equiv -(2\pi)^2\langle\mathbf{k}'|T|\mathbf{k}\rangle. \tag{4.112b}$$

Equn. (4.112b) defines the T matrix for potential scattering. It is the operator that gives the amplitude for the transition from the initial state $|\mathbf{k}\rangle$ to the final state $|\mathbf{k}'\rangle$. It is the operator whose matrix elements are primarily calculated by scattering theory.

$$\langle\mathbf{k}'|T|\mathbf{k}\rangle = \langle\mathbf{k}'|V|\chi^{(+)}(\mathbf{k})\rangle. \tag{4.113}$$

It is useful to compare the partial-wave T-matrix elements with the other scattering parameters discussed in section 4.4.4. The partial-wave expansion of (4.112b) is

$$f(\theta) = -4\pi^2\Sigma_{LM}\langle\hat{\mathbf{k}}'|LM\rangle T_L(k)\langle LM|\hat{\mathbf{k}}\rangle$$

$$= -\pi\Sigma_L(2L+1)T_L(k)P_L(\cos\theta). \tag{4.114}$$

Comparing (4.114) with (4.88) we have

$$T_L = -\frac{1}{2i\pi k}(S_L - 1) = -\frac{1}{\pi k}e^{i\delta_L}\sin\delta_L = -\frac{1}{\pi k}C_L. \tag{4.115}$$

The momentum-space integral equation is the Lippmann–Schwinger equation. It is an equation for the T matrix. We multiply (4.101) on the left by V, take the matrix element for the eigenstate $\langle\mathbf{k}'|$ of the final momentum, and introduce the spectral representation of K.

$$\langle\mathbf{k}'|V|\chi^{(+)}(\mathbf{k})\rangle = \langle\mathbf{k}'|V|\mathbf{k}\rangle$$

$$+ \int d^3k''\langle\mathbf{k}'|V|\mathbf{k}''\rangle\frac{1}{E^{(+)} - \frac{1}{2}k''^2}\langle\mathbf{k}''|V|\chi^{(+)}(\mathbf{k})\rangle. \tag{4.116}$$

The definition (4.113) of the T matrix gives

$$\langle \mathbf{k}'|T|\mathbf{k}\rangle = \langle \mathbf{k}'|V|\mathbf{k}\rangle + \int d^3k'' \langle \mathbf{k}'|V|\mathbf{k}''\rangle \frac{1}{E^{(+)} - \frac{1}{2}k''^2} \langle \mathbf{k}''|T|\mathbf{k}\rangle. \quad (4.117a)$$

We can write (4.117a) in a representation-free operator formalism as

$$T(E^{(+)}) = V + VG_0(E^{(+)})T(E^{(+)}). \quad (4.117b)$$

Much insight into the scattering process is given by (4.117a,b). For small values of the potential T is approximated by V. This is the Born approximation. In fact the T-matrix element can be calculated knowing only the potential matrix elements, given a method of solving the integral equation. Note that k'' in (4.117a) takes all possible values including $k'' \neq k$. The T-matrix element $\langle \mathbf{k}''|T|\mathbf{k}\rangle$ is a generalisation of the corresponding quantity in (4.112b). The amplitude $\langle \mathbf{k}'|T|\mathbf{k}\rangle$ for the energy-conserving scattering process is on the energy shell, or simply on shell. $\langle \mathbf{k}''|T|\mathbf{k}\rangle$ in (4.117a) is half off shell since $\frac{1}{2}k^2 = E$, but k'' is arbitrary.

Introducing the momentum representation into the representation-free form (4.117b) defines the fully-off-shell T-matrix element $\langle \mathbf{k}''|T(E^{(+)})|\mathbf{k}'\rangle$ which comes from solving the Schrödinger equation for energy $E^{(+)}$ and forming the T-matrix element for momenta k'' and k', which are unrelated to each other and to E.

The solution of the integral equation (4.117a) is accomplished by first making a partial-wave expansion to reduce the problem to integral equations in the radial dimension. The partial-wave expansion of the T- or V-matrix element is illustrated for V.

$$\langle \mathbf{k}'|V|\mathbf{k}''\rangle = \int d^3r \langle \mathbf{k}'|\mathbf{r}\rangle V(r) \langle \mathbf{r}|\mathbf{k}''\rangle$$

$$= \int dr\, r^2 \int d\hat{\mathbf{r}}$$

$$\times \left(\frac{2}{\pi}\right)^{1/2} \frac{1}{k'r} \sum_{L'M'} i^{-L'} U_{L'}(k'r) \langle \hat{\mathbf{k}}'|L'M'\rangle \langle L'M'|\hat{\mathbf{r}}\rangle$$

$$\times V(r)$$

$$\times \left(\frac{2}{\pi}\right)^{1/2} \frac{1}{k''r} \Sigma_{LM} i^{L} U_{L}(k''r) \langle \hat{\mathbf{r}}|LM\rangle \langle LM|\hat{\mathbf{k}}''\rangle$$

$$= \Sigma_{LM} \langle \hat{\mathbf{k}}'|LM\rangle \langle k'||V_L||k''\rangle \langle LM|\hat{\mathbf{k}}''\rangle, \quad (4.118)$$

where the reduced (M-independent) potential matrix element is

$$\langle k'||V_L||k''\rangle = \frac{2}{\pi k'k''} \int dr\, U_L(k'r) V(r) U_L(k''r). \quad (4.119)$$

The integral term on the right-hand side of (4.117a) is expanded thus

$$\int dk'' k''^2 \int d\widehat{\mathbf{k}}'' \sum_{L'M'LM} \langle \widehat{\mathbf{k}}'|L'M'\rangle \langle k'||V_{L'}||k''\rangle \langle L'M'|\widehat{\mathbf{k}}''\rangle \frac{1}{E^{(+)} - \frac{1}{2}k''^2}$$

$$\times \langle \widehat{\mathbf{k}}''|LM\rangle \langle k''||T_L||k\rangle \langle LM|\widehat{\mathbf{k}}\rangle. \tag{4.120}$$

We use the orthonormality of the spherical harmonics in (4.120) and write the integral equations for the coefficients of $\langle \widehat{\mathbf{k}}'|LM\rangle \langle LM|\widehat{\mathbf{k}}\rangle$ in the expanded form of (4.117a).

$$\langle k'||T_L||k\rangle = \langle k'||V_L||k\rangle + \int dk'' k''^2 \langle k'||V_L||k''\rangle \frac{1}{E^{(+)} - \frac{1}{2}k''^2} \langle k''||T_L||k\rangle. \tag{4.121}$$

The radial integral equations (4.121) are solved for each partial wave L and the half-off-shell solutions substituted in the equivalent of (4.118) for the T matrix. The on-shell solutions are in fact the T_L of (4.115), from which the scattering amplitude and cross sections can be calculated.

In solving (4.121) we use the result for integration through a pole.

$$\int dE'' \frac{f(E'')}{E^{(+)} - E''} = P \int dE'' \frac{f(E)''}{E - E''} - i\pi f(E), \tag{4.122}$$

where P denotes the Cauchy principal value. We eliminate the numerical difficulty of the principal value integration by using the result

$$P \int dE'' \frac{1}{E - E''} = 0 \tag{4.123}$$

to subtract off the pole.

We replace the integration by a quadrature rule with points k_i and weights w_i, $i = 1, N$. We add the on-shell value of k to the set of quadrature points, calling it k_0.

$$\langle k_i||T_L||k_0\rangle = \langle k_i||V_L||k_0\rangle + \sum_{j=1}^{N} w_j [k_j^2 \langle k_i||V_L||k_j\rangle \langle k_j||T_L||k_0\rangle$$

$$- k_0^2 \langle k_i||V_L||k_0\rangle \langle k_0||T_L||k_0\rangle][\tfrac{1}{2}(k_0^2 - k_j^2)]^{-1}$$

$$- i\pi k_0 \langle k_i||V_L||k_0\rangle \langle k_0||T_L||k_0\rangle; \quad i = 0, N. \tag{4.124}$$

The computational form (4.124) of the partial-wave integral equation (4.121) is a finite matrix equation of the form

$$(1 - K)T = V, \tag{4.125}$$

which is solved by inverting the kernel matrix $\delta_{ij} - K_{ij}$ (Anderson et al., 1992). The on-shell element $\langle k_0||T_L||k_0\rangle$ of the solution vector $\langle k_i||T_L||k_0\rangle$ is the partial T-matrix element T_L of (4.115).

Equn. (4.119) for the partial-wave potential matrix element shows why only a finite number of partial-wave T-matrix elements contribute to the scattering. For very large L the centrifugal barrier means that $U_L(kr)$ is appreciably greater than zero only for values of r greater than r_0, beyond which $V(r)$ is effectively zero. Note also that there is a range of L for which $\langle k'||V_L||k''\rangle$ is so small that the Born approximation is valid

$$\langle k_0||T_L||k_0\rangle = \langle k_0||V_L||k_0\rangle. \tag{4.126}$$

For the solution of the integral equation $V(r)$ is necessarily a short-range potential. It may be complex. If $V(r)$ is of the form (4.59) the Coulomb part is included with the left-hand side of the Schrödinger equation (4.100), which becomes

$$[E^{(+)} - K - V_C]|\chi^{(+)}(\mathbf{k})\rangle = U|\chi^{(+)}(\mathbf{k})\rangle, \tag{4.127}$$

where

$$V_C(r) = -z/r. \tag{4.128}$$

The integral equation is constructed from

$$|\chi^{(+)}(\mathbf{k})\rangle = |\mathbf{k}_\eta^{(+)}\rangle + \frac{1}{E^{(+)} - K - V_C} U|\chi^{(+)}(\mathbf{k})\rangle. \tag{4.129}$$

The $U = 0$ state is now a Coulomb scattering state $|\mathbf{k}_\eta^{(+)}\rangle$ (4.85). The analysis is the same as that leading to (4.117) except that the spectral representation of the inverse differential operator now has a basis of Coulomb scattering states, which is not complete until we include the hydrogenic bound states, whose radial forms are given by (4.24).

Another development of the situation where the plane-wave representation is adequate is the physically-obvious fact that the final state is the time reversal of the initial state. It is necessary to define the T-matrix elements by

$$\langle \mathbf{k}_\eta'^{(-)}|T|\mathbf{k}_\eta^{(+)}\rangle = \langle \mathbf{k}_\eta'^{(-)}|U|\chi^{(+)}(\mathbf{k})\rangle. \tag{4.130}$$

Here the physical and time-reversed Coulomb scattering functions are denoted respectively by $|\mathbf{k}_\eta^{(\pm)}\rangle$. The time-reversal operator is given by (3.111). It is simply complex conjugation. However, we must also reverse the direction of $\mathbf{k}'$, since the time-reversed function has $\mathbf{k}'$ going towards the scattering centre.

The analogue of (4.117) is

$$\langle \mathbf{k}_\eta'^{(-)}|T|\mathbf{k}_\eta^{(+)}\rangle = \langle \mathbf{k}_\eta'^{(-)}|U|\mathbf{k}_\eta^{(+)}\rangle$$
$$+ \int d^3k'' \langle \mathbf{k}_\eta'^{(-)}|U|\mathbf{k}_\eta''^{(-)}\rangle \frac{1}{E^{(+)} - \frac{1}{2}k''^2} \langle \mathbf{k}_\eta''^{(-)}|T|\mathbf{k}_\eta^{(+)}\rangle$$
$$+ \Sigma_\lambda \langle \mathbf{k}_\eta'^{(-)}|U|\lambda\rangle \frac{1}{E^{(+)} - \epsilon_\lambda} \langle \lambda|T|\mathbf{k}_\eta^{(+)}\rangle. \tag{4.131}$$

Note that the spectral representation of the Green's function has the basis states $|k_\eta^{\prime\prime(-)}\rangle$ for two reasons. First, it is necessary to have the small quantity ϵ in the definition of the Coulomb wave negative in order to have a positive-definite small quantity in the denominator of (4.131). Second, it gives a T-matrix element in the integrand that enables us to close the integral equation. The set $|k_\eta^{\prime\prime(-)}\rangle$ is not complete without the Coulomb bound states $|\lambda\rangle$. In order to close the integral equations we need more equations

$$\langle\lambda'|T|k_\eta^{(+)}\rangle = \langle\lambda'|U|k_\eta^{(+)}\rangle + \int d^3k''\langle\lambda'|U|k_\eta^{\prime\prime(-)}\rangle\frac{1}{E^{(+)}-\frac{1}{2}k''^2}\langle k_\eta^{\prime\prime(-)}|T|k_\eta^{(+)}\rangle$$

$$+ \Sigma_\lambda\langle\lambda'|U|\lambda\rangle\frac{1}{E^{(+)}-\epsilon_\lambda}\langle\lambda|T|k_\eta^{(+)}\rangle. \tag{4.132}$$

To compute the potential matrix elements we use the partial-wave expansions of the Coulomb scattering functions in the analogue of (4.118).

$$\langle r|k_\eta^{(+)}\rangle = (2/\pi)^{1/2}(kr)^{-1}\Sigma_{LM}i^Le^{i\sigma_L}F_L(\eta,kr)\langle\hat{r}|LM\rangle\langle LM|\hat{k}\rangle$$
$$\langle r|k_\eta^{(-)}\rangle = (2/\pi)^{1/2}(kr)^{-1}\Sigma_{LM}i^Le^{-i\sigma_L}F_L^*(\eta,kr)\langle\hat{r}|LM\rangle\langle LM|\hat{k}\rangle$$
$$\langle k_\eta^{(-)}|r\rangle = (2/\pi)^{1/2}(kr)^{-1}\Sigma_{LM}i^{-L}e^{i\sigma_L}F_L(\eta,kr)\langle\hat{k}|LM\rangle\langle LM|\hat{r}\rangle.$$

$$\tag{4.133}$$

In fact only a finite number of bound-state functions are needed for convergence. The role of each bound state in the numerical solution is as another quadrature point for the appropriate value of L.

The complete T-matrix element for the full potential V is not (4.130), since for $U = 0$ we still have scattering by the Coulomb potential. We must add the T-matrix element for Coulomb scattering.

$$\langle k'|V|\chi^{(+)}(k)\rangle = \langle k'|V_C|k_\eta^{(+)}\rangle + \langle k_\eta^{\prime(-)}|T|k_\eta^{(+)}\rangle. \tag{4.134}$$

A formal derivation of (4.134) is given in chapter 6 on formal scattering theory.

4.6 Resonances

Scattering cross sections may be quite smooth functions of the incident energy but sometimes, particularly at low energy, a rapid variation is observed over a small range centred at ϵ_r and of width Γ_r, before and after which the cross section has its smooth value. In such a case it may be observed by fitting phase shifts to the cross-section data that the phase shift δ_L for a particular partial wave L is responsible for the rapid variation and the others vary smoothly. The phenomenon is called a resonance with orbital angular momentum L at energy ϵ_r with width

Γ_r. Overlapping resonances in different partial waves or even in the same partial wave may be responsible for more-extensive energy fluctuations.

We first consider a single resonance in isolation. Equn. (4.89b) shows that the contribution to the total cross section from the partial wave L (the partial cross section σ_L) has its maximum value at $E = \epsilon_r$, where

$$\delta_L(\epsilon_r) = \pi/2. \tag{4.135}$$

The partial T-matrix element (4.115) has the resonance value

$$T_L(\epsilon_r) = -\frac{i}{\pi k_r}, \tag{4.136}$$

where k_r is the momentum corresponding to ϵ_r.

The partial-wave Lippmann–Schwinger equation is (4.121). We retain the convention that k is on shell, that is

$$\tfrac{1}{2}k^2 = E. \tag{4.137}$$

According to the residue theorem applied to the k'' integral the scattering is determined by the poles of the partial T-matrix element in the complex k'' plane. The existence and positions of the poles are of course determined by the details of the potential V, but we will assume that there is a pole corresponding to complex energy $\epsilon_r - \tfrac{1}{2}i\Gamma_r$. The magnitude of the partial T-matrix element varies rapidly with values of E near the pole and we can consider ϵ_r as the resonance energy. For the cross section we need only consider the on-shell partial T-matrix element

$$T_L(E) = \langle k || T_L || k \rangle. \tag{4.138}$$

We assume that it has the form near $E = \epsilon_r$

$$T_L(E) = \frac{R}{E - \epsilon_r + \tfrac{1}{2}i\Gamma_r}, \tag{4.139}$$

at least for values of E somewhere near the pole. For physical scattering E is of course on the positive real axis.

For $E = \epsilon_r$ we can compare (4.139) with (4.136)

$$T_L(\epsilon_r) = -i\frac{2R_r}{\Gamma_r} = -i\frac{1}{\pi k_r}. \tag{4.140}$$

The residue R_r is given by

$$R_r = \frac{\Gamma_r}{2\pi k_r}. \tag{4.141}$$

The on-shell partial T-matrix element is

$$T_L(E) = \frac{1}{2\pi k_r}\frac{\Gamma_r}{E - \epsilon_r + \tfrac{1}{2}i\Gamma_r}. \tag{4.142}$$

The partial cross section is given by (4.49,4.114,4.115).

$$\sigma_L = (2L+1)\frac{\pi}{k_r^2}\frac{\Gamma_r^2}{(E-\epsilon_r)^2 + \frac{1}{4}\Gamma_r^2}. \tag{4.143}$$

This is the Breit–Wigner single resonance formula (Breit and Wigner 1936). The resonance is centred at ϵ_r. The full width at half maximum is Γ_r.

The differential cross section is given by (4.48,4.112b,4.138) and the T matrix equivalent of (4.118). The scattering amplitude $f(E,\theta)$ is

$$f(E,\theta) = -\pi\Sigma_L(2L+1)T_L(E)P_L(\cos\theta). \tag{4.144}$$

It may be written as the sum of a resonant term and a slowly-varying or direct term.

$$f(E,\theta) = -\frac{2L+1}{k_r}\frac{\frac{1}{2}\Gamma_r}{E-\epsilon_r - \frac{1}{2}i\Gamma_r}P_L(\cos\theta) - \pi\sum_{L'\neq L}(2L'+1)T_{L'}(E)P_{L'}(\cos\theta)$$

$$= f_r(E,\theta) + f_D(E,\theta). \tag{4.145}$$

The complex factor in f_r has a large negative real part near resonance for $E < \epsilon_r$ and a large positive real part for $E > \epsilon_r$. The difference between the differential cross section and its non-resonant value has a rapid sign change at resonance.

The differential cross section at 33° near an elastic resonance is illustrated in fig. 4.5 for a calculation of electron scattering by the hydrogen atom. This resonance has $L = 1$ with the electrons in a state of total spin $S = 1$.

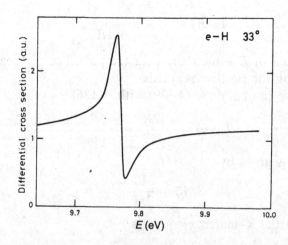

Fig. 4.5. The differential cross section at 33° near a resonance E_0=9.77 eV, Γ=8.9 $\times 10^{-3}$eV, $L = 1$, $S = 1$ in a calculation of electron–hydrogen elastic scattering.

It is interesting now to give some physical interpretation to resonances and poles of the T matrix. A resonance may be considered as a state of the compound electron–target system that we are representing in this section by our potential scattering problem. The state has a definite orbital angular momentum L. Its energy is not quite definite. The probability of finding the compound system with a certain energy E is maximum at ϵ_r but has a Lorentzian distribution (4.143) of full width at half maximum Γ_r.

We can apply the uncertainty principle, which is an approximate scattering theory, to a resonant state. Corresponding to the uncertainty Γ_r in energy it has an uncertainty τ_r in time, given by

$$\Gamma_r \tau_r \sim \hbar. \qquad (4.146)$$

The difference between a resonant state and a bound state is that a bound state has negative energy. Therefore it cannot decay. The corresponding pole is on the negative real E axis. A resonant state has enough energy to decay at least into the target system and an electron with the initial kinetic energy. The lifetime is τ_r. In a real system there may be excited target states into which decay is also allowed.

The residue theorem applied to the k'' integral in (4.121) tells us that scattering is entirely due to poles in the T matrix. Poles that are far below the positive real E axis correspond to resonances with large widths. There may be a set of poles with mean widths much larger than their mean spacing in the real E dimension. The corresponding scattering amplitude has a slow energy variation. The compound system has a correspondingly short lifetime. The minimum lifetime is the time it takes the electron to cross the target region. For a 1 eV electron this is of order 10^{-15}s. In these circumstances the scattering is called direct.

The resonance illustrated in fig. 4.5 has a width 0.008 91 eV, corresponding to a lifetime of 7.39×10^{-14}s. This is much longer than the direct lifetime, justifying the concept of a resonance as a compound state with a characteristic lifetime. The physical constants that enable these calculations to be done easily are

$$mc^2 = 0.511 \times 10^6 \text{eV} \quad \text{(mass of the electron)},$$
$$\hbar = 6.58 \times 10^{-16} \text{eV s},$$
$$c = 3 \times 10^{10} \text{cm/s},$$
$$a_0 = 0.529 \times 10^{-8} \text{cm} \quad \text{(Bohr radius of the hydrogen atom)}. \qquad (4.147)$$

4.6.1 Wave packet scattering

In this section we consider a thought experiment that can be realised in practice only in a limiting case, but gives valuable insight into the scattering process. We assume that there is some imperfect time and energy resolution in the experiment so that we write the time-dependent beam amplitude as the time development of the wave-packet form (3.53), which we rewrite for the purposes of this section as

$$\xi_0(\mathbf{r}, t) = (2\pi)^{-3/2} \int d^3k F(\mathbf{k}, \mathbf{k}_0, W) e^{i\mathbf{k}\cdot(\mathbf{r}-\mathbf{r}_0)} e^{-iEt/\hbar}. \qquad (4.148)$$

We have used the expression (3.47) for the time development of a state vector, choosing $t_0 = 0$. E is the kinetic energy corresponding to k. In this section we keep $\hbar$ and the projectile mass m explicitly in the notation for clarity.

The beam starts at the point $\mathbf{r}_0$. Its wave packet factor $F(\mathbf{k}, \mathbf{k}_0, W)$ is centred in momentum at $\mathbf{k}_0$ and characterised by width W. We assume that it travels in the direction

$$\hat{\mathbf{k}}_0 = -\hat{\mathbf{r}}_0 \qquad (4.149)$$

with velocity

$$v_0 = \hbar k_0/m. \qquad (4.150)$$

With the notation (θ_k, ϕ_k) and $(0, 0)$ for $\hat{\mathbf{k}}$ and $\hat{\mathbf{k}}_0$ respectively we choose the wave-packet factor to have the form

$$F(\mathbf{k}, \mathbf{k}_0, W) = (2\pi k^2 \sin\theta_k)^{-1} \delta(\theta_k) \Phi(k, k_0, W). \qquad (4.151)$$

With this form the wave packet is

$$\xi_0(\mathbf{r}, t) = (2\pi)^{-3/2} \int_0^\infty dk \Phi(k, k_0, W) e^{ik(r_0+r)} e^{-iEt/\hbar}. \qquad (4.152)$$

We now transform the integration variable in (4.152) from momentum k to kinetic energy E. In order that this transformation be linear we make a simplifying approximation. We drop the final term in the following expansion of E

$$E = (\hbar^2/2m)[2\mathbf{k} \cdot \mathbf{k}_0 - k_0^2 + |\mathbf{k} - \mathbf{k}_0|^2]. \qquad (4.153)$$

The nonlinear relationship due to the final term causes the wave packet to spread as it propagates. Dropping it assumes that W is so small that the detector can be placed close enough to the scattering target to neglect the spread. Note that only for a photon wave packet is E strictly proportional to k : $E = \hbar ck$. The physical situation that we will ultimately consider is that W tends to zero. In section 3.2.2 we showed that the absence of time resolution in an experiment results in the experiment being equivalent to an incoherent superposition of independent experiments, each with an incident plane wave, i.e. an incident wave packet of zero width.

We transform the integration variable in (4.152) to E using

$$k = E/\hbar v_0 + \tfrac{1}{2}k_0, \quad dk/dE = (\hbar v_0)^{-1}, \tag{4.154}$$

obtained from our approximation to (4.153). Defining the wave-packet factor in energy by

$$\Delta(E, E_0, \delta) = (\hbar v_0)^{-1}\Phi(k, k_0, W) \tag{4.155}$$

the incident wave packet becomes

$$\xi_0(\mathbf{r}, t) = (2\pi)^{-3/2}e^{ik_0(r_0+r)/2}\int_0^\infty dE\,\Delta(E, E_0, \delta)e^{iEX}, \tag{4.156}$$

where

$$X = [(r_0 + r)/v_0 - t]/\hbar. \tag{4.157}$$

There is an explicit shape for the wave-packet factor $\Delta(E, E_0, \delta)$ which leads to very easy evaluation of the integral in (4.156). This is the Lorentzian form

$$\Delta(E, E_0, \delta) = \frac{\delta/2\pi}{(E - E_0)^2 + \delta^2/4} = \frac{\delta/2\pi}{(E - E_\delta)(E - E_\delta^*)}, \tag{4.158}$$

where

$$E_\delta = E_0 - i\delta/2. \tag{4.159}$$

We extend the lower limit of integration to $-\infty$, an approximation that is justified in the ultimate limit $\delta \to 0$.

For $X < 0$ the integrand of (4.156) vanishes for large imaginary energies if we choose a contour in the lower-half E-plane, which encloses the pole at E_δ. For $X > 0$ the contour is in the upper-half E-plane enclosing the pole at E_δ^*.

$$\xi_0(\mathbf{r}, t) = (2\pi)^{-3/2}e^{ik_0(r_0+r)/2}e^{iE_\delta X}, \quad X < 0,$$
$$= -(2\pi)^{-3/2}e^{ik_0(r_0+r)/2}e^{iE_\delta^* X}, \quad X > 0. \tag{4.160}$$

Recalling the definition (4.157) of X we see that the wave packet propagates with velocity v_0. The probability of finding a beam particle at a point r is

$$|\xi_0(\mathbf{r}, t)|^2 = e^{\delta(r_0+r)/\hbar v_0}e^{-\delta t/\hbar}, \quad t > (r_0 + r)/v_0,$$
$$= e^{-\delta(r_0+r)/\hbar v_0}e^{\delta t/\hbar}, \quad t < (r_0 + r)/v_0. \tag{4.161}$$

The probability builds up exponentially in time to $t = (r_0 + r)/v_0$, after which it decays exponentially. The decay-time constant is $\tau = \hbar/\delta$. For the Lorentzian wave-packet shape (4.158) the uncertainty principle is an exact relationship if the energy uncertainty is the full width at half maximum δ and the time uncertainty is the decay time τ.

According to (4.46) the scattered wave packet $\xi(\mathbf{r}, t)$ is a linear super-position of $f(E, \theta)r^{-1}e^{ikr}$ for different k with the same coefficients as those of $e^{i\mathbf{k}\cdot\mathbf{r}}$ in (4.148). Again using (4.149,4.151,4.154,4.155,4.157) we have

$$\xi(\mathbf{r}, t) = (2\pi)^{-3/2}r^{-1}e^{ik_0(r_0+r)/2}\int_0^\infty dE\,\Delta(E, E_0, \delta)f(E, \delta)e^{iEX}. \quad (4.162)$$

From (4.162) we see again that the scattered wave depends entirely on the poles in the scattering amplitude. We consider a single pole at

$$E_r = \epsilon_r - i\Gamma_r/2 \quad (4.163)$$

for which $f(E, \theta)$ is given by (4.142,4.144).

$$f(E, \theta) = -\frac{2L+1}{k_r}\frac{\Gamma_r/2}{E - E_r}P_L(\cos\theta). \quad (4.164)$$

The scattered wave packet is obtained by using the explicit form (4.158) in (4.162).

$$\xi(\mathbf{r}, t) = -(2\pi)^{-3/2}(2L+1)(k_r r)^{-1}P_L(\cos\theta)\int_{-\infty}^\infty dE\,\frac{\delta/2\pi}{(E - E_\delta)(E - E_\delta^*)}$$

$$\times \frac{\Gamma_r/2}{E - E_r}e^{iEX}. \quad (4.165)$$

For $X > 0$ our integration contour in the upper-half E-plane encloses only the pole at E_δ^*. The scattered wave packet is proportional to $e^{iE_\delta^* X}$ as in the unscattered case (4.160). At times before the wave-packet centre reaches the scattering centre the wave packet is propagated without change of shape. Another name for this condition is causality. Disturbances due to the target appear after the centre of the wave packet has reached the target. Causality requires the scattering amplitude to be analytic in the upper-half E-plane.

Analytic properties of the scattering amplitude are better discussed in terms of the variable k rather than E, which has the square-root ambiguity. The variable in the Lippmann–Schwinger equation is k. We have used E because of its formal simplicity, but we could have made the whole argument starting with (4.152) and using k as the integration variable. We would have found that the scattering amplitude is analytic in the upper-half k-plane and the poles are on the second energy sheet.

For $X < 0$, i.e. after the centre of the wave packet has reached the target, the scattered wave packet is

$$\xi(\mathbf{r}, t) = \frac{R_{Lr}}{E_\delta - E_r}e^{iE_\delta X} - i\delta\frac{R_{Lr}}{(E_0 - \epsilon_r)^2 + \Gamma^2/4}e^{iE_r X}. \quad (4.166)$$

The residue R_{Lr} in (4.166) is obtained by comparing (4.165,4.166).

Much can be learnt from (4.166). First we can use it to interpret the resonance at E_r physically. The limit $\delta \to \infty$ corresponds to a sudden excitation of the target in comparison with times characteristic of the target, in this case

$$\tau_r = \hbar/\Gamma_r. \tag{4.167}$$

The time spectrum of the scattered electron in this limit has the shape $e^{-\Gamma_r t/\hbar}$. The lifetime of the compound electron–target system in the resonant state is τ_r. We have derived the same result for a detailed scattering theory that we knew already from the uncertainty principle. The uncertainty relation (4.146) becomes an exact equality (4.167) if Γ_r is the full width at half maximum and τ_r is the lifetime.

The differential cross section, obtained by integrating $|\xi(\mathbf{r}, t)|^2$, given by (4.166), over time, is

$$\frac{d\sigma}{d\Omega} = \frac{R_{Lr} R_{Lr}^* [(E_0 - \epsilon_r)^2 \Gamma_r/2 + (\Gamma_r + \delta)^2 (\Gamma_r/2 + \delta)/4]}{2\Gamma_r [(E_0 - \epsilon_r)^2 + (\Gamma_r + \delta)^2/4]^2}. \tag{4.168}$$

In the usual scattering experiment $\delta \to 0$ and (4.168) reduces to the Breit–Wigner form (4.143). Equn. (4.168) also tells us under what conditions the wave-packet width is significant in the experiment. We must have experimental time resolution $\hbar/\delta$ such that δ is comparable to the resonance width Γ_r. The width in this case is the sum of Γ_r and δ.

While we have developed the theory of wave-packet scattering and resonances in the context of potential scattering of electrons it is easy to generalise. In particular there is no reason why the scattered particle should not be a photon. In this case the wave packet does not spread and the formalism is valid for general values of δ. Wave packets are known whose widths correspond to a lifetime of order 10^{-7}s, which is easily resolved with nanosecond electronics. Such wave packets arise in the photon decay of many atomic states. The time spectrum of detected photons is given by $|\xi(\mathbf{r}, t)|^2$ for $X < 0$. We see from (4.166) that this involves an interference between a term whose lifetime is $\hbar/\delta$ and one whose lifetime is τ_r. The resulting time oscillations have been observed experimentally. They are called quantum beats.

4.7 Relativistic electron in a local, central potential

In the previous part of the chapter we expressed the problem of an electron in a local, central potential in terms of radial equations and eigenstates of orbital angular momentum. In generalising to the case where the electron obeys the Dirac equation (3.154) we remember that spin and orbital angular momentum are coupled.

The spin–orbital eigenstates (3.92) are conveniently written as eigenstates of an operator k.

$$k = \boldsymbol{\sigma} \cdot \mathbf{L} + 1. \tag{4.169}$$

Since

$$\boldsymbol{\sigma} \cdot \mathbf{L} = J^2 - L^2 - S^2, \tag{4.170}$$

the eigenvalues of k are

$$\kappa = \pm(j + \tfrac{1}{2}), \quad \text{for} \quad \ell = j \pm \tfrac{1}{2}. \tag{4.171}$$

It is useful to use the notation $\bar{\ell}$.

$$\ell + \bar{\ell} = 2j. \tag{4.172}$$

ℓ and $\bar{\ell}$ differ by 1. The eigenstates of k are written in the coordinate–spin representation as

$$\chi_{\kappa m}(\hat{\mathbf{r}}, \boldsymbol{\sigma}) = \langle \hat{\mathbf{r}} \boldsymbol{\sigma} | \ell j m \rangle, \tag{4.173}$$

where

$$\chi_{\kappa m}(\hat{\mathbf{r}}, \boldsymbol{\sigma}) = \sum_{\mu\nu} \langle \ell \tfrac{1}{2} \mu\nu | jm \rangle Y_{\ell\mu}(\hat{\mathbf{r}}) \chi_\nu^{1/2}(\boldsymbol{\sigma}),$$

$$\chi_{-\kappa m}(\hat{\mathbf{r}}, \boldsymbol{\sigma}) = \sum_{\mu\nu} \langle \bar{\ell} \tfrac{1}{2} \mu\nu | jm \rangle Y_{\bar{\ell}\mu}(\hat{\mathbf{r}}) \chi_\nu^{1/2}(\boldsymbol{\sigma}). \tag{4.174}$$

Besides the eigenvalue equation for k

$$k\chi_{\kappa m} = -\kappa\chi_{\kappa m} \tag{4.175}$$

we have the relationship

$$\boldsymbol{\sigma} \cdot \hat{\mathbf{r}} \chi_{\kappa m} = -\chi_{-\kappa m}. \tag{4.176}$$

To understand (4.176) we observe that $\boldsymbol{\sigma} \cdot \hat{\mathbf{r}}$ is a pseudoscalar. It does not change j and m since it is scalar under rotation, but it is odd under inversion so it changes the parity. We see from (4.172, 4.174) that $\chi_{-\kappa m}$ has the opposite parity to $\chi_{\kappa m}$ since the corresponding values of ℓ differ by 1. Our understanding is completed by applying the operator $(\boldsymbol{\sigma} \cdot \hat{\mathbf{r}})^2 = 1$ to $\chi_{\kappa m}$.

We are now in a position to express the Dirac Hamiltonian in terms of radial operators and a four-component spin–orbital operator K defined by

$$K = \begin{pmatrix} k & 0 \\ 0 & -k \end{pmatrix} \tag{4.177}$$

$$= \beta(\boldsymbol{\Sigma} \cdot \mathbf{L} + 1). \tag{4.178}$$

We recall the definitions (3.158, 3.163) of α, β and Σ. Making use of (3.32, 3.59, 3.157) and the vector operator identity

$$\nabla = \hat{\mathbf{r}}\frac{\partial}{\partial r} - i\frac{\hat{\mathbf{r}}}{r} \times \mathbf{L} \qquad (4.179)$$

we find

$$\boldsymbol{\alpha} \cdot \mathbf{p} = -i(\boldsymbol{\alpha} \cdot \hat{\mathbf{r}})\left[\frac{\partial}{\partial r} + \frac{1}{r}(1 - \beta K)\right] \qquad (4.180)$$

and write the free-electron Hamiltonian (3.153) as

$$H = -ic(\boldsymbol{\alpha} \cdot \hat{\mathbf{r}})\left[\frac{\partial}{\partial r} + \frac{1}{r}(1 - \beta K)\right] + \beta c^2. \qquad (4.181)$$

The one-electron state $|\psi\rangle$ is written in the coordinate–spin representation as two two-component spinors.

$$\langle \mathbf{r}\boldsymbol{\sigma}|\psi\rangle = r^{-1}\begin{pmatrix} g(r)\chi_{\kappa m}(\hat{\mathbf{r}}, \boldsymbol{\sigma}) \\ if(r)\chi_{-\kappa m}(\hat{\mathbf{r}}, \boldsymbol{\sigma}) \end{pmatrix}. \qquad (4.182)$$

$|\psi\rangle$ is a simultaneous eigenstate of $H + V(r)$ and K. The latter is seen by applying (4.175, 4.177) to (4.182), obtaining

$$K|\psi\rangle = -\kappa|\psi\rangle. \qquad (4.183)$$

The Dirac equation in two-component form is

$$\left[E - V + ic\begin{pmatrix} 0 & \boldsymbol{\sigma} \cdot \hat{\mathbf{r}} \\ \boldsymbol{\sigma} \cdot \hat{\mathbf{r}} & 0 \end{pmatrix}\frac{d}{dr} - ic\begin{pmatrix} 0 & -\boldsymbol{\sigma} \cdot \hat{\mathbf{r}} \\ \boldsymbol{\sigma} \cdot \hat{\mathbf{r}} & 0 \end{pmatrix}\frac{K}{r} + \begin{pmatrix} 0 & 0 \\ 0 & 2 \end{pmatrix}c^2\right]$$

$$\times \begin{pmatrix} g(r)\chi_{\kappa m} \\ if(r)\chi_{-\kappa m} \end{pmatrix} = 0. \qquad (4.184)$$

Applying (4.176, 4.183) to (4.184) we have the two coupled equations

$$(E - V)g(r) + c\left(\frac{d}{dr} - \frac{\kappa}{r}\right)f(r) = 0,$$

$$(E - V + 2c^2)f(r) - c\left(\frac{d}{dr} + \frac{\kappa}{r}\right)g(r) = 0. \qquad (4.185)$$

The coupled radial equations (4.185) are the relativistic analogue of (4.19) for bound states and (4.57) for scattering states. In order to set up partial-wave integral equations corresponding to (4.121) we need the partial-wave form of the free-electron state (3.170). This is set up by generalising (4.56) to include the spin and using it in the partial-wave expansion of (3.170), which becomes a four-component spinor.

$$\langle \mathbf{r}\boldsymbol{\sigma}|\nu\mathbf{p}\rangle = \frac{(2/\pi)^{1/2}}{pr}\Sigma_{L\mu}i^L U_L(pr)Y_{L\mu}^*(\hat{\mathbf{r}})\begin{pmatrix} Y_{L\mu}(\hat{\mathbf{p}})\chi_\nu^{1/2}(\boldsymbol{\sigma}) \\ \omega\boldsymbol{\sigma} \cdot \hat{\mathbf{p}}Y_{L\mu}(\hat{\mathbf{p}})\chi_\nu^{1/2}(\boldsymbol{\sigma}) \end{pmatrix}, \qquad (4.186)$$

where

$$\omega = \frac{cp}{E + 2c^2}. \tag{4.187}$$

Inverting the momentum–spin form of (4.174) and using the momentum form of (4.176) yields

$$\langle \mathbf{r}\sigma | v\mathbf{p} \rangle = \frac{(2/\pi)^{1/2}}{pr} \sum_{LJM\mu} i^L U_L(pr) Y^*_{L\mu}(\hat{\mathbf{r}}) \langle L\tfrac{1}{2}\mu v | JM \rangle$$

$$\times \begin{pmatrix} \chi_{\kappa M}(\hat{\mathbf{p}}, \boldsymbol{\sigma}) \\ -\omega \chi_{-\kappa M}(\hat{\mathbf{p}}, \boldsymbol{\sigma}) \end{pmatrix}. \tag{4.188}$$

The integral equations for relativistic potential scattering are conveniently written in terms of a four-dimensional notation for the four-component spinor $|v\mathbf{p}\rangle$.

$$|v\mathbf{p}\rangle \equiv |\mathbf{p}_\alpha\rangle, \quad \alpha = 1, 4. \tag{4.189}$$

The Dirac equation (3.159) is written in analogy to (4.100) as

$$(E + c^2 - H)|\chi^{(+)}_\beta(\mathbf{p})\rangle = V|\chi^{(+)}_\beta(\mathbf{p})\rangle, \tag{4.190}$$

where H is given by (3.153). The T-matrix element is

$$\langle \mathbf{p}'_\alpha | T | \mathbf{p}_\beta \rangle = \langle \mathbf{p}'_\alpha | V | \chi^{(+)}_\beta(\mathbf{p}) \rangle. \tag{4.191}$$

The coupled integral equations analogous to (4.117) are

$$\langle \mathbf{p}'_\alpha | T | \mathbf{p}_\beta \rangle = \langle \mathbf{p}'_\alpha | V | \mathbf{p}_\beta \rangle + \Sigma_\gamma \int d^3q \langle \mathbf{p}'_\alpha | V | \mathbf{q}_\gamma \rangle \frac{1}{E^{(+)} - E(q)} \langle \mathbf{q}_\gamma | T | \mathbf{p}_\beta \rangle, \tag{4.192}$$

where $E(q)$ is the relativistic kinetic energy for momentum q. The partial-wave form of the integral equations is obtained from (4.192) by generalising (4.118) using (4.188).

5

Theory of atomic bound states

The problem of N bound electrons interacting under the Coulomb attraction of a single nucleus is the basis of the extensive field of atomic spectroscopy. For many years experimental information about the bound eigenstates of an atom or ion was obtained mainly from the photons emitted after random excitations by collisions in a gas. Energy-level differences are measured very accurately. We also have experimental data for the transition rates (oscillator strengths) of the photons from many transitions. Photon spectroscopy has the advantage that the photon interacts relatively weakly with the atom so that the emission mechanism is described very accurately by first-order perturbation theory. One disadvantage is that the accessibility of states to observation is restricted by the dipole selection rule.

Photon spectroscopy associates two numbers with the pair of states involved in a transition, the energy-level difference and the transition rate. The correlated emission directions of photons in successive transitions are determined trivially by the dipole selection rule. In most cases it is impossible to solve the many-body problem accurately enough to reproduce spectroscopic data within experimental error and we are left wondering how good our theoretical methods really are.

Because our description of differential cross sections for momentum transfer in a reaction initiated by an electron beam depends on our ability to describe both the structure and the reaction mechanism, scattering provides much more information about bound states. This is even more true of ionisation. The information is less accurate than from photon spectroscopy and is obtained only after a thorough understanding of reactions, the subject of this book, is achieved. The understanding of structure and reactions is of course achieved iteratively. A theoretical description of a reaction is completely tested only when we know the structure of the relevant target states with accuracy that is at least commensurate with that of the reaction calculation. The hydrogen atom is the prototype target from this point of view, although important relativistic effects can be tested only in larger atoms.

We are concerned here with the use of the theoretical descriptions of atomic eigenstates in the calculation of a reaction. It is necessary to know in principle how the structure calculations are done and to know the detail of the different forms that can be adapted to reaction calculations.

5.1 The Hartree—Fock problem

Many-body structure calculations are done in terms of one-electron states $|\alpha\rangle$, which we call 'orbitals' to distinguish them from the states of the N-electron system. One-electron states are discussed in chapter 4. The simplest states in the N-electron space are independent-particle configurations $|\rho\rangle$ whose coordinate—spin representation consists of antisymmetric products (determinants) of orbitals. The coordinate—spin representation of a normalised configuration $|\rho\rangle$ is

$$\langle \mathbf{r}_1\sigma_1\cdots\mathbf{r}_N\sigma_N|\rho\rangle = (N!)^{-1/2}\det\langle\mathbf{r}_i\sigma_i|\alpha\rangle. \tag{5.1}$$

All the eigenstates $|f_{\ell j}\rangle$ of an atomic Hamiltonian H that have the same values of the quantum numbers j (total angular momentum) and ℓ (parity), collectively called the symmetry, belong to the symmetry manifold ℓj. The use of the quantum number ℓ to represent the parity is related to its use to represent the total orbital angular momentum if spin—orbit coupling is neglected. In general, atomic states are not eigenstates of total orbital angular momentum but if it is even/odd their parity is even/odd. Useful values of ℓ are 0/1 for even/odd parity. If we consider the solution of the structure problem as the diagonalisation of the Hamiltonian in a matrix representation using a basis of states with various symmetries, the problem reduces to independent problems for each symmetry manifold, since states with different symmetries are orthogonal. We consider a particular symmetry manifold and omit the indices ℓj.

The Hartree—Fock problem in its simplest form (Hartree, 1927; Fock, 1930) consists in finding the best orbitals $|\alpha\rangle$ so that the configuration $|\rho\rangle$ approximates as closely as possible the lowest-energy eigenstate of H in the symmetry manifold ℓj. This is done using the variation theorem.

Consider a state $|f\rangle$ in the space spanned by the eigenstates of a Hamiltonian H, which belong to the symmetry manifold ℓj. If the variational energy

$$E = \frac{\langle f|H|f\rangle}{\langle f|f\rangle} \tag{5.2}$$

is a minimum, then $|f\rangle$ is an eigenstate of H belonging to the minimum eigenvalue. The theorem is proved by considering the minimum value E_0

of E.

$$E_0 = \frac{\langle f_0|H|f_0\rangle}{\langle f_0|f_0\rangle}. \tag{5.3}$$

If $|f_0\rangle$ is varied by adding $\pm\epsilon|g\rangle$, where $|g\rangle$ is a state in the same space and ϵ is a small positive number, then E_0 changes to $E_0 + \delta E_0$, where

$$\delta E_0 \geq 0. \tag{5.4}$$

Substituting $|f_0\rangle \pm \epsilon|g\rangle$ for $|f\rangle$ in (5.2), the inequality (5.4) becomes

$$\mp\epsilon(\langle g|E_0 - H|f_0\rangle + \langle f_0|E_0 - H|g\rangle \pm \epsilon\langle g|E_0 - H|g\rangle) \geq 0. \tag{5.5}$$

Since E_0 is the minimum value of E,

$$\langle g|E_0 - H|g\rangle \leq 0. \tag{5.6}$$

Using the Hermitian property of H we have

$$\epsilon\langle g|E_0 - H|g\rangle \leq 2\langle g|E_0 - H|f_0\rangle \leq -\epsilon\langle g|E_0 - H|g\rangle. \tag{5.7}$$

Since $|g\rangle$ is arbitrary we choose

$$|g\rangle = (E_0 - H)|f_0\rangle. \tag{5.8}$$

Allowing ϵ to tend to zero we find

$$\langle f_0|(E_0 - H)(E_0 - H)|f_0\rangle = 0 \tag{5.9}$$

and hence

$$(E_0 - H)|f_0\rangle = 0. \tag{5.10}$$

The theorem is the basis of the variational method of approximating the lowest eigenstate of a particular symmetry manifold. We choose a trial form of $|f\rangle$, which is varied to minimise $\langle f|H|f\rangle$ with the constraint that $\langle f|f\rangle = 1$. The form of $|f\rangle$ that gives the minimum is the best approximation.

We now apply the variational method to the N-electron problem. In order to ensure antisymmetry we express the N-electron Hamiltonian in the occupation-number representation.

$$H = \sum_{\alpha\beta} a_\alpha^\dagger a_\beta \langle\alpha|t|\beta\rangle + \tfrac{1}{2}\sum_{\alpha\beta\gamma\delta} a_\alpha^\dagger a_\beta^\dagger a_\gamma a_\delta \langle\alpha\beta|v|\gamma\delta\rangle, \tag{5.11}$$

where t is the Hamiltonian $K + V$ for the interaction of one electron with the nucleus and v is the electron–electron potential.

The best configuration $|\rho\rangle$ of the form (5.1) is found variationally. We vary $|\rho\rangle$ by adding $\epsilon|\delta\rho\rangle$, where ϵ is a small positive number. The problem is to choose orbitals $|\alpha\rangle$ such that

$$\langle\delta\rho|H|\rho\rangle = 0. \tag{5.12}$$

The configuration $|\delta\rho\rangle$ is chosen by annihilating an electron in an occupied orbital $|\eta\rangle$ ($\eta \leq N$) of $|\rho\rangle$ and creating one in an unoccupied orbital $|\xi\rangle$ ($\xi > N$). We choose the set of orbitals to be orthonormal. This ensures that $\langle\delta\rho|\rho\rangle$ is zero so that $|\delta\rho\rangle$ is normalised to first order in ϵ.

$$|\delta\rho\rangle = \epsilon a_\xi^\dagger a_\eta |\rho\rangle. \tag{5.13}$$

Using (5.11,5.13) the condition (5.12) becomes

$$\sum_{\alpha\beta} \langle\rho|a_\eta^\dagger a_\xi a_\alpha^\dagger a_\beta|\rho\rangle\langle\alpha|t|\beta\rangle$$

$$+ \tfrac{1}{2}\sum_{\alpha\beta\gamma\delta} \langle\rho|a_\eta^\dagger a_\xi a_\alpha^\dagger a_\beta^\dagger a_\gamma a_\delta|\rho\rangle\langle\alpha\beta|v|\gamma\delta\rangle = 0. \tag{5.14}$$

We evaluate the matrix elements of (5.14) using the commutation rules (3.137–3.139), remembering that $\xi > N$ and $\eta \leq N$.

A systematic way of evaluating matrix elements of a string of creation and annihilation operators is to choose an operator that gives zero when operating on the ket and move it to the right until it operates on the ket. In the case of (5.14) we move the annihilation operator a_ξ, which gives zero when operating on $|\rho\rangle$ since the orbital $|\xi\rangle$ is unoccupied in $|\rho\rangle$. Each time we exchange a_ξ with an operator $a_\zeta^\dagger$ we produce a new term involving $\pm\delta_{\xi\zeta}$. A term resulting from an odd/even number of exchanges is positive/negative.

The variational condition (5.14) becomes

$$\langle\xi|t|\eta\rangle + \tfrac{1}{2}\sum_{\zeta=1}^N (\langle\xi\zeta|v|\zeta\eta\rangle - \langle\xi\zeta|v|\eta\zeta\rangle - \langle\zeta\xi|v|\zeta\eta\rangle + \langle\zeta\xi|v|\eta\zeta\rangle)$$
$$= 0, \ \xi > N, \eta \leq N. \tag{5.15}$$

We now define a one-electron operator, the Hartree–Fock Hamiltonian H_{HF}.

$$H_{HF} = \sum_{\xi\eta}[\langle\xi|t|\eta\rangle + \sum_{\zeta=1}^N(\langle\xi\zeta|v|\zeta\eta\rangle - \langle\xi\zeta|v|\eta\zeta\rangle)]a_\xi^\dagger a_\eta. \tag{5.16}$$

In obtaining (5.16) from (5.15) we have used the fact that we can exchange the orbitals in both the bra and ket of a two-electron matrix element, which represents an integration over dummy coordinate–spin variables. H_{HF} is a one-electron operator since we sum over orbitals $|\zeta\rangle$. Equation (5.16) shows that we can diagonalise H_{HF} in the space of occupied orbitals only, since it has no matrix elements connecting occupied and unoccupied orbitals. Performing the diagonalisation we have

$$\langle\xi|t|\eta\rangle + \sum_{\zeta=1}^N(\langle\xi\zeta|v|\zeta\eta\rangle - \langle\xi\zeta|v|\eta\zeta\rangle) = \epsilon_\eta\langle\xi|\eta\rangle, \ \xi,\eta \leq N. \tag{5.17}$$

The ϵ_η are eigenvalues of H_{HF}. We interpret them as the one-electron energies of the orbitals $|\eta\rangle$ in the independent-particle model.

An alternative way of understanding (5.17) is that it results from the minimisation of $\langle \rho|H|\rho \rangle$ subject to the constraints

$$\langle \eta|\eta \rangle = 1 \quad \text{for all } |\eta\rangle. \tag{5.18}$$

The ϵ_η are the Lagrange multipliers for the constraints.

It is important that the orbitals $|\eta\rangle$ that satisfy (5.17) are orthogonal for different eigenvalues ϵ_η. We can use them to construct an orthonormal set with which to express the many-electron problem. We have assumed orthogonality in deriving (5.17). We now show that this is consistent by considering (5.17) as an equation for the matrix elements formed for the bra orbital $\langle \xi|$ from a Schrödinger equation for $|\eta\rangle$. The set of such equations for N different $|\eta\rangle$ is called the Hartree–Fock equations.

$$(\epsilon_\eta - t)|\eta\rangle = \tilde{V}|\eta\rangle, \tag{5.19}$$

where the Hartree–Fock potential is

$$\tilde{V}|\eta\rangle = \sum_{\zeta=1}^{N} \langle \zeta|v(1-P)|\zeta\rangle|\eta\rangle. \tag{5.20}$$

The operator P exchanges the electrons in the orbitals $|\zeta\rangle$ and $|\eta\rangle$. The Hermitian conjugate equation of (5.19), replacing $|\eta\rangle$ by $|\xi\rangle$, is

$$\langle \xi|(\epsilon_\xi - t) = \langle \xi|\tilde{V}. \tag{5.21}$$

This equation is true because P may operate either on the bra or ket vectors of a two-electron matrix element with the same result since only a redefinition of dummy integration coordinates is involved. From (5.19,5.21) we form the matrix equations

$$\langle \xi|\epsilon_\eta - t|\eta\rangle = \langle \xi|\tilde{V}|\eta\rangle,$$
$$\langle \xi|\epsilon_\xi - t|\eta\rangle = \langle \xi|\tilde{V}|\eta\rangle, \tag{5.22}$$

which are subtracted to yield the desired result

$$(\epsilon_\xi - \epsilon_\eta)\langle \xi|\eta\rangle = 0. \tag{5.23}$$

Equn. (5.17) is a formal statement of the Hartree–Fock problem in the simplest, single determinant, form. We must find orbitals that satisfy the equation.

A configuration in the independent-particle model may be of either the closed-shell or open-shell type. In the former the N electrons occupy all the orbitals of the lowest-energy sets with the same symmetry and principal quantum number n, called shells. In the latter some orbitals with particular values of the projection quantum numbers are unoccupied. The symmetry of a state that is either a closed shell or has one electron in an open shell outside a closed-shell core (such as in the simplest model of an alkali-metal atom) is sufficiently determined by a single configuration. For more-general symmetries linear combinations of configurations with different sets of projection quantum numbers are necessary to give the symmetry. We will not discuss these cases, since we are interested in the Hartree–Fock problem as a means of generating orbitals to obtain a basis of configurations for diagonalising the Hamiltonian, rather than as a model for open-shell states.

Hartree–Fock calculations are done using the coordinate–spin representation for the orbitals. For the relativistic Hamiltonian this is written in two-component form (4.182) as

$$\phi_{n\kappa m}(x) = r^{-1} \begin{pmatrix} g_{n\kappa}(r)\chi_{\kappa m}(\hat{\mathbf{r}}, \sigma) \\ if_{n\kappa}(r)\chi_{-\kappa m}(\hat{\mathbf{r}}, \sigma) \end{pmatrix}, \tag{5.24}$$

where x stands for the set of position and spin coordinates $\mathbf{r}$, σ. The radial functions are called radial orbitals. If it is sufficient to use a nonrelativistic Hamiltonian with a spin–orbit potential (3.174), we use the large component of (5.24) in the jj-coupling description. It may be appropriate to use the orbital- and total-angular-momentum quantum numbers ℓ and j, which are equivalent to κ (4.171). If the spin plays no part except in the Pauli exclusion principle it is more economical to use LS coupling, for which the one-electron function is

$$\phi_{n\ell m\nu}(x) = r^{-1}u_{n\ell}(r)Y_{\ell m}(\hat{\mathbf{r}})\chi_{\nu}^{1/2}(\sigma). \tag{5.25}$$

The object of solving the Hartree–Fock equations is to determine a set of orbitals that are self-consistent in the sense that they are solutions of the set of equations (5.19) when the same orbitals are used to construct the potential (5.20), called the self-consistent field.

5.2 Numerically-specified orbitals

The coordinate–spin representation of the Hartree–Fock equation (5.19) is

$$(\epsilon_{\eta} - t)\phi_{\eta}(x) = \int dx' \tilde{V}(x', x)\phi_{\eta}(x'). \tag{5.26}$$

This is an extension of the Schrödinger equation for a bound state to include the nonlocal potential

$$\tilde{V}(x', x) = \sum_{\zeta}\left[\int dx''\phi_{\zeta}^{\dagger}(x'')v(x'', x')\phi_{\zeta}(x'')\delta(x' - x) - \phi_{\zeta}^{\dagger}(x')v(x', x)\right.$$

$$\left. \times \phi_{\zeta}(x)\right]. \tag{5.27}$$

The notations for integration and the corresponding δ-function refer to the set x of coordinate—spin variables. The first term of (5.27) is the direct potential due to the screening of the nucleus for the electron in orbital $|\eta\rangle$ by the other electrons. The second term is the exchange potential. Note that the term for $|\zeta\rangle = |\eta\rangle$ cancels between the direct and exchange potentials. The potential $v(x',x)$ is the two-electron Coulomb potential $|\mathbf{r}' - \mathbf{r}|^{-1}$. In the relativistic case there are additional magnetic terms $-\boldsymbol{\alpha}' \cdot \boldsymbol{\alpha}|\mathbf{r}' - \mathbf{r}|^{-1}$ that are small and make little difference. The relativistic treatment of electron—electron interaction can only be approximate. It is discussed by Brown (1952).

5.2.1 Nonrelativistic Hamiltonian

For the nonrelativistic case with neglect of spin—orbit coupling we separate the space and spin parts of the coordinate—spin representation of the orbital

$$\phi_{n\ell m\nu}(x) = \psi_{n\ell m}(\mathbf{r})\chi_\nu^{1/2}(\sigma). \tag{5.28}$$

The spin functions vanish from the formalism due to their orthonormality. The exchange term however has a factor $\langle v_\zeta | v_\eta \rangle$ restricting its effect to pairs of electrons with the same spin projection. Equn. (5.26) becomes an integrodifferential equation in coordinate space, which is reduced to an equation in the radial variable by the methods of sections 3.3 and 4.3. The coordinate-space Hartree—Fock equation for a closed-shell structure is

$$(\epsilon_\eta + \tfrac{1}{2}\nabla^2 + Z/r)\psi_\eta(\mathbf{r})$$
$$= \sum_\zeta \int d^3r' \Big[2\psi_\zeta^*(\mathbf{r}')|\mathbf{r}' - \mathbf{r}|^{-1}\psi_\zeta(\mathbf{r}')\psi_\eta(\mathbf{r}) - \psi_\zeta^*(\mathbf{r}')|\mathbf{r}' - \mathbf{r}|^{-1}$$

$$\times \psi_\zeta(\mathbf{r})\psi_\eta(\mathbf{r}') \Big], \tag{5.29}$$

where Z is the charge of the nucleus. The factor 2 in the screening potential is due to the fact that there are two spin states (singlet and triplet) for each pair of electrons with position-space quantum numbers n, ℓ, m.

We do not pursue the reduction of (5.29) to radial integrodifferential equations, beyond noting that it is analogous to (4.10). The angular-momentum reduction of two-electron matrix elements is given essentially by (3.109). Two-electron matrix elements are treated fully for scattering in chapter 7. A numerical method for solving integrodifferential equations is given by Sams and Kouri (1969). Details of the numerical Hartree—Fock calculation are given, for example, by Froese-Fischer (1977). The resultant numerical quantities used in electron—atom reaction calculations are the radial orbitals $u_{n\ell}(r)$ of (5.25), which are specified by the computation on a radial grid. A computer program for Hartree–Fock calculations is part of the basic equipment of an atomic theory group. A good example is the program of Chernysheva, Cherepkov and Radojevic (1976).

5.2.2 *Relativistic Hamiltonian*

The Hartree–Fock problem with the Dirac Hamiltonian (3.153) is called Dirac–Fock. The coordinate–spin representation of the orbital $|\eta\rangle$ is

$$\phi_\eta(x) = \begin{pmatrix} P_\eta(x) \\ Q_\eta(x) \end{pmatrix}. \tag{5.30}$$

This is substituted into (5.19) to obtain

$$
(\epsilon_\eta - t)\begin{pmatrix} P_\eta(x) \\ Q_\eta(x) \end{pmatrix} = \sum_\zeta \int dx' \left(P_\zeta^\dagger(x') \quad Q_\zeta^\dagger(x') \right) v(x', x) \begin{pmatrix} P_\zeta(x') \\ Q_\zeta(x') \end{pmatrix} \begin{pmatrix} P_\eta(x) \\ Q_\eta(x) \end{pmatrix}
$$
$$
- \sum_\zeta \int dx' \left(P_\zeta^\dagger(x') \quad Q_\zeta^\dagger(x') \right) v(x', x) \begin{pmatrix} P_\zeta(x) \\ Q_\zeta(x) \end{pmatrix} \begin{pmatrix} P_\eta(x') \\ Q_\eta(x') \end{pmatrix}. \tag{5.31}
$$

We substitute the Dirac Hamiltonian (3.153) into (5.31), obtaining

$$
[\epsilon_\eta - V_N(x) - V(x)]P_\eta(x) + \int dx' V_{PP}(x', x)P_\eta(x')
$$
$$
+ \int dx' V_{QP}(x', x)Q_\eta(x') - ic\boldsymbol{\sigma} \cdot \boldsymbol{\nabla} Q_\eta(x) = 0,
$$
$$
[\epsilon_\eta + 2c^2 - V_N(x) - V(x)]Q_\eta(x) + \int dx' V_{PQ}(x', x)P_\eta(x')
$$
$$
+ \int dx' V_{QQ}(x', x)Q_\eta(x') - ic\boldsymbol{\sigma} \cdot \boldsymbol{\nabla} P_\eta(x) = 0. \tag{5.32}
$$

Here the local screening potential is

$$V(x) = \sum_\zeta \int dx' \left[P_\zeta^\dagger(x')v(x', x)P_\zeta(x') + Q_\zeta^\dagger(x')v(x', x)Q_\zeta(x') \right]. \tag{5.33}$$

There are four nonlocal exchange potentials

$$V_{PP}(x', x) = \sum_\zeta P_\zeta^\dagger(x')v(x', x)P_\zeta(x),$$

$$V_{QP}(x', x) = \sum_\zeta Q_\zeta^\dagger(x')v(x', x)P_\zeta(x),$$

$$V_{PQ}(x', x) = \sum_\zeta P_\zeta^\dagger(x')v(x', x)Q_\zeta(x),$$

$$V_{QQ}(x', x) = \sum_\zeta Q_\zeta^\dagger(x')v(x', x)Q_\zeta(x). \tag{5.34}$$

The electron—nucleus potential is not quite trivial. For larger atoms the radial functions that are large at the nucleus are affected by the finite charge distribution of the nucleus. It is sufficient to use the potential for a uniform charge distribution of radius R

$$V_N(x) = -\frac{Z}{2R}\left(3 - \frac{r^2}{R^2}\right), \quad r \le R,$$
$$= -\frac{Z}{r}, \qquad\qquad r > R. \tag{5.35}$$

Equns. (5.32) are reduced to coupled integrodifferential equations in the radial variable by techniques analogous to the derivation of (4.184) with the use of the angular-momentum algebra of section 3.3, which will be applied in detail to two-electron matrix elements in chapter 7. Details of the Dirac—Fock theory are given by Grant (1970). The resulting quantities used in electron—atom reaction theory are the large- and small-component radial orbitals $g_{n\kappa}(r)$ and $f_{n\kappa}(r)$ of (5.24), which are specified by the computation on a numerical radial grid.

5.3 Analytic orbitals

The extension of the matrix solution of section 4.3 for one-electron bound states to the Hartree—Fock problem has many advantages. It results in radial orbitals specified as linear combinations of analytic functions, usually normalised Slater-type orbitals (4.38). This is a very convenient form for the computation of potential matrix elements in reaction theory. The method has been described by Roothaan (1960) for a closed-shell or single-open-shell structure.

We illustrate the method by applying it to the simplest closed-shell form (5.29) of the Hartree—Fock problem. The radial orbitals $u_{n\ell}(r)$ are expressed as a linear combination of basis radial orbitals $f_{i\ell}(r)$.

$$u_{n\ell}(r) = \sum_i c_{i\ell n} f_{i\ell}(r). \tag{5.36}$$

The basis one-electron wave functions are

$$\chi_{i\ell m}(\mathbf{r}) = r^{-1} f_{i\ell}(r) Y_{\ell m}(\hat{\mathbf{r}}). \tag{5.37}$$

We define a matrix notation, omitting explicit reference to the coordinate **r**.

$$\psi_{n\ell m} = \sum_i \chi_{i\ell m} c_{i\ell n} \equiv \mathbf{\chi}_{\ell m} \mathbf{c}_{\ell n}, \tag{5.38}$$

where $\mathbf{\chi}_{\ell m}$ is a row vector whose components are all the basis functions $\chi_{i\ell m}$, and $\mathbf{c}_{\ell n}$ is a column vector whose components are the corresponding coefficients $c_{i\ell n}$. The overlap matrix $\mathbf{S}_\ell$ is defined by

$$S_{\ell ij} = N_\ell^{-1} \sum_m \langle \chi_{i\ell m} | \chi_{j\ell m} \rangle, \tag{5.39}$$

where N_ℓ is half the number of electrons in the shell ℓ.

$$N_\ell = 2\ell + 1. \tag{5.40}$$

The vectors $\mathbf{c}_{\ell n}$ are normalised by

$$\mathbf{c}_{\ell n'}^\dagger \mathbf{S}_\ell \mathbf{c}_{\ell n} = N_\ell \delta_{n'n}. \tag{5.41}$$

The matrices $\mathbf{t}_\ell$ for the bare-nucleus Hamiltonian t are defined by

$$t_{\ell ij} = N^{-1} \Sigma_m \langle \chi_{i\ell m} | t | \chi_{j\ell m} \rangle. \tag{5.42}$$

In order to take into account the two-electron interaction terms we define a supermatrix notation that we illustrate for $\mathbf{t}_\ell$. The components of a supervector $\mathbf{t}$ are the matrix elements $t_{\ell ij}$ in dictionary order. Supervectors are transformed by supermatrices. We need the supermatrix $\mathbf{V}$, defined by

$$V_{\ell'i'j',\ell ij} = (N_{\ell'}N_\ell)^{-1} \sum_{m'm} \int d^3r' \int d^3r \chi_{i'\ell'm'}^*(\mathbf{r})\chi_{i\ell m}^*(\mathbf{r}')|\mathbf{r}' - \mathbf{r}|^{-1}$$
$$\times \left[2\chi_{j'\ell'm'}(\mathbf{r}')\chi_{j\ell m}(\mathbf{r}) - \chi_{j'\ell'm'}(\mathbf{r})\chi_{j\ell m}(\mathbf{r}') \right]. \tag{5.43}$$

We apply the variational method to the total energy in order to obtain linear equations for the vectors $\mathbf{c}_{\ell n}$. From a set of trial vectors $\mathbf{c}_{\ell n}$ that satisfy the normalisation constraint (5.41) we compute the density matrix

$$\mathbf{D}_\ell = \Sigma_n \mathbf{c}_{\ell n} \mathbf{c}_{\ell n}^\dagger \tag{5.44}$$

and form the supermatrix $\mathbf{D}$, analogous to $\mathbf{V}$. The potential supermatrix $\mathbf{P}$ is defined by

$$\mathbf{P} = \mathbf{V}\mathbf{D}. \tag{5.45}$$

We now revert to considering the supervectors as a collection of matrices for ℓ. The eigenvalue equations for each ℓ, corresponding to (5.29), are

$$(\epsilon_{\ell n}\mathbf{S}_\ell - \mathbf{t}_\ell - \mathbf{P}_\ell)\mathbf{c}_{\ell n} = 0. \tag{5.46}$$

These equations are solved, yielding a new set of trial vectors. The process is repeated until the new set agrees with the previous set to a prescribed limit of accuracy. In this way we achieve a self-consistent solution.

The simple closed-shell problem is generalised to open-shell cases by Roothaan (1960) and Roothaan and Bagus (1963). The present formalism gives an introduction to the generalisations.

If the basis radial orbitals $f_{i\ell}(r)$ are Slater-type orbitals (4.38), each is characterised by nonlinear parameters, an exponent $\zeta_{i\ell}$ and a power

$n_{i\ell}$. These are chosen to span the space required to describe the occupied states. The choice can be optimised to lower the total energy E, but this is a laborious process requiring a complete Hartree—Fock calculation for each variation. An extensive investigation of this has resulted in the tables of Clementi and Roetti (1974) for the low-lying states of neutral atoms, positive ions and isoelectronic series of ions up to $Z=54$. These eigenvectors have been very sensitively verified by the (e,2e) reaction (chapter 11) and form an excellent start for structure calculations.

It is of course possible to solve the Dirac—Fock problem with a linear combination of analytic orbitals. However, owing to the rapid variation of the orbitals near the nucleus it requires an awkwardly-large basis. If an analytic representation is convenient for a reaction calculation it may be obtained by a least-squares fit to a numerical orbital.

5.4 Frozen-core Hartree—Fock calculations

Hartree—Fock calculations may be performed to find sets of orbitals describing the lowest-lying states of different symmetry manifolds of an atom. It is found that each different state has a closed-shell core whose orbitals are closely independent of the state.

A frozen-core calculation involves choosing a particular state (for example the one lowest in energy), performing a Hartree—Fock calculation to find the best orbitals, then using the orbitals of the core to generate a nonlocal potential (5.27), which is taken to represent the core in calculations of further states.

A good example is provided by the alkali-metal atoms, which consist of one electron outside a closed-shell core in the single-configuration model. If the frozen-core approximation is valid a frozen-core calculation of the orbital occupied by one electron will give the same result as a Hartree—Fock calculation and the core orbitals will not depend on the state.

Table 5.1 illustrates the frozen-core approximation for the case of sodium using a simple Slater (4.38) basis in the analytic-orbital representation. The core ($1s^2\ 2s^2\ 2p^6$) is first calculated by Hartree—Fock for the state characterised by the $3s$ one-electron orbital, which we call the $3s$ state. The frozen-core calculation for the $3p$ state uses the same core orbitals and solves the $3p$ one-electron problem in the nonlocal potential (5.27) of the core. Comparison with the core and $3p$ orbitals from a $3p$ Hartree—Fock calculation illustrates the approximation. The overwhelming component of the $3p$ orbital agrees to almost five significant figures.

The main use of the frozen-core approximation is to generate an orthonormal set of orbitals for use as a basis in structure or reaction

Table 5.1. *Comparison of a frozen-core (FC) calculation for the sodium 3p orbital with a Hartree—Fock (HF) calculation of the same state. The basis column gives the parameters $n_{i\ell}$ and $\zeta_{i\ell}$ of the basis Slater orbitals (4.38). The other columns give the coefficients $c_{i\ell n}$ (5.36). The frozen core is the 3s Hartree—Fock core*

S-BASIS		1s(FC)	1s(HF)	2s(FC)	2s(HF)
1s	12.584 30	-0.375 898 51	-0.375 885 65	-0.023 176 36	-0.023 078 14
1s	9.438 84	-0.631 352 16	-0.631 354 66	0.329 536 58	0.329 265 09
2s	3.859 28	0.001 277 65	0.001 233 73	-0.767 448 66	-0.766 253 48
2s	2.394 34	-0.002 406 05	-0.002 378 09	-0.327 025 79	-0.328 350 68
3s	1.252 77	0.000 840 05	0.000 832 00	0.013 388 39	0.013 815 58
3s	0.746 08	-0.000 402 36	-0.000 399 05	-0.005 085 80	-0.005 014 98

P-BASIS		2p(FC)	2p(HF)	3p(FC)	3p(HF)
2p	5.477 23	0.335 386 93	0.336 135 70	-0.034 383 71	-0.031 534 00
2p	2.562 67	0.734 614 11	0.734 664 86	-0.061 018 47	-0.055 104 17
3p	0.600 21	0.008 238 16	0.000 076 73	1.002 827 21	1.002 869 53

calculations, rather than as a model for an atomic state. The example of table 5.1 is rather extreme, since the 3s and 3p states are orthogonal because of symmetry and independent Hartree—Fock calculations are appropriate.

5.5 Multiconfiguration Hartree—Fock

Up to this stage we have been concerned with the use of the variational method for calculating atomic orbitals to provide an independent-particle model that is as realistic as possible. We now turn to a variational method of approximating the lower-energy eigenstates in the spectrum of an atom.

The variational foundation for the approximation is understood by varying the state $|f\rangle$ in the N-electron space by $\epsilon|g\rangle$, where ϵ is a small positive number. The variational energy E (5.2) becomes $E + \delta E$. We consider only quantities of first order in the small quantities ϵ and δE.

$$\langle f|f\rangle^2 \delta E = \epsilon \langle f|f\rangle (\langle f|H|g\rangle + \langle g|H|f\rangle) - \epsilon \langle f|H|f\rangle (\langle f|g\rangle + \langle g|f\rangle) + O(\epsilon^2). \tag{5.47}$$

Since H is Hermitian, any state $|i\rangle$ that makes δE zero to first order satisfies the eigenvalue equation for H

$$(\epsilon_i - H)|i\rangle = 0. \tag{5.48}$$

Conversely any eigenstate of H makes the first-order variation δE of E zero.

The eigenstate $|i\rangle$ is characterised by the quantum numbers n, ℓ, j, m. The total-angular-momentum quantum number is j and its projection is m. The parity is denoted by a number ℓ that is 0 or 1 for even or odd parity respectively. The principal quantum number n specifies which state of the ℓj manifold we are concerned with. If spin—orbit coupling is neglected it is convenient to use n, ℓ, s, where ℓ and s are the total-orbital-angular-momentum and spin quantum numbers, and the corresponding projections. We use the generally-valid total-angular-momentum and parity specification in general discussions.

The multiconfiguration Hartree—Fock procedure is concerned with a particular symmetry manifold ℓj. It is therefore necessary to specify an eigenstate only by the principal quantum number n. The eigenstate $|n\rangle$ is expanded in a set of N_r symmetry configurations $|r\rangle$ that belong to the same manifold. That is they are eigenstates of parity and total angular momentum with quantum numbers ℓ, j, m.

$$|n\rangle = \Sigma_r c_{nr} |r\rangle. \tag{5.49}$$

The symmetry configurations are linear combinations of single determinant configurations $|\rho\rangle$ that have the required quantum numbers. They are formed by coupling the angular momenta of the configurations. Normally all symmetry configurations are formed from a common set of orbitals.

The computational procedure is to solve the equations that give $\delta E = 0$, subject to the constraints

$$\langle r'|r\rangle = \delta_{r'r},$$
$$\Sigma_r c_{nr}^2 = 1, \tag{5.50}$$

which are represented by Lagrange multipliers. The eigenstate $|n\rangle$ is represented in coordinate—spin space by a set of orbitals, defined by (5.24) or (5.25), and the coefficients $c_{nr}, r = 1, N_r$.

An optimal-level calculation optimises on the variational energy of a particular eigenstate $|n\rangle$. The calculation must be repeated for each eigenstate. Faster computations, in which individual eigenstates are not completely optimised, involve optimising on different linear combinations of the variational energies for the different eigenstates.

For the relativistic Hamiltonian the procedure is called multiconfiguration Dirac—Fock. A computer program for structure calculations in this approximation has been described by Grant *et al.* (1980). The non-relativistic procedure has been described by Froese-Fischer (1977) and implemented by the same author (Froese-Fischer, 1978).

5.6 Configuration interaction

The coordinate–spin representation of the states $|n\ell jm\rangle$ of an N-electron atom or ion can be expanded in an M-dimensional linear combination of single-determinant configurations $|\rho_k\rangle$ (5.1). This is the configuration-interaction expansion. The orbitals $|\alpha\rangle$ forming the determinants are represented as orthonormal square-integrable functions $\phi_\alpha(x)$.

We denote the atomic eigenstate $|n\ell jm\rangle$ by $|i\rangle$ and the configuration-interaction expansion by

$$|i\rangle = \Sigma_k |\rho_k\rangle\langle\rho_k|i\rangle. \tag{5.51}$$

The configuration-interaction approximation to $|i\rangle$ results from diagonalising the atomic Hamiltonian H in the M-dimensional basis $|\rho_k\rangle$.

$$\sum_{k'k} \langle i'|\rho_{k'}\rangle\langle\rho_{k'}|\epsilon_i - H|\rho_k\rangle\langle\rho_k|i\rangle = 0. \tag{5.52}$$

If the states $|i'\rangle$ and $|i\rangle$ belong to different symmetry manifolds, characterised by the quantum numbers ℓ, j, then the Hamiltonian matrix element is zero. It is economical to consider the diagonalisation in a particular symmetry manifold and we will begin our discussion in this way. The basis states $|r_k\rangle$ are now symmetry configurations consisting of linear combinations of configurations which have the symmetry ℓj of the manifold.

5.6.1 The hydrogen atom

In one sense the hydrogen atom is a trivial case since the symmetry configurations are one-orbital determinants and in any case the exact eigenstates are known. However, we use it to illustrate the answer to a nontrivial question. How well can the lower-energy eigenstates of an atomic system be represented by an M_ℓ-dimensional square-integrable basis for each symmetry manifold? We remember that a complete set of atomic states includes the ionisation continuum.

The diagonalisation problem for the symmetry manifold ℓ (neglecting spin–orbit coupling) is

$$\langle i'|\bar{\epsilon}_i - H|i\rangle = 0, \tag{5.53}$$

where

$$\langle \mathbf{r}|i\rangle = \langle \mathbf{r}|\bar{n}\ell m\rangle = r^{-1}u_{\bar{n}\ell}(r)Y_{\ell m}(\hat{\mathbf{r}}) \tag{5.54}$$

and

$$u_{\bar{n}\ell}(r) = \sum_{k=1}^{M_\ell} C_{\bar{n}\ell k}\xi_{k\ell}(r). \tag{5.55}$$

We choose the Laguerre basis

$$\xi_{k\ell}(r) = \left[\frac{\lambda(k-1)!}{(2\ell+1+k)!} \right]^{1/2} (\lambda r)^{\ell+1} \exp(-\lambda r/2) L_{k-1}^{2\ell+2}(\lambda r), \tag{5.56}$$

where the associated Laguerre polynomial $L_{k-1}^{2\ell+2}(\lambda r)$ is defined by (4.26). We choose $\lambda = 2$ so that $\xi_{10}(r)$ is the radial orbital $u_{10}(r)$ of the ground state (4.27). The radial orbitals $u_{\bar{n}\ell}(r)$ are known as the Sturmians (Rotenberg, 1962). They are sometimes called 'pseudostates'.

The eigenvalue $\bar{\epsilon}_i$ of the M_ℓ-dimensional diagonalisation problem is of course not equal to any of the hydrogen bound-state eigenvalues ϵ_i in general, but for the states of lower energy it comes closer as M_ℓ is increased. We use the closeness of $\bar{\epsilon}_i$ to the corresponding ϵ_i as a measure of the quality of the approximation to the eigenstate of hydrogen.

Table 5.2 shows the number of hydrogen states for each ℓ, represented by the principal quantum number n of the highest state, whose eigenvalues are given by the Sturmian expansion of dimension M within different relative-error tolerances δ. (It turns out that the dimension of the required expansion is independent of ℓ.) The table confirms that the expansion of the lower atomic states in a square-integrable basis converges with the basis size and gives an idea of the rapidity of convergence.

5.6.2 The Hartree–Fock basis

It is sensible to choose the basis orbitals $|\alpha\rangle$ that form the basis configurations in an optimal way. The symmetry configuration that most-closely approximates the lowest-energy state of the ℓj manifold is the Hartree–Fock configuration $|r_0\rangle$, in which the lowest-energy orbitals are occupied. The basis orbitals include ones that are unoccupied in $|r_0\rangle$. They may be calculated as eigenstates of the Hartree–Fock equation (5.26).

Table 5.2. *The highest principal quantum number for an eigenstate of hydrogen whose eigenvalue is approximated within the relative-error tolerance δ by a Sturmian expansion of dimension M (adapted from Bray, Konovalov and McCarthy, 1991a)*

M	$\delta = 10^{-14}$	$\delta = 10^{-7}$	$\delta = 10^{-4}$
7	1	1	2
10	1	2	–
20	2	3	–
100	6	7	8

The Hartree—Fock basis has a particular significance. The eigenstate $|n\rangle$ of the ℓj manifold is represented by the configuration sum

$$|n\rangle = \Sigma_k |r_k\rangle\langle r_k|n\rangle. \tag{5.57}$$

We sort the terms of (5.57) into partial sums of configurations involving μ one-electron excitations, that is μ electrons are annihilated in $|r_0\rangle$ and μ electrons are created in unoccupied orbitals. The partial sum for μ in $|n\rangle$ is denoted $|n_\mu\rangle$ and the expansion (5.57) becomes

$$|n\rangle = \Sigma_\mu |n_\mu\rangle. \tag{5.58}$$

The Hamiltonian matrix elements are

$$\langle n'|H|n\rangle = \sum_{\mu'\mu}\langle n'_{\mu'}|H|n_\mu\rangle, \tag{5.59}$$

where

$$|n_0\rangle = |n'_0\rangle = |r_0\rangle. \tag{5.60}$$

The Hartree—Fock basis is defined by the relations (5.12,5.13), which become in the present notation

$$\langle r_0|H|n_1\rangle = \langle n'_1|H|r_0\rangle = 0. \tag{5.61}$$

Therefore there are no terms of first order in the number $\mu' + \mu$ of excitations in the representation (5.59) of the Hamiltonian.

5.6.3 Practical calculations, natural orbitals

The configuration-interaction representation of the lower-energy states of an atom is the N-electron analogue of the Sturmians in the hydrogen-atom problem. We choose an orbital basis of dimension P, form from them a subset of all possible N-electron determinants $|\rho_k\rangle, k = 0, M_P$, and use these determinants as a basis for diagonalising the N-electron Hamiltonian. It may be convenient first to form symmetry configurations $|r_k\rangle$ from the $|\rho_k\rangle$.

There are many ways of obtaining an orthonormal set of orbitals. The solution of the Hartree—Fock equations (5.26) for the ground state produces self-consistent occupied orbitals and a set of unoccupied orbitals into which electrons are excited to form the basic configurations $|r_k\rangle, k \neq 0$. The unoccupied orbitals are not optimised in any sense for the description of excited states and in fact provide a bad description since they have boundary conditions that result in their extension too far from the main matter distribution in coordinate space.

The unoccupied orbitals that give the best convergence of the configuration-interaction problem with increasing number M of configurations are the natural orbitals, first discussed by Löwdin (1955). They are obtained by considering the quantity

$$\gamma_{\mu\nu}(0,0) = \langle 0|a_\mu^\dagger a_\nu|0\rangle. \tag{5.62}$$

This is the one-electron density matrix in orbital space for the ground state $|0\rangle$. For the present purpose we call it the density matrix. It is manifestly Hermitian. Therefore it is possible to find a unitary transformation U of the orbitals $|\mu\rangle$ that diagonalises it. The new orbitals $|\alpha\rangle$ are the natural orbitals

$$|\alpha\rangle = U|\mu\rangle, \tag{5.63}$$

for which

$$\gamma_{\alpha\beta}(0,0) = \langle 0|a_\alpha^\dagger a_\alpha|0\rangle \delta_{\alpha\beta}. \tag{5.64}$$

In practice it is necessary to truncate the configuration basis $|r_k\rangle$ to a dimension $M < M_P$. Satisfactory results are achieved by including all single and double excitations to unoccupied orbitals $|\alpha\rangle$ from the highest-energy major shell in the Hartree–Fock configuration. Examples of LS-coupled Hartree–Fock configurations $|\rho_0\rangle$ are

$$\begin{aligned}
\text{magnesium} &: \quad 1s^2 2s^2 2p^6 3s^2, \\
\text{argon} &: \quad 1s^2 2s^2 2p^6 3s^2 3p^6.
\end{aligned} \tag{5.65}$$

In these cases we consider configuration-interaction calculations that allow excitations from the $n = 3$ orbitals. Electrons involved in the excitations are called active electrons.

Satisfactory convergence is achieved by using the natural orbitals for the $(ns)^2$ electron pair in the highest-energy major Hartree–Fock shell. These orbitals are found by modelling the atomic ground state as a two-electron pair in the frozen-orbital self-consistent field of the other electrons. The ground state is a spin-singlet, $\ell = 0$, even-parity state, denoted $^1S^e$ (the symmetry nomenclature is described in section 5.8). The two-electron basis configurations are denoted for orbitals $|m\rangle$, $|n\rangle$ in the ℓ_α manifold by $|mn\rangle_{\ell_\alpha}$. The configuration-interaction expansion of the ground state is

$$|0\rangle = \sum_{\ell_\alpha} \left[\sum_{mn} c_{mn} |mn\rangle_{\ell_\alpha} \right]. \tag{5.66}$$

We consider the sum for each ℓ_α separately, since orbitals for different ℓ_α are orthogonal, and drop the subscript ℓ_α temporarily.

The density matrix is

$$\gamma_{\alpha\beta}(0,0) = \sum_{m'n'mn} c_{m'n'} c_{mn} \langle m'n'|a_\alpha^\dagger a_\beta|nm\rangle. \tag{5.67}$$

The result of annihilating the electrons in different orbitals α, β is

$$\alpha = m, \beta = m : \; \gamma_{\alpha\beta}(0,0) = \sum_{n'n} c_{mn'} c_{mn} \langle n'|n \rangle = \sum_{n} c_{mn}^2,$$

$$\alpha = m, \beta = n : \; \gamma_{\alpha\beta}(0,0) = \sum_{n'm} c_{mn'} c_{mn} \langle n'|m \rangle = \sum_{m} c_{mm} c_{mn}. \quad (5.68)$$

Therefore if the orbitals are chosen so that the off-diagonal density-matrix elements are zero, then c_{mn} is diagonal and the natural-orbital expansion is

$$|0\rangle = \sum_{\ell_\alpha} \left[\sum_{\overline{m}} c_{\overline{m}\,\overline{m}} |\overline{m}\,\overline{m}\rangle_{\ell_\alpha} \right]. \quad (5.69)$$

Natural orbitals are indicated by a bar. There are considerably fewer configurations in (5.69) than in (5.66).

While natural orbitals chosen in this way do not diagonalise the density matrix simultaneously for all electron pairs in the major shell or for excited atomic eigenstates $|i\rangle$, it is found in practice that they occupy the part of coordinate space one would expect of excited one-electron states in a singly-charged potential, off-diagonal coefficients c_{mn} are small, and the configuration-interaction calculation is sufficiently convergent in the dimension M of the configuration basis. The natural orbitals for one-electron states that are occupied in $|r_0\rangle$ are very similar to the Hartree–Fock orbitals.

In finding the natural orbitals numerically the coefficients c_{mn} must be known. They are of course found by diagonalising the Hamiltonian in a configuration basis of dimension M, so nothing has been gained. In practice the natural orbitals are found for a smaller configuration basis and the most important of these are used in the full-scale calculation.

5.6.4 Methods of diagonalisation

Much economy in the size of the Hamiltonian matrices to be diagonalised is achieved by diagonalising independent blocks of the same symmetry. Here the basis configurations are the symmetry configurations $|r_k\rangle$. They are formed from the determinants $|\rho_k\rangle$ by angular-momentum coupling techniques (Racah, 1942), assisted by the coefficients of fractional parentage for atomic $n\ell j$ shells. For a reasonably-large number of active electrons and shells over which they are distributed the coupling becomes very complicated and the computational algebra proliferates. A computer program for configuration-interaction calculations using a symmetry-configuration basis has been described by Hibbert (1975).

An alternative approach that is well-adapted to modern computing is simply to use a basis of determinants $|\rho_k\rangle$ of the same parity. Since determinants are eigenstates of J_z (or L_z and S_z) as well as J^2 (or L^2 and

S^2) this is called the *m*-scheme. The diagonalisation of the Hamiltonian of course generates eigenstates of definite symmetry. The inefficiency is due to the fact that matrix elements connecting basis states of different symmetry are zero so that the Hamiltonian matrix is sparse. However, the lower-energy eigenstates of large, sparse matrices may be calculated using the Lanczos (1950) algorithm, which can handle matrices of dimension $M=50$ 000 quite easily. This technique for configuration-interaction calculations was pioneered by Whitehead *et al.* (1977) for nuclear structure and adapted for atomic structure by Mitroy (Mitroy, Amos and Morrison, 1979; Mitroy, 1983). It removes the need for calculating angular-momentum coupling coefficients, which are absorbed in the eigenvector elements.

5.7 Perturbation theory

While the multiconfiguration methods lead to large and accurate descriptions of atomic states, formal insight that can lead to a productive understanding of structure-related reaction problems can be obtained from first-order perturbation theory. We consider the atomic states as perturbed frozen-orbital Hartree–Fock states. It is shown in chapter 11 on electron momentum spectroscopy that the perturbation is quite small, so it is sensible to consider the first order. Here the term 'Hartree–Fock' is used to describe the procedure for obtaining the unperturbed determinantal configurations $|\rho_k\rangle$. The orbitals may be those obtained from a Hartree–Fock calculation of the ground state. A refinement would be to use natural orbitals.

The Hartree–Fock problem is written as

$$(\tilde{E}_k - K)|\rho_k\rangle = 0. \tag{5.70}$$

The Hartree–Fock Hamiltonian is written as a sum over electrons with coordinate–spin labels s.

$$K = \Sigma_s(K_s + V_s + U_s), \tag{5.71}$$

where K_s is the one-electron kinetic energy, V_s is the electron–nucleus potential and U_s is the (possibly refined) Hartree–Fock potential for the ground state.

The N-electron Hamiltonian is

$$H = K + V, \tag{5.72}$$

where we subtract the Hartree–Fock potential from the two-electron potential to obtain

$$V = \sum_{s<t} v_{st} - \sum_s U_s \equiv \sum_{s<t} \tilde{v}_{st}. \tag{5.73}$$

The symmetric two-electron residual potential is defined by

$$\tilde{v}_{st} = v_{st} - (U_s + U_t)/(N - 1). \tag{5.74}$$

The Schrödinger equation for the atomic state $|i\rangle$ is written in the form

$$(\tilde{E}_i - K)|i\rangle = (\tilde{E}_i - E_i + V)|i\rangle. \tag{5.75}$$

If the perturbation V is reduced to zero then E_i is the same as the Hartree–Fock energy $\tilde{E}_i$. This is used to define the label i for a Hartree–Fock configuration $|\rho_i\rangle$. It is the configuration of the same symmetry as $|i\rangle$ whose energy $\tilde{E}_i$ is nearest to E_i. The residual electron–electron potential V splits the Hartree–Fock energy levels so that there is more than one atomic state for every Hartree–Fock state.

We define projection operators P_i and Q_i by

$$P_i|i\rangle = |\rho_i\rangle, \tag{5.76}$$

$$P_i + Q_i = 1. \tag{5.77}$$

This gives the state $|i\rangle$ in the form

$$
\begin{aligned}
|i\rangle &= |\rho_i\rangle + Q_i|i\rangle \\
&= |\rho_i\rangle + \sum_{k \neq i} |\rho_k\rangle\langle\rho_k|i\rangle.
\end{aligned} \tag{5.78}
$$

We multiply the Schrödinger equation (5.75) on the left by the inverse of the Hartree–Fock operator to obtain

$$|i\rangle = \frac{1}{\tilde{E}_i - K}(\tilde{E}_i - E_i + V)|i\rangle. \tag{5.79}$$

This is substituted into (5.78) to obtain, with the aid of the function theorem (3.19),

$$|i\rangle = |\rho_i\rangle + \sum_{k \neq i} |\rho_k\rangle \frac{1}{\tilde{E}_i - \tilde{E}_k}\langle\rho_k|\tilde{E}_i - E_i + V|i\rangle. \tag{5.80}$$

Equn. (5.80) can be formed into a set of equations to be solved for $\langle\rho_k|\tilde{E}_i - E_i + V|i\rangle$ in analogy to (4.101,4.116), but a close approximation is given by the first iteration, which we write using the second-quantised form (3.149) of the symmetric two-electron operator V as

$$|i\rangle = |\rho_i\rangle + \sum_{k \neq i} |\rho_k\rangle \sum_{\alpha\beta\gamma\delta} \frac{\langle\alpha\beta|\tilde{v}|\gamma\delta\rangle}{\epsilon_\gamma + \epsilon_\delta - \epsilon_\alpha - \epsilon_\beta}\langle\rho_k|a_\alpha a_\beta a_\gamma^\dagger a_\delta^\dagger|\rho_i\rangle. \tag{5.81}$$

In practice we eliminate the problem of the possible degeneracy of one-electron orbitals with different angular-momentum projections by summing over the angular-momentum states to isolate the reduced matrix elements. This is simple in relevant special cases.

It is significant that the relevant two-electron matrix elements are those for double excitations, which are of the same order in the number of excitations as those usually considered in the configuration basis of a configuration-interaction calculation.

5.8 Comparison with spectroscopic data

Since the description of a reaction is based on the description of the atomic states involved, it is important to see how well our structure methods match the data of photon spectroscopy. The symmetry of an atomic state for which LS coupling is a good approximation is denoted

$$^{2S+1}X_j^\pi,$$

where X is a capital letter of the old spectroscopic notation $S, P, D, ..$ denoting the total orbital angular momentum ℓ, the left superscript $2S+1$ is the spin multiplicity, the right superscript π is either e or o, indicating even or odd parity, and the right subscript j is the total angular momentum. The indices π and/or j are sometimes omitted when no confusion results. The state is sometimes further identified by a principal quantum number n placed before the symmetry symbols. If spin—orbit coupling is significant we use the notation

$$j\pm,$$

where j is the total angular momentum and the parity is indicated by $\pm$.

Table 5.3. *One-electron separation energies for the lower-energy states of sodium (units eV). Experimental data (EXP) are from Moore (1949). The calculations are FCHF, frozen-core Hartree—Fock; and POL, frozen-core Hartree—Fock with the phenomenological core-polarisation potential (5.82)*

State	EXP	FCHF	POL
3^2S	5.139	4.956	5.140
4^2S	1.948	1.910	1.947
5^2S	1.024	1.009	1.022
3^2P	3.038	2.980	3.040
4^2P	1.387	1.370	1.387
5^2P	0.795	0.787	0.795
3^2D	1.523	1.515	1.522
4^2D	0.856	0.852	0.856
4^2F	0.851	0.850	0.851

For atoms with more than a few electrons, practicable calculations involve freezing some of the orbitals to form a frozen core. Sodium is a prototype for this approximation since the calculation of the state of one electron with a frozen $1s^2 2s^2 2p^6$ core does not require configuration interaction. Table 5.3 shows the one-electron separation energy calculated in the Hartree–Fock approximation for the $3^2 S$ ground state (where the active electron is in the $3s$ orbital) and in the frozen-core Hartree–Fock approximation for the excited states. Quite large errors occur in comparison with experiment for the lower states. These must be due to the neglect of excitations from the core (core polarisation). It will be shown for scattering in section 7.5.4 that polarisation may be described by an attractive potential whose long-range form is r^{-4}. We therefore show the separation energies obtained by including a phenomenological potential of the form (Zhou *et al.*, 1990)

$$V_{\text{pol}}(r) = -\frac{\alpha}{2r^4}\left[1 - e^{-(r/r_0)^6}\right], \quad \alpha = 0.99, r_0 = 1.439, \quad (5.82)$$

where all quantities are expressed in atomic units. We expect the inclusion of this potential to give a structure model that provides a good test of reaction calculations.

Magnesium requires at least a two-electron configuration-interaction calculation with a frozen $1s^2 2s^2 2p^6$ core. We give a detailed example of a large calculation (Mitroy, 1983) and the results of other calculations to show how well structure calculations can describe spectroscopic data.

The calculation of Mitroy started by calculating the Hartree–Fock approximation to the ground state $|3s^2 \, ^1S^e\rangle$, where we denote the states by the orbitals of the two active electrons in the configuration with the largest coefficient, in addition to the symmetry notation. The calculation used the analytic method with the basis set of Clementi and Roetti (1974) augmented by further Slater-type orbitals in order to give flexibility for the description of unoccupied orbitals. The total energy calculated by this method was $-199.614\ 61$, which should be compared with the result of a numerical Hartree–Fock calculation, $-199.614\ 64$.

The $n=1$ and 2 shells were frozen at the ground-state Hartree–Fock values. The orbital set included the $\overline{4s}, \overline{3p}, \overline{3d}, \overline{4f}$ and $\overline{5g}$ natural orbitals and $3p, 3d, 4s, 4p, 4d, 4f, 5s, 5p, 5d, 5f$ orbitals from frozen-core Hartree–Fock calculations to provide representations for states whose dominant configuration is $|3s \, n\ell\rangle$. This set was again augmented by extra *ad hoc* orbitals to increase flexibility. The full set contained 24 orbitals (6 s-type, 7 p-type, 6 d-type, 3 f-type, 2 g-type) which were all orthogonalised using the prescription for two orbitals $|a\rangle$ and $|b\rangle$

$$\langle b \left[|a\rangle - |b\rangle\langle b|a\rangle \right] = 0. \quad (5.83)$$

We first consider the ground state. Table 5.4 shows the coefficients of some of the more-important configurations. This gives an idea of the importance of correlations in lower-energy atomic states.

Table 5.5 illustrates the accuracy available for energy calculations by showing the one-electron separation energy calculated in the detailed example (Mitroy, 1983), the multiconfiguration Hartree–Fock method of Froese-Fischer (1979) and the natural-orbital configuration-interaction method of Weiss (1967), compared with experimental values. The status of structure calculations is shown, for example, by the ground state where calculations agree within a factor 10^{-3} but differ from experiment by 0.11 eV, an energy comparable with the resolution of a scattering experiment.

In addition to energy levels, transition rates are available experimentally for pairs of states where the transition satisfies the dipole selection rule. The status of atomic spectroscopy is illustrated in table 5.6 for the lowest

Table 5.4. *Expansion coefficients of the dominant configurations for the ground state of magnesium in the calculation of Mitroy (1983)*

Configuration	Coefficient
$3s^2$	0.9646
$3s\overline{4s}$	0.0271
$\overline{4s}^2$	-0.0405
$\overline{3p}^2$	0.2563
$\overline{3d}^2$	-0.0280

Table 5.5. *The one-electron separation energies of the lower-energy states of magnesium. EXAMPLE, Mitroy (1983); MCHF, Froese-Fischer (1979); NOCI, Weiss (1967); EXP, experimental data compiled by Moore (1949)*

STATE	EXAMPLE	MCHF	NOCI	EXP
$3s^2\,^1S^e$	0.276 83	0.276 60	0.2765	0.280 99
$3s3p\,^3P^o$	0.181 38	0.181 34	0.1815	0.181 25
$3s3p\,^1P^o$	0.119 61	0.119 49	0.1192	0.121 28
$3s4s\,^3S^e$	0.092 49	0.092 53	0.0924	0.093 28
$3s4s\,^1S^e$	0.082 18	0.082 13	0.0821	0.082 78
$3s3d\,^1D^e$	0.070 64	0.071 13	0.0668	0.069 56
$3s3d\,^3D^e$	0.062 25	0.062 61	0.0622	0.062 48

Table 5.6. *Transition rate (optical oscillator strength) for the lowest two dipole transitions in magnesium. EXP: experimental data of Kelly and Mathur (1980)[1], Liljeby et al. (1980)[2], Lundin et al. (1973)[3], Lurio (1964)[4], Smith and Gallagher (1966)[5], Smith and Liszt (1971)[6], Mitchell (1975)[7]. EXAMPLE: Mitroy (1983). MCHF: Froese-Fischer (1979). NOCI: Weiss (1974)*

Upper state	Lower state	EXP		EXAMPLE	MCHF	NOCI
$3s3p\,^1P$	$3s^2\,^1S$	1.83	±0.08 [1]	1.74	1.76	1.75
		1.83	±0.18 [2]			
		1.75	±0.07 [3]			
		1.85	±0.07 [4]			
		1.80	±0.05 [5]			
		1.86	±0.03 [6]			
$3s4p\,^1P$	$3s^2\,^1S$	0.19	±0.02 [1]	0.104	0.117	–
		0.102	±0.22 [2]			
		0.18	±0.04 [6]			
		0.109	±0.008 [7]			

two dipole transitions in magnesium. Not only do experiments sometimes disagree within the quoted errors but there may be significant differences between structure calculations. The calculated transition rate is sensitive to the representation of the states. Here it is calculated by the dipole-length method where the transition strength is

$$S = \sum_{m'm} |\langle n'\ell' j'm'| \sum_{k=1}^{N} L_k |n\ell jm\rangle|^2. \tag{5.84}$$

The primed quantum numbers represent the final state. The coordinate representation of the length operator L_k for the kth electron is

$$L_k = \mathbf{r}_k. \tag{5.85}$$

In summary, structure calculations can obtain 1 or 2% agreement with accurate optical data. A broader perspective is given in chapter 11 by electron momentum spectroscopy. Hartree–Fock calculations agree with one-electron momentum densities within experimental error, but configuration-interaction calculations agree only qualitatively with detailed data on correlations.

6

Formal scattering theory

We have considered the measurement of observables in electron–atom collisions and the description of the structure of the target and residual atomic states. We are now in a position to develop the formal theory of the reaction mechanism. Our understanding of potential scattering serves as a useful example of the concepts involved.

Reactions are understood in terms of channels. A channel is a quantum state of the projectile–target system when the projectile and target are so far apart that they do not interact. It is specified by the incident energy and spin projection of the projectile and the quantum state of the N-electron target, which may be bound or ionised.

The reaction mechanism is studied by considering targets whose description is simple and, at least from the spectroscopic point of view, believable within an accuracy appropriate to the scattering experiment. Hydrogen is the obvious example, although experiments are difficult because of the need to make the atomic target by dissociating molecules. Sodium is a target for which a large quantity of experimental data is available and whose structure can be quite well described for the lower-energy states. When the reaction mechanism is sufficiently understood the reaction may be used as a probe for the structure of more-complicated target or residual systems.

6.1 Formulation of the problem

Scattering theory concerns a collision of two bodies, that may change the state of one or both of the bodies. In our application one body (the projectile) is an electron, whose internal state is specified by its spin-projection quantum number v. The other body (the target) is an atom or an atomic ion, whose internal bound state is specified by the principal quantum number n and quantum numbers j, m and ℓ for the total angular momentum, its projection and the parity respectively. We are also concerned with singly-ionised states of the target, specified by the internal quantum numbers of the separated electron and the residual ion and their relative momentum. The state of relative motion of the projectile and the target is specified by the relative momentum $\mathbf{k}$. We assume that the nucleus is so massive that its kinetic energy may be neglected.

In a collision experiment we have an incident beam of electrons of kinetic energy E_0. If one spin projection predominates then the beam is polarised, the polarisation P being given in terms of the intensities I_v of electrons with projection v. The intensity is the number of particles per second incident on unit area normal to the beam direction.

$$P = \frac{I_{1/2} - I_{-1/2}}{I_{1/2} + I_{-1/2}}. \tag{6.1}$$

The target atoms are usually in a beam whose kinetic energy is negligible.

In the final state one or two electrons are detected with specified kinetic energies and in specified directions so that their momenta are known. The numbers of electrons per second in a particular range of energies and solid angles are recorded. The polarisation of the final-state electron beams (or related observables) may or may not be observed.

The total Hamiltonian H of the projectile–target system is partitioned into a projectile–target potential V, whose range must be short compared with the Coulomb potential, and channel Hamiltonian K. In the case of a charged target there is a residual Coulomb potential V_C, which is subtracted from V and added to K.

$$H = (K + V_C) + (V - V_C). \tag{6.2}$$

The channel Hamiltonian governs the system at macroscopic separation distances, in particular the injection and detection distances of the experiment. It consists of the kinetic-energy operator K_0 of the electron and the Hamiltonian H_T of the target.

$$K = K_0 + H_T. \tag{6.3}$$

We develop scattering theory using the ideas of Gell-Mann and Goldberger (1953), starting with the case $V_C = 0$. The case of a charged target will be considered later. Modifications to the notation are trivial.

The sytem is observed before and after the collision in time-dependent channel states $|\Phi_i(t)\rangle$. The channel index i stands not only for the channel quantum numbers n, j, m, ℓ, v but also for the relative momentum $\mathbf{k}_i$. The entrance channel is denoted by $i = 0$. The Schrödinger equation of motion for the channel i is, in atomic units,

$$i\frac{\partial}{\partial t}|\Phi_i(t)\rangle = K|\Phi_i(t)\rangle. \tag{6.4}$$

The normalised stationary channel state corresponding to an energy eigen-value E_i is $|\Phi_i\rangle$, where

$$|\Phi_i(t)\rangle = e^{-iE_it}|\Phi_i\rangle. \tag{6.5}$$

The channel Schrödinger equation is

$$(E - K)|\Phi_i\rangle = 0. \tag{6.6}$$

A necessary condition for K is that it is separable in the translational coordinates of the projectile and all other coordinates. We may therefore write the more-explicit form for the stationary channel state

$$|\Phi_i\rangle = |i\mathbf{k}_i\rangle, \tag{6.7}$$

which applies if i is a discrete (bound target state) channel. If i is a continuum (ionised target state) channel we write

$$|\Phi_i\rangle = |\mu\mathbf{k}_\mu\mathbf{k}_i\rangle, \tag{6.8}$$

where μ denotes the quantum state of the residual ion and the spin projections of the continuum electrons, and $\mathbf{k}_\mu$ is the relative momentum of the residual ion and the target electron. At this stage we distinguish projectile and target electrons for illustration. We must keep in mind that the states are antisymmetric in the coordinates and spins of all electrons.

The Schrödinger equation of motion for the collision problem is

$$i\frac{\partial}{\partial t}|\Psi_j(t)\rangle = (K + V)|\Psi_j(t)\rangle. \tag{6.9}$$

The stationary state for the collision is $|\Psi_j\rangle$, where

$$|\Psi_j(t)\rangle = e^{-iE_jt}|\Psi_j\rangle. \tag{6.10}$$

The Schrödinger equation for the collision is

$$(E - H)|\Psi_j\rangle = (E - K - V)|\Psi_j\rangle = 0. \tag{6.11}$$

The collision state $|\Psi_j\rangle$ is in a one-to-one correspondence with the channel state $|\Phi_j\rangle$, the entrance channel for the collision. The physical collision state is $|\Psi_0\rangle$, but the index j is needed for some formal purposes when we use the spectral representation of H.

It is useful to keep in mind a simple problem for illustration. This is potential scattering (section 4.4), i.e. scattering of a spinless particle by a short-range potential V. The coordinate representation of the scattering quantities is, in atomic units,

$$\langle\mathbf{r}'|K|\mathbf{r}\rangle = -\tfrac{1}{2}\nabla^2\delta(\mathbf{r}' - \mathbf{r}),$$
$$\langle\mathbf{r}'|V|\mathbf{r}\rangle = V(r)\delta(\mathbf{r}' - \mathbf{r}),$$
$$\langle\mathbf{r}|\Phi_i\rangle = (2\pi)^{-3/2}e^{i\mathbf{k}_i\cdot\mathbf{r}},$$
$$\langle\mathbf{r}|\Psi_0\rangle = (2\pi)^{-3/2}\chi^{(+)}(\mathbf{k}_0,\mathbf{r}). \tag{6.12}$$

The relationship of the mathematical constructions to the physical situation is given by the interpretation of section 3.2. The amplitude for detecting the channel state $|\Phi_i(t)\rangle$ at time t for the collision state $|\Psi_0(t)\rangle$ is

$$f_{i0}(t) = \langle \Phi_i(t)|\Psi_0(t)\rangle. \tag{6.13}$$

The corresponding probability is

$$w_{i0}(t) = |f_{i0}(t)|^2 N_0^{-1}, \tag{6.14}$$

where the normalisation is

$$N_0 = \langle \Psi_0(t)|\Psi_0(t)\rangle. \tag{6.15}$$

Since probability is conserved N_0 is constant in time.

At this stage we encounter a mathematical difficulty. Continuum states are not normalisable in the conventional sense $N_0=1$. In fact we have no algebra for continuum states. The algebra of Hilbert space, where a basis has discrete eigenvalues, is applied by modifying the physical system we are considering. We enclose the whole system in a cubic box (section 4.1) of side L. Eigenvalues of states in the box are discrete. At the end of the analysis we take the continuum limit $L \to \infty$. The experimental observables turn out to be independent of L. For box-normalised states the index i stands for a countable set of quantum numbers representing the internal states of the target and projectile and the relative motion in the box.

6.2 Box-normalised wave-packet states

We are interested in the rate of transition from the entrance channel $|\Phi_0(t)\rangle$ to the exit channel $|\Phi_i(t)\rangle$ caused by the interaction V. We choose $t = 0$ to be the time at which the collision occurs and consider the mathematical representation of the dependence of the collision state $|\Psi_0(t)\rangle$ on $|\Phi_0(T)\rangle$, its state at time T in the distant past (on an atomic scale) when the system was so well separated that the potential V did not act. Applying the time-development operator (3.45) we have

$$|\Psi_0(t)\rangle = e^{-iH(t-T)}|\Phi_0(T)\rangle. \tag{6.16}$$

However, this does not represent the physical situation. The system does not start suddenly at T. The beam is switched on over a finite period τ. We therefore represent $|\Psi_0(t)\rangle$ as the time development of a wave packet prepared with an exponentially-increasing weight $e^{\epsilon T}$, where

$$\epsilon = 1/\tau. \tag{6.17}$$

The exponential form is chosen for mathematical convenience, since it leads to an integral that can be evaluated. The exact form of the wave packet is physically unimportant. We have seen in our discussion of the wave-packet description of beams (section 3.2) that a scattering experiment should be described by an initial eigenstate of momentum. In order to achieve this we will take the limit $\epsilon \to 0+$ at the end of the analysis. The wave-packet collision state is

$$|\Psi_0^{(\epsilon)}(t)\rangle = \epsilon \int_{-\infty}^{0} dT e^{\epsilon T} e^{-iH(t-T)} |\Phi_0(t)\rangle. \qquad (6.18)$$

Note that in the time-reversed situation the beam is switched off over a finite period τ and the weight factor of the wave packet is $e^{-\epsilon T}$.

The experimental situation is that times characteristic of an experiment are of the order 10^{-9}s, while the time characteristic of an atomic collision is, for example, the time it takes an electron projectile to traverse an atom. This is of the order 10^{-16}s. It is therefore physically reasonable to consider the limit $\tau \to \infty$ or $\epsilon \to 0+$. Experiments involving time resolution have been devised with resonant states whose lifetimes are greater than 10^{-9}s. Such experiments must be described by explicit wave packets rather than the formalism of the present section. An example is given in section 4.6.

The double limit $\epsilon \to 0+$, $L \to \infty$ must be taken in such a way that the whole system is inside the normalising box at all times. τ is the length of the incident train of particles divided by their velocity v. We require

$$\tau < Lv^{-1}. \qquad (6.19)$$

When τ and L both tend to infinity

$$\epsilon^{-1} L^{-3} \longrightarrow 0. \qquad (6.20)$$

This is the quantity that is relevant to our derivation.

6.3 Integral equation for the box-normalised collision state

The integration over T in (6.18) may be performed formally to obtain

$$|\Psi_0^{(\epsilon)}(t)\rangle = e^{-iHt} \frac{i\epsilon}{E_0 - H + i\epsilon} |\Phi_0\rangle. \qquad (6.21)$$

We have used (6.5) to introduce the stationary channel state. From equations (6.2,6.6) we have

$$(E_0 - H)|\Phi_0\rangle = -V|\Phi_0\rangle. \qquad (6.22)$$

We use (6.22) to obtain from (6.21) the box-normalised wave-packet collision state at $t = 0$.

$$|\Psi_0^{(\epsilon)}(0)\rangle = |\Phi_0\rangle + \frac{1}{E_0 - H + i\epsilon} V|\Phi_0\rangle. \qquad (6.23)$$

This is an explicit expression for the collision state in terms of the corresponding channel state. The difficulty is in the integral operator. We can turn this into numbers by introducing the spectral representation of H, but a knowledge of this requires a solution of the problem. To obtain a form that leads to a solution we use the operator identity

$$\frac{1}{A} - \frac{1}{B} = \frac{1}{B}(B - A)\frac{1}{A} \tag{6.24}$$

on the last term of (6.23), with

$$A = E_0 - H + i\epsilon, \quad B = E_0 - K + i\epsilon, \tag{6.25}$$

to obtain

$$\frac{1}{E_0 - H + i\epsilon} V|\Phi_0\rangle = \frac{1}{E_0 - K + i\epsilon} V\left[1 + \frac{1}{E_0 - H + i\epsilon} V\right]|\Phi_0\rangle$$

$$= \frac{1}{E_0 - K + i\epsilon} V|\Psi_0^{(\epsilon)}(0)\rangle.$$

This is substituted into (6.23) to obtain the integral equation for the collision state

$$|\Psi_0^{(\epsilon)}(0)\rangle = |\Phi_0\rangle + \frac{1}{E_0 - K + i\epsilon} V|\Psi_0^{(\epsilon)}(0)\rangle. \tag{6.26}$$

Here the spectral representation of the integral operator is known. The difficulty is in solving the integral equation.

6.4 The physical limiting procedure : normalisation

The key quantity in the calculation of experimental observables is the collision amplitude (6.13). The box-normalised collision amplitude for the wave packet is given for $t = 0$ by using (6.26).

$$f_{i0}^{(\epsilon)}(0) = \langle\Phi_i|\Phi_0\rangle + \langle\Phi_i|\frac{1}{E_0 - K + i\epsilon} V|\Psi_0^{(\epsilon)}(0)\rangle$$

$$= \delta_{i0} + \frac{1}{E_0 - E_i + i\epsilon} R_{i0}^{(\epsilon)}, \tag{6.27}$$

where

$$R_{i0}^{(\epsilon)} = \langle\Phi_i|V|\Psi_0^{(\epsilon)}(0)\rangle. \tag{6.28}$$

The significance of this expression is that the divergent integral (6.13) over all space has been replaced by an integral over the finite volume in which V is nonzero, and other quantities that are easily handled.

We first consider the limit $L \to \infty$, applied to $R_{i0}^{(\epsilon)}$. We normalise both the channel and collision states of (6.28) in the same way by considering the part of $|\Phi_i\rangle$ that describes the relative motion of the projectile and the target. After taking the limit this is an eigenstate of momentum $|\mathbf{k}_i\rangle$, according to (6.7). Its coordinate representation is a plane wave (3.25), i.e. $(2\pi)^{-3/2} e^{i\mathbf{k}_i \cdot \mathbf{r}}$. Since the probability of the state is uniform over the whole box, the box normalisation is $L^{-3/2}$. The properly-normalised wave-packet quantity that is independent of L is

$$T_{i0}^{(\epsilon)} = \lim_{L\to\infty} (L/2\pi)^3 R_{i0}^{(\epsilon)}. \tag{6.29}$$

Next we turn to the normalisation N_0 of the collision amplitude. Introducing the unit operator for the channel space into the definition (6.15) in the wave-packet case we have

$$N_0 = \Sigma_i \langle \Psi_0^{(\epsilon)}(t)|\Phi_i(t)\rangle \langle \Phi_i(t)|\Psi_0^{(\epsilon)}(t)\rangle = \Sigma_i |f_{i0}^{(\epsilon)}(t)|^2. \tag{6.30}$$

Using (6.27) in (6.30) we find, after some complex-number arithmetic,

$$N_0 = 1 + \frac{2}{\epsilon} \operatorname{Im} R_{00}^{(\epsilon)} + \Sigma_i \frac{1}{(E_0 - E_i)^2 + \epsilon^2} |R_{i0}^{(\epsilon)}|^2. \tag{6.31}$$

Since $R_{i0}^{(\epsilon)}$ is of order L^{-3} and $\epsilon^{-1} L^{-3} \to 0$ (6.20) we have

$$\lim_{L\to\infty} N_0 = 1. \tag{6.32}$$

The limit $\epsilon \to 0+$ is kept in the formalism. We introduce a notation for it below. The important quantity in this limit is the T-matrix element

$$T_{i0} = \lim_{\epsilon\to 0+} T_{i0}^{(\epsilon)}. \tag{6.33}$$

We continue the formal development in terms of the box-normalised quantities.

6.5 Transition rate and differential cross section

We first consider the transition rate to a particular final state i. The transition rate at $t = 0$ is given by (6.14,6.27).

$$\dot{w}_{i0} = \frac{\partial}{\partial t} |f_{i0}^{(\epsilon)}(0)|^2. \tag{6.34}$$

In order to evaluate it we need an identity obtained by writing the time-dependent collision state in (6.13) as the time-development of the collision state at $t = 0$ and using (6.5).

$$f_{i0}^{(\epsilon)}(t) = \langle \Phi_i|e^{i(E_i - H)t}|\Psi_0^{(\epsilon)}(0)\rangle. \tag{6.35}$$

We differentiate (6.35), use (6.2,6,28) and set $t = 0$.

$$\frac{\partial}{\partial t} f_{i0}^{(\epsilon)}(0) = -i R_{i0}^{(\epsilon)}. \tag{6.36}$$

After some complex-number arithmetic we find for the box-normalised quantities

$$\dot{w}_{i0} = \lim_{\epsilon \to 0+} \left[2\delta_{i0} \mathrm{Im} R_{00}^{(\epsilon)} + \frac{2\epsilon}{(E_0 - E_i)^2 + \epsilon^2} |R_{i0}^{(\epsilon)}|^2 \right]. \tag{6.37}$$

The case of forward scattering $i = 0$ is special. It is treated in the next section. Defining R_{i0} by

$$R_{i0} = \lim_{\epsilon \to 0+} R_{i0}^{(\epsilon)} = (2\pi/L)^3 T_{i0}, \tag{6.38}$$

we have, in view of (3.14),

$$\begin{aligned} \dot{w}_{i0} &= \lim_{\epsilon \to 0+} \frac{2\epsilon}{(E_0 - E_i)^2 + \epsilon^2} |R_{i0}|^2 \\ &= 2\pi\delta(E_0 - E_i)|R_{i0}|^2. \end{aligned} \tag{6.39}$$

Now the finite value of ϵ corresponds to a finite uncertainty in the energy of the initial state. We may consider the transition as involving a collection of initial states in which the number of states in the interval dE_0 is $\rho(E_0)dE_0$, where $\rho(E_0)$ is the density of states. From this point of view the transition rate is given by

$$\begin{aligned} \dot{w}_{i0} &= 2\pi|R_{i0}|^2 \int dE_0 \delta(E_0 - E_i)\rho(E_0) \\ &= 2\pi|R_{i0}|^2 \rho(E_i). \end{aligned} \tag{6.40}$$

This expression is the exact form of Fermi's Golden Rule, familiar in time-dependent perturbation theory where $|\Psi_0^{(\epsilon)}(0)\rangle$ is approximated by $|\Phi_0\rangle$ (Merzbacher, 1970). $\rho(E_i)$ is the density of final states.

In the discrete notation for box-normalised channels the differential cross section σ_{i0} for the transition $0 \to i$ is defined as the transition rate (6.39) divided by the incident flux vL^{-3}.

$$\sigma_{i0} = \dot{w}_{i0}L^3 v^{-1}. \tag{6.41}$$

We will see in sections 6.7 and 6.8 that this is independent of L in particular cases so that the limit $L \to \infty$ is well defined.

6.6 The optical theorem

The transition $0 \to 0$ is forward scattering. The total cross section is

$$\sigma_T = \sum_{i \neq 0} \sigma_{i0}. \tag{6.42}$$

The optical theorem expresses the fact that the normalisation N_0 is constant in time. This means that the rate of depletion of the entrance channel $\dot{w}_{00}$ is equal to the sum of the transition rates to the other channels.

Formally this is seen by differentiating (6.30) and using (6.34).

$$\dot{N}_0 = \Sigma_i \frac{\partial}{\partial t}|f_{i0}^{(\epsilon)}(0)|^2 = \Sigma_i \dot{w}_{i0} = 0. \tag{6.43}$$

Hence

$$-\dot{w}_{00} = \sum_{i \neq 0} \dot{w}_{i0}. \tag{6.44}$$

We calculate $\dot{w}_{00}$ by substituting (6.27) in (6.34) and using (6.36).

$$\dot{w}_{00} = 2\mathrm{Im}R_{00}^{(\epsilon)} + \frac{2}{\epsilon}|R_{00}^{(\epsilon)}|^2. \tag{6.45}$$

The second term vanishes by (6.20) in the double limit and we have in the limit $\epsilon \to 0+$

$$-2\mathrm{Im}R_{00} = \sum_{i \neq 0} \dot{w}_{i0}. \tag{6.46}$$

We multiply by $L^3 v^{-1}$ and use (6.41,6.42) and the definition (6.29,6.33) of T_{00}.

$$\sigma_T = -(2/v)(2\pi)^3 \mathrm{Im}T_{00}. \tag{6.47}$$

The total cross section is proportional to the imaginary part of the forward scattering amplitude.

6.7 Differential cross section for scattering

In the case of scattering the channel states of relative motion are defined by the momentum of one electron relative to the collision centre of mass, which is at the nucleus if we neglect the kinetic energy of the nucleus. To obtain the differential cross section we use the form (6.40) for $\dot{w}_{i0}$ in the definition (6.41).

We must first find the density of final states, which we characterise in terms of the relative momentum k_i. The permitted values of $\mathbf{k}_i$ in the normalisation box are given by (4.7).

$$k_{ix} = 2\pi n_x / L \quad \text{and similarly for } y \text{ and } z. \tag{6.48}$$

There are $(L/2\pi)^3 d^3k_i$ states in the range d^3k_i about $\mathbf{k}_i$. Many different final states belong to the same energy channel, defined by the channel kinetic energy E_i, but have different directions given by the polar angles θ, ϕ.

$$\rho(k_i)dE_i = (L/2\pi)^3 d^3k_i = (L/2\pi)^3 k_i^2 dk_i \sin\theta d\theta d\phi. \tag{6.49}$$

We transform the integration variable from E_i to k_i.

$$dE_i = k_i dk_i. \tag{6.50}$$

The density in momentum of final states is

$$\rho(k_i) = (L/2\pi)^3 k_i d\Omega, \tag{6.51}$$

where the element of solid angle $d\Omega$ is defined by

$$d\Omega = \sin\theta d\theta d\phi. \tag{6.52}$$

The experiment measures the rate of transition into a solid angle $d\Omega$ subtended by the detector at scattering angles θ, ϕ. The energy channels $i, 0$ are defined by energy resolution. In atomic units the relative velocity v is k_0. We use the notation $d\sigma_i(\theta, \phi)$ for the differential cross section in this experiment. The definition (6.41) becomes

$$d\sigma_i(\theta, \phi) = \dot{w}_{i0} L^3 k_0^{-1}. \tag{6.53}$$

We obtain $\dot{w}_{i0}$ by using (6.29,6.33,6.51) in (6.40) and find

$$d\sigma_i(\theta, \phi) = 2\pi(\frac{2\pi}{L})^6 |T_{i0}|^2 (\frac{L}{2\pi})^3 k_i d\Omega L^3 k_0^{-1}. \tag{6.54}$$

The differential cross section is independent of L, the size of the normalisation box.

The calculation of T_{i0} for potential scattering has been discussed in great detail in section 4.5. This is a particular case of the present discussion which is useful for illustration. The differential cross section in this case is given by (4.48,4.112b).

The differential cross section is defined for experiments that do not resolve angular-momentum projections or observe polarisations. States with different values of these observables are degenerate. We average over initial-state degeneracies and sum over final-state degeneracies. In the absence of details of the states this is denoted by Σ_{av}. The final form of the differential cross section is

$$\frac{d\sigma_i}{d\Omega} = (2\pi)^4 \frac{k_i}{k_0} \Sigma_{av} |T_{i0}|^2. \tag{6.55}$$

6.8 Differential cross section for ionisation

In the case of ionisation the channel states of relative motion are defined by the momenta $\mathbf{k}_\mu, \mathbf{k}_i$ of the two electrons. Our derivation parallels closely the derivation for scattering in section 6.7.

It is convenient at this stage to redefine the notation for energy. We drop the channel index from the total energy E, since it is the same in all channels. This is expressed for example by the energy-conserving delta function in (6.39). We now use a subscripted energy variable to refer to the kinetic energy of an electron, for example in channel 0 or i. The density of final states is the energy density of states specified by the electron kinetic energies E_μ, E_i with the energy-conservation condition

$$E = E_0 + \epsilon_0 = E_\mu + E_i. \tag{6.56}$$

The number of final states in the range $d^3k_\mu d^3k_i$ about $\mathbf{k}_\mu, \mathbf{k}_i$ is

$$
\begin{aligned}
\rho(k_\mu, k_i)dE_\mu dE_i &= (L/2\pi)^6 d^3k_\mu d^3k_i \delta(E - E_\mu - E_i) \\
&= (L/2\pi)^6 k_\mu k_i dE_\mu dE_i d\Omega_\mu d\Omega_i \delta(E - E_\mu - E_i).
\end{aligned} \tag{6.57}
$$

The differential cross section is again obtained from the expression (6.40) for the transition rate $\dot{w}_{i0}$. However, in the case of ionisation the continuum limit of $|\Phi_i\rangle$ in the definition (6.28) of $R_{i0}^{(\epsilon)}$ is (6.8), which has two plane waves. The analogue of (6.29,6.33) for the limit as $\epsilon \to 0+$ of the properly-normalised quantity independent of L is

$$T_{\mu 0}(\mathbf{k}_0, \mathbf{k}_\mu, \mathbf{k}_i) = \lim_{L\to\infty} (L/2\pi)^{9/2} R_{i0}. \tag{6.58}$$

We therefore have the following expression for the differential cross section

$$
\begin{aligned}
d^5\sigma_\mu(\mathbf{k}_0, \mathbf{k}_\mu, \mathbf{k}_i) &= 2\pi \left(\frac{2\pi}{L}\right)^9 |T_{\mu 0}(\mathbf{k}_0, \mathbf{k}_\mu, \mathbf{k}_i)|^2 \left(\frac{L}{2\pi}\right)^6 k_\mu k_i \\
&\quad \times dE_\mu dE_i d\Omega_\mu d\Omega_i \delta(E - E_\mu - E_i) L^3 k_0^{-1}.
\end{aligned} \tag{6.59}
$$

Accounting for degeneracies in the initial and final states the differential cross section becomes

$$\frac{d^5\sigma_\mu}{d\Omega_\mu d\Omega_i dE_i} = (2\pi)^4 \frac{k_\mu k_i}{k_0} \Sigma_{av} |T_{\mu 0}(\mathbf{k}_0, \mathbf{k}_\mu, \mathbf{k}_i)|^2. \tag{6.60}$$

6.9 The continuum limit : Lippmann–Schwinger equation

In this section we first summarise the meaning of the notation for the channel and collision states with box normalisation and in the continuum limit $L \to \infty$. We then define notation for the limit $\epsilon \to 0+$ and write the corresponding integral equations.

With box normalisation the channel states $|\Phi_i\rangle$ are countable. For discrete target states the index i stands for the internal quantum numbers n, j, m, ℓ, v of the target and projectile and the box quantum numbers n_{ix}, n_{iy}, n_{iz} characterising the relative motion. When $L \to \infty$ the box quantum number set is replaced by the momentum continuum $\mathbf{k}_i$. The limiting procedure is summarised as follows

$$|\Phi_i\rangle = |in_{ix}n_{iy}n_{iz}\rangle \longrightarrow |i\mathbf{k}_i\rangle, \quad L \longrightarrow \infty. \tag{6.61}$$

For a continuum target state the internal quantum numbers i include the set μ for the residual ion state and electron-spin projections and the box quantum numbers $n_{\mu x}, n_{\mu y}, n_{\mu z}$ for the motion of the freed electron. The limiting procedure is summarised by

$$|\Phi_i\rangle = |\mu n_{\mu x} n_{\mu y} n_{\mu z} n_{ix} n_{iy} n_{iz}\rangle \longrightarrow |\mu \mathbf{k}_\mu \mathbf{k}_i\rangle, \quad L \longrightarrow \infty. \tag{6.62}$$

The same meaning is given to the relative motion described by the collision state $|\Psi_0^{(\epsilon)}(0)\rangle$. For this state we must also consider the limit $\epsilon \to 0+$. Introduction of the energy width ϵ enabled us to write an integral equation (6.26) for the collision state. The limiting procedure is represented by

$$\lim_{\epsilon \to 0+} |\Psi_0^{(\epsilon)}(0)\rangle \equiv |\Psi_0^{(+)}\rangle. \tag{6.63}$$

The channel subscript 0 indicates that the entrance channel for the collision is $|\Phi_0\rangle$.

The $\epsilon \to 0+$ limit of the integral operator of (6.26) is represented by

$$\lim_{\epsilon \to 0+} \frac{1}{E_0 - K + i\epsilon} \equiv \frac{1}{E^{(+)} - K}. \tag{6.64}$$

Note that (6.64) expresses the redefinition of energy (6.56). We drop the channel subscript from the total energy E since it is the same in all channels in the limit $\epsilon \to 0+$.

The operator (6.64) is called the resolvent or Green's function operator

$$G_0(E^{(+)}) \equiv \frac{1}{E^{(+)} - K}. \tag{6.65}$$

In the illustrative case of potential scattering equn. (4.107) shows that the coordinate representation of $G_0(E^{(+)})$ is a spherical outgoing wave. We may generalise this to the coordinate representation for channel i, using (6.3) with $V_C = 0$.

$$\langle \mathbf{r}'i | \frac{1}{E^{(+)} - K_0 - H_T} | i\mathbf{r} \rangle = -(2\pi)^{-1} \frac{e^{ik_i|\mathbf{r}' - \mathbf{r}|}}{|\mathbf{r}' - \mathbf{r}|}, \tag{6.66}$$

where

$$\tfrac{1}{2}k_i^2 = E - \epsilon_i. \tag{6.67}$$

This is an outgoing spherical wave in channel i.

The time-reversed collision state and Green's function operator are denoted by

$$\lim_{\epsilon \to 0+} |\Psi_0^{(-\epsilon)}(0)\rangle \equiv |\Psi_0^{(-)}\rangle, \tag{6.68}$$

$$\lim_{\epsilon \to 0+} \frac{1}{E_0 - K - i\epsilon} \equiv \frac{1}{E^{(-)} - K}. \tag{6.69}$$

The coordinate representation of (6.69) for channel i is an ingoing spherical wave, generalising (4.108).

The $\epsilon \to 0\pm$ limit of the integral equation (6.26) is the Lippmann—Schwinger equation

$$|\Psi_0^{(\pm)}\rangle = |\Phi_0\rangle + \frac{1}{E^{(\pm)} - K} V |\Psi_0^{(\pm)}\rangle. \tag{6.70}$$

The T-matrix element is given by (6.28,6.29,6.33).

$$T_{i0} \equiv \langle\Phi_i| T |\Phi_0\rangle = \langle\Phi_i| V |\Psi_0^{(+)}\rangle. \tag{6.71}$$

In the limit $L \to \infty$ the index i has become a convenient discrete notation including the projectile continuum for channel i, defined by (6.7), or including the projectile—target continuum when the notation is defined by (6.8). We will retain this notation for formal convenience, but use the more-explicit forms (6.7,6.8) when it is necessary to specify electron momenta. The more-explicit form for the T-matrix element is

$$\langle \mathbf{k}_i i| T |0\mathbf{k}_0\rangle = \langle \mathbf{k}_i i| V |\Psi_0^{(+)}(\mathbf{k}_0)\rangle. \tag{6.72}$$

We form the T-matrix element (6.72) in the integral equation (6.70) and expand in the complete set of channel states $|j\mathbf{k}'\rangle$ to obtain the Lippmann—Schwinger equation for the T-matrix element.

$$\langle \mathbf{k}i| T |0\mathbf{k}_0\rangle = \langle \mathbf{k}i| V |0\mathbf{k}_0\rangle + \Sigma_j \int d^3k' \langle \mathbf{k}i| V |j\mathbf{k}'\rangle$$
$$\times \frac{1}{E^{(+)} - \epsilon_j - \frac{1}{2}k'^2} \langle \mathbf{k}'j| T |0\mathbf{k}_0\rangle. \tag{6.73}$$

Note that the integration over $\mathbf{k}'$ involves all momentum values since $|j\mathbf{k}'\rangle$ form a complete set of eigenstates of K. Not all momentum values conserve energy. The corresponding V- and T-matrix elements are half off shell, while the V-matrix element in the kernel of (6.73) is fully off shell. We use the channel subscript j to denote the physical (on-shell) momentum $\mathbf{k}_j$ of the projectile in channel j (6.67). The solution of (6.73) is the complete set of half-off-shell T-matrix elements, which includes the physical (on-shell) T-matrix element T_{i0} (6.71) that is used to calculate the differential cross section (6.55). For continuum target states the channel index j is a discrete notation for the target continuum and the notation Σ_j represents the corresponding integral. The potential-scattering analogue of (6.73) is (4.117).

Some of the early calculations of electron—atom scattering assumed that the potential was small compared with the total energy so that it is a good approximation to iterate the Lippmann—Schwinger equation (6.73). Using the notation (6.65,6.71) the resultant series

$$\langle\Phi_i| T |\Phi_0\rangle = \langle\Phi_i| V + V G_0 V + V G_0 V G_0 V + \cdots |\Phi_0\rangle. \tag{6.74}$$

is the Born or perturbation series. The first order is the Born approximation. In electron-atom scattering the Born series is strictly divergent (Stelbovics, 1990) but it can sometimes give useful results.

A significant improvement is achieved by the unitarised Born approximation, which neglects the real part of the resolvent in (6.73), but keeps the imaginary part, given by the identity (4.122) for integration through a pole.

$$\langle \mathbf{k}_i i | T | 0\mathbf{k}_0 \rangle = \langle \mathbf{k}_i i | V | 0\mathbf{k}_0 \rangle - i\pi \Sigma_j \int d\hat{\mathbf{k}}_j \langle \mathbf{k}_i i | V | j\mathbf{k}_j \rangle \langle \mathbf{k}_j j | T | 0\mathbf{k}_0 \rangle. \qquad (6.75)$$

This approximation relates on-shell amplitudes for all the channels, since the imaginary part of (4.122) conserves energy.

Subsequent chapters will be concerned with non-perturbative solutions of the Lippmann–Schwinger equation.

6.10 The distorted-wave transformation

It would be convenient for solving the Lippmann–Schwinger equation (6.73) if we could make the potential matrix elements as small as possible. For example, we could hope to find a transformed equation whose iteration would converge much more quickly. This is achieved by a judicious choice of a local, central potential U, which is called the distorting potential since the problem is reformulated in terms of the distorted-wave eigenstates of U rather than the plane waves of (6.73). An important particular case of U is the Coulomb potential V_C in the case where the target is charged. The Hamiltonian (6.2) is repartitioned as follows

$$H = (K + U) + (V - U). \qquad (6.76)$$

By multiplying (6.70) on the left by the channel Schrödinger operator $E^{(+)} - K$ we obtain the form of the Schrödinger equation that embodies the correct collision boundary conditions. Using (6.76) this becomes

$$(E^{(\pm)} - K - U)|\Psi_0^{(\pm)}\rangle = (V - U)|\Psi_0^{(\pm)}\rangle. \qquad (6.77)$$

The homogeneous equation is written using (6.3) as

$$(E^{(\pm)} - K_0 - U - H_T)|\chi_i^{(\pm)}\rangle = 0. \qquad (6.78)$$

It is separable in projectile and target operators

$$(E^{(\pm)} - K_0 - U)|\mathbf{k}_i^{(\pm)}\rangle = 0,$$
$$(\epsilon_i - H_T)|i\rangle = 0. \qquad (6.79)$$

The solution $|\chi_i^{(\pm)}\rangle$ of (6.78) is the distorted-wave channel state. Its more-explicit form is written in analogy to equn. (6.7) for the channel state, with the distorted waves $|\mathbf{k}_i^{(\pm)}\rangle$ replacing the plane waves $|\mathbf{k}_i\rangle$. It is convenient to use the same notation for the more-general distorted waves as we used for Coulomb waves in (4.129).

$$|\chi_i^{(\pm)}\rangle = |i\mathbf{k}_i^{(\pm)}\rangle. \tag{6.80}$$

The transformed integral equation, corresponding to (6.77), is given for physical boundary conditions indicated by the superscript (+).

$$|\Psi_0^{(+)}\rangle = |\chi_0^{(+)}\rangle + \frac{1}{E^{(+)} - K - U}(V - U)|\Psi_0^{(+)}\rangle. \tag{6.81}$$

We may arrive at this equation by repeating the derivation of this chapter for the Hamiltonian (6.76).

We also have an integral equation corresponding to (6.78). We write it for the time-reversed bra vector $\langle\chi_i^{(-)}|$ using analogous arguments. Note that the resolvent for the bra vector is the complex conjugate of the resolvent analogous to (6.70).

$$\langle\chi_i^{(-)}| = \langle\Phi_i| + \langle\chi_i^{(-)}|U\frac{1}{E^{(+)} - K}. \tag{6.82}$$

Using the operator identity (6.24) we write the corresponding explicit expression for $\langle\chi_i^{(-)}|$.

$$\langle\chi_i^{(-)}| = \langle\Phi_i| + \langle\Phi_i|U\frac{1}{E^{(+)} - K - U}. \tag{6.83}$$

In the distorted-wave formalism we denote the T matrix by $\tilde{T}$.

$$\langle\Phi_i|\tilde{T}|\Phi_0\rangle \equiv \langle\Phi_i|V|\Psi_0^{(+)}\rangle. \tag{6.84}$$

We substitute (6.81) into (6.84) and use (6.83) to obtain

$$\langle\Phi_i|\tilde{T}|\Phi_0\rangle = \langle\Phi_i|U|\chi_0^{(+)}\rangle + \langle\chi_i^{(-)}|V - U|\Psi_0^{(+)}\rangle. \tag{6.85}$$

The first term of (6.85) is the T-matrix element for elastic scattering by the potential U. If U is the Coulomb potential V_C it is the Rutherford-scattering T-matrix element. The second term is the distorted-wave T-matrix element for which we solve the distorted-wave Lippmann–Schwinger equation formed from (6.81). Its explicit form is written by expanding in the complete set of eigenstates of $K + U$. This may include projectile bound states $|\lambda\rangle$ defined by

$$(\epsilon_\lambda - K_0 - U)|\lambda\rangle = 0. \tag{6.86}$$

We use time-reversed scattering eigenstates so that the corresponding small imaginary energy adds to the one signified by $E^{(+)}$ in the resolvent to make the limit well-defined. The distorted-wave Lippmann–Schwinger equation is

$$\langle \mathbf{k}^{(-)}i| T |0\mathbf{k}_0^{(+)} \rangle = \langle \mathbf{k}^{(-)}i| V - U |0\mathbf{k}_0^{(+)} \rangle$$

$$+ \Sigma_j \int d^3k' \langle \mathbf{k}^{(-)}i| V - U |j\mathbf{k}''^{(-)} \rangle \frac{1}{E^{(+)} - \epsilon_j - \frac{1}{2}k'^2} \langle \mathbf{k}''^{(-)}j| T |0\mathbf{k}_0^{(+)} \rangle$$

$$+ \Sigma_j \Sigma_{\lambda'} \langle \mathbf{k}^{(-)}i| V - U |j\lambda' \rangle \frac{1}{E^{(+)} - \epsilon_j - \epsilon_{\lambda'}} \langle \lambda'j| T |0\mathbf{k}_0^{(+)} \rangle,$$

$$\langle \lambda i| T |0\mathbf{k}_0^{(+)} \rangle = \langle \lambda i| V - U |0\mathbf{k}_0^{(+)} \rangle$$

$$+ \Sigma_j \int d^3k' \langle \lambda i| V - U |j\mathbf{k}''^{(-)} \rangle \frac{1}{E^{(+)} - \epsilon_j - \frac{1}{2}k'^2} \langle \mathbf{k}''^{(-)}j| T |0\mathbf{k}_0^{(+)} \rangle$$

$$+ \Sigma_j \Sigma_{\lambda'} \langle \lambda i| V - U |j\lambda' \rangle \frac{1}{E^{(+)} - \epsilon_j - \epsilon_{\lambda'}} \langle \lambda'j| T |0\mathbf{k}_0^{(+)} \rangle. \tag{6.87}$$

The driving term of (6.87) is the distorted-wave Born approximation (DWBA). If $i \neq 0$ the T-matrix element in the DWBA is

$$\langle \mathbf{k}_i i| T |0\mathbf{k}_0 \rangle = \langle \mathbf{k}_i^{(-)}i| V - U |0\mathbf{k}_0^{(+)} \rangle, \tag{6.88}$$

where the distorted waves are calculated in the potential U according to the first of equations (6.79). Their calculation has been discussed in detail in section 4.4.

Up to this stage the distorting potential U has been arbitrary. We now derive an optimum form for it. According to (6.85) the exact explicit form for the T-matrix element in the case $i \neq 0$ is

$$\langle \mathbf{k}_i i| T |0\mathbf{k}_0 \rangle = \langle \mathbf{k}_i^{(-)}i| V - U |\Psi_0^{(+)}(\mathbf{k}_0) \rangle. \tag{6.89}$$

We choose U so that $|0\mathbf{k}_0^{(+)} \rangle$ is as close to $|\Psi_0^{(+)}(\mathbf{k}_0) \rangle$ as possible. To accomplish this we first project the explicit form of (6.81) onto the target ground state $|0\rangle$ and expand in the complete set of target eigenstates $|j\rangle$.

$$\langle 0|\Psi_0^{(+)}(\mathbf{k}_0) \rangle = |\mathbf{k}_0^{(+)} \rangle$$

$$+ \Sigma_j \langle 0| \frac{1}{E^{(+)} - H_T - K_0 - U} |j\rangle \langle j| V - U |\Psi_0^{(+)}(\mathbf{k}_0) \rangle. \tag{6.90}$$

The approximation requires the second term of (6.90) to be small. Making the approximation the second term becomes

$$\frac{1}{E^{(+)} - \epsilon_0 - K_0 - U} \langle 0| V - U |0\rangle |\mathbf{k}_0^{(+)} \rangle.$$

The approximation is consistent if we choose

$$U = \langle 0|V|0\rangle. \tag{6.91}$$

The optimum choice of U is the ground-state average of the projectile–target potential V. Note that the set of integral equations (6.87) can only be closed if the choice of U is unique. However, the choice of the ground-state average gives a DWBA T-matrix element (6.88) that is not time-reversal invariant. If we want to use the DWBA as an approximation to the T-matrix element, as distinct from using the distorted-wave representation as a numerical aid in solving the integral equations, it is advisable to satisfy time-reversal invariance by calculating $|k_0^{(+)}\rangle$ in the average potential for the target ground state $|0\rangle$ and $|k_i^{(-)}\rangle$ in the average potential for the excited state $|i\rangle$.

7

Calculation of scattering amplitudes

The background for the details of multichannel scattering calculations has now been established. We consider methods based on the integral-equation formulation of chapter 6. These momentum-space methods have proved accurate at all energies in a sufficient variety of situations to justify the belief that they can be generally applied. In some situations sufficient accuracy is achieved without resorting to the full power of the integral-equation solution. The methods used in these situations are distorted-wave methods. Their relationship to the full solutions will be examined in a simplified illustrative case. A brief outline will be given of alternative methods based on a coordinate-space formulation of the multichannel problem.

There are two characteristic difficulties of multichannel many-fermion problems. The first is that computational methods can of course directly address only a finite number of channels whereas the physical problem has an infinite number of discrete channels and the ionisation continuum. The second is that the electrons are identical so that the formulation in terms of one-electron states must be explicitly antisymmetric in the position (or momentum) and spin coordinates.

We first show how to set up the problem within the framework of formal scattering theory using antisymmetric products of one-electron states. The problem is then formulated in terms of the calculation of reduced T-matrix elements relating the absolute values of initial and final momenta in different angular-momentum states. This depends on a knowledge of the corresponding potential matrix elements, whose calculation we treat in detail. We then show how the target continuum is accounted for in the scattering formalism.

7.1 Antisymmetrisation

The scattering problem is formulated in terms of one-electron states, which we call orbitals to distinguish them from the N-electron target states and the $(N+1)$-electron collision and channel states of scattering theory. The space of collision states is spanned by products of $N+1$ orbitals, which we explicitly antisymmetrise in this section.

In formulating the problem antisymmetrically we describe it by the Schrödinger equation equivalent to (6.70).

$$(E - K - V)|\Psi_0^{(+)}(\mathbf{k})\rangle = 0. \tag{7.1}$$

This equation formally includes the scattering boundary conditions.

We introduce the antisymmetric multichannel expansion

$$|\Psi_0^{(+)}(\mathbf{k})\rangle = \Sigma_j |j u_j^{(+)}(\mathbf{k})\rangle. \tag{7.2}$$

We are not interested in the spin projection v of the projectile at this stage and will drop it from the notation until it is needed.

7.1.1 Reduction to direct and exchange amplitudes

The antisymmetric multichannel expansion is

$$|\Psi_0^{(+)}(\mathbf{k})\rangle = \Sigma_j a_{\mathbf{k}}^\dagger |j\rangle, \tag{7.3}$$

where $a_{\mathbf{k}}^\dagger$ creates an electron in the continuum orbital $|u_j^{(+)}(\mathbf{k})\rangle$. The antisymmetrised coupled differential equations corresponding to (7.1) are

$$\Sigma_j \langle \mathbf{k}'i|(E - H)a_{\mathbf{k}}^\dagger|j\rangle = 0. \tag{7.4}$$

The coordinate–spin representation of the $(N+1)$-electron state $a_{\mathbf{k}}^\dagger|j\rangle$ is a linear combination of determinants whose first column consists of elements $\langle x_r|u_j^{(+)}(\mathbf{k})\rangle$. Subsequent columns have elements $\langle x_r|\alpha\rangle$ for orbitals $|\alpha\rangle$ occupied in the corresponding configuration. Rows are characterised by x_r, $r = 0, N$. It is written in the notation of (3.141) as an expansion over products of the elements $\langle x_r|u_j^{(+)}(\mathbf{k})\rangle$ of the first column and their cofactors $\langle \bar{x}_r|j\rangle$ in the determinants comprising $a_{\mathbf{k}}^\dagger|j\rangle$.

$$\langle x|a_{\mathbf{k}}^\dagger|j\rangle = (N+1)^{-1/2} \sum_{r=0}^{N} (-1)^r \langle \bar{x}_r|j\rangle \langle x_r|u_j^{(+)}(\mathbf{k})\rangle. \tag{7.5}$$

In terms of this notation the bra channel state in (7.4) is

$$\langle \mathbf{k}'i|x\rangle = \langle \mathbf{k}'|x_0\rangle \langle i|\bar{x}_0\rangle. \tag{7.6}$$

It is not necessary to antisymmetrise the bra state. Since $E-H$ is symmetric the amplitude (7.4) with an explicitly-antisymmetrised bra simply involves a redefinition of dummy integration coordinates.

The coupled equations (7.4) are equivalent to

$$\int d\bar{x}_0 \langle i|\bar{x}_0\rangle (E-H)\langle \bar{x}_0|j\rangle \langle x_0|u_j^{(+)}(\mathbf{k})\rangle$$

$$+ \sum_{r=1}^{N}(-1)^r \int d\bar{x}_0 \langle i|\bar{x}_0\rangle (E-H)\langle \bar{x}_r|j\rangle \langle x_r|u_j^{(+)}(\mathbf{k})\rangle = 0, \qquad (7.7)$$

where the sum in (7.5) has been substituted into (7.4) in two parts. In the first part the coordinate of the projectile is the same (x_0) in the representations of the channel and collision states. This is the direct term. The N remaining terms are exchange terms, the $r=1$ term being

$$X_1 = -\int d\bar{x}_0 \langle i|\bar{x}_0\rangle (E-H)\langle \bar{x}_1|j\rangle \langle x_1|u_j^{(+)}(\mathbf{k})\rangle. \qquad (7.8)$$

For the rth term in the sum we interchange the coordinates x_1 and x_r.

$$X_r = (-1)^r \int d\bar{x}_0 \langle i|x_r \cdots x_1 \cdots x_N\rangle (E-H)\langle x_0 x_r \cdots x_N|j\rangle \langle x_1|u_j^{(+)}(\mathbf{k})\rangle. \qquad (7.9)$$

To restore cyclic order we take a permutation of order 1 in the coordinate ket and a permutation of order $r-2$ in the bra. Since $|i\rangle$ and $|j\rangle$ are antisymmetric

$$X_r = X_1 \qquad (7.10)$$

and (7.7) becomes

$$\int d\bar{x}_0 \langle i|\bar{x}_0\rangle (E-H)\langle \bar{x}_0|j\rangle \langle x_0|u_j^{(+)}(\mathbf{k})\rangle$$

$$- N \int d\bar{x}_0 \langle i|\bar{x}_0\rangle (E-H)\langle \bar{x}_1|j\rangle \langle x_1|u_j^{(+)}(\mathbf{k})\rangle = 0. \qquad (7.11)$$

We now partition the Hamiltonian as in (6.2,6.3), in order to have a convenient form for expressing (7.1) as a Lippmann–Schwinger equation (6.73).

$$H = \sum_{r=0}^{N}(K_r + V_r) + \sum_{r<s} v_{rs} = (K_0 + H_T) + \left(V_0 + \sum_{r=1}^{N} v_{0r}\right), \qquad (7.12)$$

where V_r is the electron–nucleus potential for electron r and the target Schrödinger equation is

$$(\epsilon_j - H_T)|j\rangle = 0. \qquad (7.13)$$

Using (7.11) the direct amplitude of (7.4) becomes

$$\langle \mathbf{k}'i|E - \epsilon_j - K_0|ju_j^{(+)}(\mathbf{k})\rangle - \sum_{r=1}^{N}\langle \mathbf{k}'i|N^{-1}V_0 + v_{0r}|ju_j^{(+)}(\mathbf{k})\rangle. \qquad (7.14)$$

The first term of (7.14) is the inverse of the Green's function of (6.73). We will express the potential term using orbitals $|\alpha\rangle$ for $|i\rangle$ and $|\beta\rangle$ for $|j\rangle$ that are occupied in the configuration-interaction basis of determinants. This is done by expanding the target states $|i\rangle$ and $|j\rangle$ in the orbital elements $|\alpha\rangle$ and $|\beta\rangle$ of the first row and their cofactors $a_\alpha|i\rangle$ and $a_\beta|j\rangle$. The annihilation operator takes care of the possibility that an orbital is not occupied in a particular determinant by giving zero for such a term. Using the fact that all terms of the r sum are equal the direct amplitude is

$$\langle \mathbf{k}'i|E - \epsilon_j - K_0|ju_j^{(+)}(\mathbf{k})\rangle - \sum_{\alpha\beta}\langle i|a_\alpha^\dagger a_\beta|j\rangle\langle \mathbf{k}'\alpha|N^{-1}V_0 + v_{01}|\beta u_j^{(+)}(\mathbf{k})\rangle. \quad (7.15)$$

The detail of the derivation of the potential term is given below for the exchange case. ·

Note that the potential (7.15) is uncharged, since there is one nuclear charge for each electron charge. For electron–ion scattering V_0 is replaced by $V_0 - U_0$, where the distorting potential U_0 is charged so that

$$V_0(r_0) - U_0(r_0) = Nr_0^{-1}. \qquad (7.16)$$

The exchange amplitude is written using (7.11) in (7.4). It is

$$-N\int dx_0 \int d\bar{x}_0\langle \mathbf{k}'|x_0\rangle\langle i|\bar{x}_0\rangle(E - H)\langle \bar{x}_1|j\rangle\langle x_1|u_j^{(+)}(\mathbf{k})\rangle. \qquad (7.17)$$

This is a matrix element of a nonlocal potential $V_X(x_0, x_1)$, which we express by splitting H into two terms.

$$V_X(x_0, x_1) =$$
$$- N\int d\bar{x}_{01}\langle i|\bar{x}_0\rangle(E - H_T - v_{01})\langle \bar{x}_1|j\rangle$$
$$+ N\int d\bar{x}_{01}\langle i|\bar{x}_0\rangle(K_0 + V_0 + \sum_{r=2}^{N}v_{0r})\langle \bar{x}_1|j\rangle. \qquad (7.18)$$

We first apply the target Hamiltonian symmetrically to $|i\rangle$ and $|j\rangle$ to replace it by $\frac{1}{2}(\epsilon_i + \epsilon_j)$. V_X is thus manifestly hermitian. We then apply the orbital-cofactor expansion to the first term, which becomes

$$-\sum_{\alpha\beta}\langle i|a_\alpha^\dagger a_\beta|j\rangle\langle\alpha|x_1\rangle\left[E - \frac{1}{2}(\epsilon_i + \epsilon_j) - v_{01}\right]\langle x_0|\beta\rangle. \qquad (7.19)$$

For the second term of (7.18) we treat the part in $K_0 + V_0$ similarly. The sum over r is treated by applying the same argument that resulted in

(7.10) to find that each term is equal to the v_{02} term. This time we make two successive orbital-cofactor expansions obtaining

$$N(N-1) \int d\bar{x}_{01} \langle i | \bar{x}_0 \rangle v_{02} \langle \bar{x}_1 | j \rangle$$

$$= \sum_{\alpha\beta\gamma\delta} \int d\bar{x}_{012} \int dx_2 \langle i | a_\alpha^\dagger a_\gamma^\dagger | \bar{x}_{012} \rangle \langle \alpha | x_1 \rangle \langle \gamma | x_2 \rangle v_{02}$$

$$\times \langle x_2 | \delta \rangle \langle x_0 | \beta \rangle \langle \bar{x}_{012} | a_\delta a_\beta | j \rangle$$

$$= \sum_{\alpha\beta\gamma\delta} \langle i | a_\alpha^\dagger a_\gamma^\dagger a_\delta a_\beta | j \rangle \langle \alpha | x_1 \rangle \langle \gamma | v_{02} | \delta \rangle \langle x_0 | \beta \rangle. \qquad (7.20)$$

Note that the indices on $\bar{x}$ denote the coordinates that are omitted from the set.

Applying the commutation rules (3.137–3.139) we obtain

$$\langle i | a_\alpha^\dagger a_\gamma^\dagger a_\delta a_\beta | j \rangle = \langle i | a_\alpha^\dagger a_\beta | j \rangle \delta_{\gamma\delta} - \langle i | a_\alpha^\dagger a_\delta | j \rangle \delta_{\beta\gamma}. \qquad (7.21)$$

Here we have assumed that $a_\gamma^\dagger | j \rangle$ is always zero. This is strictly true only if $| \gamma \rangle$ is occupied in $| j \rangle$ as it would be if $| i \rangle = | j \rangle$ or if it is an orbital of an inert core. This approximation is further discussed below equation (7.23).

The nonlocal potential (7.20) is written, after interchanging the sum indices β and δ in the last term of (7.21), in terms of the operator $\tilde{H}_0$.

$$\sum_{\alpha\beta} \langle i | a_\alpha^\dagger a_\beta | j \rangle \langle \alpha | x_1 \rangle \tilde{H}_0 \langle x_0 | \beta \rangle \equiv \sum_{\alpha\beta} \langle i | a_\alpha^\dagger a_\beta | j \rangle$$

$$\times \left[\langle \alpha | x_1 \rangle (K_0 + V_0 + \Sigma_\gamma \langle \gamma | v_{02} | \gamma \rangle) \langle x_0 | \beta \rangle \right.$$

$$\left. - \langle \alpha | x_1 \rangle \Sigma_\gamma \langle \gamma | v_{02} | \beta \rangle \langle x_0 | \gamma \rangle \right]. \qquad (7.22)$$

From (5.17) we see that $\tilde{H}_0$ is formally identical to the Hartree–Fock Hamiltonian.

A useful simplification of (7.22) is achieved if we choose the orbitals $| \alpha \rangle$ and $| \beta \rangle$ to be eigenstates of $\tilde{H}_0$ with eigenvalues ϵ_α and ϵ_β.

$$(\epsilon_\beta - \tilde{H}_0) | \beta \rangle = 0. \qquad (7.23)$$

$\tilde{H}_0$ is the Hamiltonian chosen to be best suited for modelling the target states. If a single determinant is a sufficiently-accurate model for $| j \rangle$ then the definition (7.22) is self-consistent if $\tilde{H}_0$ is the Hartree–Fock Hamiltonian. However, the self-consistent potential is not the same for all target states $| j \rangle$. The one-electron potential is discussed in chapter 5.

An appropriate potential for many cases is the frozen-core Hartree–Fock potential with the addition of a core-polarisation term.

We substitute the nonlocal potential terms (7.19,7.22) in the amplitudes of (7.4), using symmetry in (7.22) to replace $\hat{H}_0$ by $\frac{1}{2}(\epsilon_\alpha + \epsilon_\beta)$. The explicitly-antisymmetric form of the coupled equations (7.4) is

$$\Sigma_j \langle \mathbf{k}'i| E - \epsilon_j - K_0 - V |j u_j^{(+)}(\mathbf{k})\rangle = 0, \tag{7.24}$$

where the potential matrix element is given by (7.15,7.19,7.22).

$$\langle \mathbf{k}'i|V|j u_j^{(+)}(\mathbf{k})\rangle = \sum_{\alpha\beta} \langle i|a_\alpha^\dagger a_\beta|j\rangle \Big[\langle \mathbf{k}'\alpha|N^{-1}V_0 + v_{01}|\beta u_j^{(+)}(\mathbf{k})\rangle$$

$$- \langle \alpha\mathbf{k}'|v_{01}|\beta u_j^{(+)}(\mathbf{k})\rangle + \langle \mathbf{k}'|\beta\rangle [E - \tfrac{1}{2}(\epsilon_i + \epsilon_j + \epsilon_\alpha + \epsilon_\beta)] \langle \alpha|u_j^{(+)}(\mathbf{k})\rangle \Big].$$

$$\tag{7.25}$$

This equation defines a nonlocal projectile–target potential V for use in the Lippmann–Schwinger equations (6.73) corresponding to (7.24). The factor $\langle i|a_\alpha^\dagger a_\beta|j\rangle$ relates the orbitals to the target states. It is an extension of the one-electron density matrix (5.62). We call it the density matrix. The first and second terms of (7.25) are the direct and two-electron-exchange terms. The third term is the one-electron-exchange term, which is nonzero because $|u_j^{(+)}(\mathbf{k})\rangle$ is not orthogonal to any of the orbitals $|\alpha\rangle$. The notation convention used for the two-electron states $|\alpha\mathbf{k}\rangle$, which are unsymmetrised, is that the outermost one-electron state belongs to the coordinate x_0, the innermost to x_1. Subscripts indicate the space to which the operators apply.

7.1.2 Validation of the exchange potential

The antisymmetric multichannel expansion (7.3) of the collision state $|\Psi_0^{(+)}(\mathbf{k})\rangle$ has a serious difficulty. It is not unique. This is seen by considering the orbital-cofactor expansion (7.5) in the elements of the $|u_j^{(+)}(\mathbf{k})\rangle$ column. If we add $\lambda\langle x_r|\alpha\rangle$ to $\langle x_r|u_j^{(+)}(\mathbf{k})\rangle$, where the orbital $|\alpha\rangle$ is occupied in all the determinants comprising $|j\rangle$, we are adding determinants with two identical columns, which are zero. Therefore $|u_j^{(+)}(\mathbf{k})\rangle$ is ambiguous, at least with respect to the addition of a linear combination of occupied orbitals.

The removal of the ambiguity in the context of formal scattering theory was first achieved by Stelbovics and Bransden (1989) in the case of a one-electron target. They give references to related considerations in a coordinate-space formulation of the problem. The method was extended to a square-integrable representation of the target (e.g. equn. (5.53)) by

Stelbovics (1991). A general unique formulation for hydrogen has been given by Stelbovics (1990). Here we generalise the formulation to the case of an N-electron target.

Using (7.5) the antisymmetric multichannel expansion for M target states is

$$\langle x|\Psi_0^{(+)}(\mathbf{k})\rangle = (N+1)^{-1/2}\sum_{j=1}^{M}\sum_r(-1)^r\langle\bar{x}_r|j\rangle\langle x_r|u_j^{(+)}(\mathbf{k})\rangle$$

$$\equiv (N+1)^{-1/2}\sum_r(-1)^r\langle x|\psi_r^{(+)}(\mathbf{k})\rangle, \tag{7.26}$$

where the unsymmetrised multichannel expansion is

$$\langle x|\psi_r^{(+)}(\mathbf{k})\rangle = \sum_{j=1}^{M}\langle\bar{x}_r|j\rangle\langle x_r|u_j^{(+)}(\mathbf{k})\rangle. \tag{7.27}$$

The ambiguity is removed by requiring that all terms $(-1)^r\langle x|\psi_r^{(+)}(\mathbf{k})\rangle$ of (7.26) be identical. This is a stronger condition than overall antisymmetry.

$$|\psi_r^{(+)}(\mathbf{k})\rangle = |\psi_0^{(+)}(\mathbf{k})\rangle \quad \text{for all } r. \tag{7.28}$$

The existence of the $|\psi_r^{(+)}(\mathbf{k})\rangle$ in (7.28) is assured if the Pauli exclusion principle applies.

The expansion (7.27) can be considered as the application of a target-state projection operator $\bar{I}_r$, whose coordinate representation is

$$\langle\bar{x}_r'|\bar{I}_r|\bar{x}_r\rangle = \sum_{j=1}^{M}\langle\bar{x}_r'|j\rangle\langle j|\bar{x}_r\rangle. \tag{7.29}$$

In order to obtain a formulation in terms of individual orbitals we make the orbital-cofactor expansion of the determinants comprising $\langle\bar{x}_r|j\rangle$ in the elements $\langle x_s|\alpha\rangle$ of the row s.

$$\langle x|\bar{I}_r|\psi_0^{(+)}(\mathbf{k})\rangle = N^{-1/2}\sum_{j\alpha}\langle\bar{x}_{rs}|a_\alpha|j\rangle\langle x_s|\alpha\rangle\langle x_r|u_j^{(+)}(\mathbf{k})\rangle. \tag{7.30}$$

The symmetry condition is expressed by making the expansion equivalent to (7.26), with s replacing r, followed by the expansion (7.30) of $\langle\bar{x}_s|j\rangle$ in the elements $\langle x_r|\alpha\rangle$ of the row r. The representation of $|\psi_0^{(+)}(\mathbf{k})\rangle$ changes sign on interchanging r and s.

$$\langle x|\bar{I}_r|\psi_0^{(+)}(\mathbf{k})\rangle = \langle x|\bar{I}_s|\psi_0^{(+)}(\mathbf{k})\rangle$$

$$= -N^{-1/2}\sum_{j\alpha}\langle\bar{x}_{rs}|a_\alpha|j\rangle\langle x_r|\alpha\rangle\langle x_s|u_j^{(+)}(\mathbf{k})\rangle. \tag{7.31}$$

The symmetry condition (7.31) is applied to the E amplitude of (7.24). For consistency with (7.18) we choose $r = 0$ and $s = 1$. We first choose the operator $\bar{I}_0$ to expand both the bra and ket, and use the definition (7.26,7.28) to obtain

$$\int dx \langle \mathbf{k}'i|\bar{I}_0|x\rangle E \langle x|\bar{I}_0|\psi_0^{(+)}(\mathbf{k})\rangle$$
$$= \sum_j \sum_{\alpha\beta} \langle i|a_\alpha^\dagger a_\beta|j\rangle \langle \mathbf{k}'\alpha|N^{-1}E|\beta u_j^{(+)}(\mathbf{k})\rangle. \tag{7.32}$$

According to (7.31), (7.32) is equivalent to the analogous expansion applying $\bar{I}_1$ to expand the bra.

$$\int dx \langle \mathbf{k}'i|\bar{I}_1|x\rangle E \langle x|\bar{I}_0|\psi_0^{(+)}(\mathbf{k})\rangle$$
$$= -\sum_j \sum_{\alpha\beta} \langle i|a_\alpha^\dagger a_\beta|j\rangle \langle \alpha\mathbf{k}'|N^{-1}E|\beta u_j^{(+)}(\mathbf{k})\rangle. \tag{7.33}$$

We avoid the difficulty of computing the overlap $\langle \mathbf{k}'|u_j^{(+)}(\mathbf{k})\rangle$ in (7.32) by multiplying E by the orbital projection operator I_0, whose coordinate–spin representation is

$$\langle x_0'|I_0|x_0\rangle = \sum_\gamma \langle x_0'|\gamma\rangle \langle \gamma|x_0\rangle. \tag{7.34}$$

We multiply (7.32) (with the inclusion of (7.34)) by an arbitrary constant $N\theta$ and add and subtract equivalent terms given by (7.32,7.33). The potential matrix element (7.25) is replaced by

$$\langle \mathbf{k}'i|V|ju_j^{(+)}(\mathbf{k})\rangle = \sum_{\alpha\beta} \langle i|a_\alpha^\dagger a_\beta|j\rangle \bigg[\langle \mathbf{k}'\alpha|N^{-1}V_0 + v_{01} - \theta E I_0|\beta u_j^{(+)}(\mathbf{k})\rangle$$
$$- \langle \alpha\mathbf{k}'|v_{01}|\beta u_j^{(+)}(\mathbf{k})\rangle$$
$$+ \langle \mathbf{k}'|\beta\rangle [(1-\theta)E - \tfrac{1}{2}(\epsilon_i + \epsilon_j + \epsilon_\alpha + \epsilon_\beta)]\langle \alpha|u_j^{(+)}(\mathbf{k})\rangle \bigg]. \tag{7.35}$$

The one-electron form of (7.35) is given by Bray and Stelbovics (1992b).

A detailed discussion of the effect of different choices of θ has been given for a hydrogen target by Stelbovics (1990). The full solution of the Lippmann–Schwinger equations (6.73) consists of half-off-shell T-matrix elements in addition to the on-shell elements that describe scattering processes. Half-off-shell T-matrix elements for a particular scattering problem may be used in distorted-wave Born approximation solutions of a larger problem. In the numerical solution of the Lippmann–Schwinger equation using a finite basis of orbitals the choice of θ is important. The choice $\theta = 0$ is a special case for which the off-shell T-matrix elements are numerically unstable although the on-shell elements are normally calculated accurately. Any other choice of θ gives stable solutions both on and off shell. A useful choice is $\theta = 1$.

Use of the potential (7.35) in solving the coupled Lippmann–Schwinger equations (6.73,6.87) corresponding to (7.24) is a unique and numerically-valid description of the electron–atom scattering problem in the context of formal scattering theory.

7.2 Reduced Lippmann–Schwinger equations

The coupled Lippmann–Schwinger equations (6.73) are reduced for computation to coupled integral equations in one radial dimension by extension of the angular-momentum techniques leading to equn. (4.121). The distorted-wave representation (6.87) is essential for charged targets and gives vital numerical simplification for targets whose description includes a closed-shell core. It is necessary to describe the reduction only for the plane-wave representation (6.73) with reference to the obvious extension to (6.87).

7.2.1 jj coupling

The jj-coupling representation is generally useful in the reduction of the Lippmann–Schwinger equations since it applies to situations where spin–orbit coupling is not negligible. The quantum numbers used in the representation are defined in table 7.1. Primed and double-primed quantities are used to distinguish different angular-momentum states.

The channel state $|i\mathbf{k}\rangle$ is described by quantities that determine the differential cross section (6.55) and other experimental observables described

Table 7.1. *Quantities describing* jj-*coupling states*

Principal quantum number of target state	n
Total angular momentum of target state	j
Corresponding projection	m
Parity of target state	ℓ
Absolute momentum of projectile	k
Spin projection of projectile	ν
Total angular momentum of projectile	J
Corresponding projection	M
Orbital angular momentum of projectile	L
Corresponding projection	μ
Total angular momentum	P
Corresponding projection	Q
Overall parity	π

in chapters 8 and 9. They are the projectile momentum **k** and spin projection v and the target-state quantum numbers n, ℓ, j, m described in section 5.1. The projectile state is represented by a partial-wave expansion (4.188). The quantum numbers of the projectile partial wave and the target state are coupled to total angular momentum P and parity π. The jj-coupling expansion of the potential matrix element is

$$\langle \mathbf{k}'v'n'\ell'j'm'|V|n\ell jmv\mathbf{k}\rangle = \sum_{L'\mu'J'M'} \sum_{L\mu JM} \sum_{PQ}$$

$$\times \langle \widehat{\mathbf{k}}'|L'\mu'\rangle\langle L'\tfrac{1}{2}\mu'v'|J'M'\rangle\langle J'j'M'm'|PQ\rangle\langle k'L'J'n'\ell'j' \parallel V_{P\pi} \parallel j\ell nJLk\rangle$$

$$\times \langle PQ|JjMm\rangle\langle JM|L\tfrac{1}{2}\mu v\rangle\langle L\mu|\widehat{\mathbf{k}}\rangle. \tag{7.36}$$

The T-matrix element is expanded similarly. The reduced V- and T-matrix elements are obtained by inverting (7.36) using the orthonormality relations (3.71) of the spherical harmonics and (3.89) of the Clebsch–Gordan coefficients.

$$\langle k'L'J'n'\ell'j' \parallel V_{P\pi} \parallel j\ell nJLk\rangle$$

$$= \int d\widehat{\mathbf{k}}' \int d\widehat{\mathbf{k}} \sum_{\mu'M'm'v'} \sum_{\mu Mmv} \langle L'\mu'|\widehat{\mathbf{k}}'\rangle\langle J'M'|L'\tfrac{1}{2}\mu'v'\rangle\langle PQ|J'j'M'm'\rangle$$

$$\times \langle \mathbf{k}'v'n'\ell'j'm'|V|n\ell jmv\mathbf{k}\rangle\langle JjMm|PQ\rangle\langle L\tfrac{1}{2}\mu v|JM\rangle\langle \widehat{\mathbf{k}}|L\mu\rangle. \tag{7.37}$$

The reduced Lippmann–Schwinger equations are obtained by expanding all the amplitudes of (6.73) according to (7.36) and again using the orthonormality relations (3.71,3.89) to eliminate the integral over $\widehat{\mathbf{k}}$ and the sum over Clebsch–Gordan coefficients in the expansion of the projection operator

$$\Sigma_j \int d\widehat{\mathbf{k}}'|j\mathbf{k}'\rangle\langle \mathbf{k}'j|.$$

They are equations for the coefficients of the product of spherical harmonics and Clebsch–Gordan coefficients in the sums analogous to (7.36).

$$\langle kL'J'n'\ell'j' \parallel T_{P\pi} \parallel j\ell nJLk_0\rangle = \langle kL'J'n'\ell'j' \parallel V_{P\pi} \parallel j\ell nJLk_0\rangle$$

$$+ \sum_{L''j''n''\ell''j''} \int dk'k'^2 \langle kL'J'n'\ell'j' \parallel V_{P\pi} \parallel j''\ell''n''J''L''k'\rangle$$

$$\times \frac{1}{E^{(+)} - \epsilon'' - \tfrac{1}{2}k'^2} \langle k'L''J''n''\ell''j'' \parallel T_{P\pi} \parallel j\ell nJLk_0\rangle. \tag{7.38}$$

There are independent sets of equations for each total angular momentum

P and parity π. The numerical solution of (7.38) is analogous to the solution of (4.121), described by (4.124).

In solving the distorted-wave integral equations (6.87) the potential matrix elements are calculated using distorted waves for the potential U (6.79), with appropriate phases (4.133), to replace the plane waves. In the channel states $|i\mathbf{k}\rangle$ the plane waves $|\mathbf{k}\rangle$ are replaced, where appropriate, by bound states $|\lambda\rangle$ of U whose quantum numbers are N, L, J. The integration over the momentum k is replaced by a sum over the bound-state index λ. If the distorting potential U is positively charged, as it must be for electron scattering by a positive ion, there are infinitely-many bound states. The sum over λ is, however, rapidly convergent and may be cut off at $N \sim 5$.

$$|j\ell nJLk\rangle \rightarrow |j\ell nJLN\rangle. \tag{7.39}$$

7.2.2 LS coupling

For small atomic targets the spin–orbit potential (3.174) may be neglected. The target states may be described by LS coupling. This introduces an economy in the number of channels to be coupled since there is no spin–orbit splitting of the target states whose orbital angular momentum is ℓ. The description of the reduced Lippmann–Schwinger equations parallels that for jj coupling. The quantum numbers used in the representation are defined in table 7.2.

Table 7.2. *Quantities describing LS-coupling states*

Principal quantum number of target state	n
Orbital angular momentum of target state	ℓ
Corresponding projection	m
Spin of target state	s
Corresponding projection	σ
Absolute momentum of projectile	k
Spin projection of projectile	v
Orbital angular momentum of projectile	L
Corresponding projection	μ
Total orbital angular momentum	K
Corresponding projection	M
Total spin	S
Corresponding projection	N
Total angular momentum	P
Corresponding projection	Q
Overall parity	π

The *LS*-coupling expansion of the potential matrix element is

$$\langle \mathbf{k}'v'n'\ell'm's'\sigma'|V|n\ell ms\sigma v\mathbf{k}\rangle$$
$$= \sum_{L'\mu'K'M'S'N'} \sum_{L\mu KMSN} \sum_{PQ} \langle \hat{\mathbf{k}}'|L'\mu'\rangle \langle L'\ell'\mu'm'|K'M'\rangle \langle \tfrac{1}{2}s'v'\sigma'|S'N'\rangle$$
$$\times \langle K'S'M'N'|PQ\rangle \langle k'L'n'\ell's'K'S' \parallel V_{P\pi} \parallel SKs\ell nLk\rangle \langle PQ|KSMN\rangle$$
$$\times \langle SN|\tfrac{1}{2}sv\sigma\rangle \langle KM|L\ell\mu m\rangle \langle L\mu|\hat{\mathbf{k}}\rangle. \tag{7.40}$$

If the potential cannot change the total spin S and orbital angular momentum K the reduced matrix element becomes

$$\langle k'L'n'\ell's'K'S' \parallel V_{P\pi} \parallel SKs\ell nLk\rangle$$
$$= \langle k'L'n'\ell's' \parallel V_{KS\pi} \parallel s\ell nLk\rangle \delta_{K'K}\delta_{S'S} \tag{7.41}$$

and (7.40) reduces, using the orthonormality of the Clebsch–Gordan coefficients (3.90), to

$$\langle \mathbf{k}'v'n'\ell'm's'\sigma'|V|n\ell ms\sigma v\mathbf{k}\rangle$$

$$= \sum_{L'\mu'} \sum_{L\mu} \sum_{KMSN} \langle \hat{\mathbf{k}}'|L'\mu'\rangle \langle L'\ell'\mu'm'|KM\rangle \langle \tfrac{1}{2}s'v'\sigma'|SN\rangle$$

$$\times \langle k'L'n'\ell's' \parallel V_{KS\pi} \parallel s\ell nLk\rangle \langle SN|\tfrac{1}{2}sv\sigma\rangle \langle KM|L\ell\mu m\rangle \langle L\mu|\hat{\mathbf{k}}\rangle. \tag{7.42}$$

The reduced potential matrix elements are obtained by inverting (7.42).

$$\langle k'L'n'\ell's' \parallel V_{KS\pi} \parallel s\ell nLk\rangle$$
$$= \int d\hat{\mathbf{k}}' \int d\hat{\mathbf{k}} \sum_{\mu'm'v'\sigma'} \sum_{\mu mv\sigma} \langle L'\mu'|\hat{\mathbf{k}}'\rangle \langle KM|L'\ell'\mu'm'\rangle \langle SN|\tfrac{1}{2}s'v'\sigma'\rangle$$
$$\times \langle \mathbf{k}'v'n'\ell'm's'\sigma'|V|n\ell ms\sigma v\mathbf{k}\rangle \langle \tfrac{1}{2}sv\sigma|SN\rangle \langle L\ell\mu m|KM\rangle \langle \hat{\mathbf{k}}|L\mu\rangle. \tag{7.43}$$

The reduced Lippmann–Schwinger equations are

$$\langle kL'n'\ell's' \parallel T_{KS\pi} \parallel s\ell nLk_0\rangle = \langle kL'n'\ell's' \parallel V_{KS\pi} \parallel s\ell nLk_0\rangle$$
$$+ \sum_{L''n''\ell''s''} \int dk'k'^2 \langle kL'n'\ell's' \parallel V_{KS\pi} \parallel s''\ell''n''L''k'\rangle$$
$$\times \frac{1}{E^{(+)} - \epsilon'' - \tfrac{1}{2}k'^2} \langle k'L''n''\ell''s'' \parallel T_{KS\pi} \parallel s\ell nLk_0\rangle. \tag{7.44}$$

7.3 Potential matrix elements

The solution of the reduced Lippmann–Schwinger equations requires the calculation of potential matrix elements that are the off-shell analogues of the Born approximation, or the distorted-wave Born approximation if the potential U of (6.76) is not zero. To derive the reduced potential matrix elements (7.37,7.43) we require the full potential matrix elements (7.36,7.40). For the two-electron states we use the more-general jj-coupling case to illustrate the derivation of the reduced potential matrix elements. For LS coupling we give the results. The derivations are analogous. We again consider the plane-wave representation in detail, noting the changes necessary to implement the distorted-wave representation. We generalise the potential-scattering case (4.118).

The coordinate–spin representation of an eigenstate of the momentum $\mathbf{k}$ of an electron with spin projection v is written in partial-wave form as

$$\langle \mathbf{r}_0 \sigma_0 | v \mathbf{k} \rangle = \left(\frac{2}{\pi}\right)^{1/2} \frac{1}{kr_0} \sum_{L\mu} i^L U_L(kr_0) \langle \hat{\mathbf{r}}_0 | L\mu \rangle \langle \sigma_0 | \tfrac{1}{2} v \rangle \langle L\mu | \hat{\mathbf{k}} \rangle. \tag{7.45}$$

This form is used in the LS-coupling representation of the potential matrix element. For jj coupling we use

$$\langle \mathbf{r}_0 \sigma_0 | v \mathbf{k} \rangle = \left(\frac{2}{\pi}\right)^{1/2} \frac{1}{kr_0} \sum_{L\mu JM} i^L U_L(kr_0) \langle \hat{\mathbf{r}}_0 \sigma_0 | LJM \rangle \langle JM | L\tfrac{1}{2}\mu v \rangle \langle L\mu | \hat{\mathbf{k}} \rangle \tag{7.46}$$

obtained from (7.45) by inverting the definition (3.92) of $\langle \hat{\mathbf{r}}_0 \sigma_0 | LJM \rangle$ using (3.90).

For inclusion of the target states it is useful to define a further reduction of the notation by separating the radial coordinate of the projectile from the projectile-spin–angle and target coordinates. The variables displayed in the state vectors indicate the context of the reduced notation. The coordinate–spin representation of an $(N+1)$-electron state is given in the notation of (7.26) as

$$\langle \mathbf{r}_0 \sigma_0 \bar{x}_0 | i v \mathbf{k} \rangle \equiv \langle \mathbf{r}_0 \sigma_0 \bar{x}_0 | n\ell j m v \mathbf{k} \rangle$$

$$= \left(\frac{2}{\pi}\right)^{1/2} \frac{1}{kr_0} \sum_{P} \sum_{L\mu JM} i^L U_L(kr_0) \langle \hat{\mathbf{r}}_0 \sigma_0 \bar{x}_0 \parallel PiJL \rangle \langle PQ | JjMm \rangle$$

$$\times \langle JM | L\tfrac{1}{2}\mu v \rangle \langle L\mu | \hat{\mathbf{k}} \rangle. \tag{7.47}$$

The reduced notation is expressed by inverting (7.47) using (3.71,3.89).

$$\langle \hat{\mathbf{r}}_0 \sigma_0 \bar{x}_0 \parallel PiJL \rangle = \sum_{Mm} \langle \bar{x}_0 | i \rangle \langle \hat{\mathbf{r}}_0 \sigma_0 | LJM \rangle \langle JjMm | PQ \rangle. \tag{7.48}$$

The target states are expressed, according to equn. (7.35) for the full potential matrix elements, in terms of orbitals $|\alpha\rangle$ and $|\beta\rangle$. The quantity that relates the orbitals to the target states $|i'\rangle$ and $|i\rangle$ is the m-scheme density matrix $\langle i'|a_\alpha^\dagger a_\beta|i\rangle$. Its transformation properties under rotations are important in finding the reduced potential matrix elements.

The notation used to represent the orbital in coordinate–spin space is

$$\langle \mathbf{r}_1\sigma_1|\alpha\rangle \equiv \langle \mathbf{r}_1\sigma_1|n_\alpha\ell_\alpha j_\alpha m_\alpha\rangle$$
$$= r_1^{-1}u_\alpha(r_1)\langle \hat{\mathbf{r}}_1\sigma_1|\ell_\alpha j_\alpha m_\alpha\rangle. \tag{7.49}$$

The reduced notation for orbitals is

$$\langle \hat{\mathbf{r}}_1\sigma_1 \parallel \alpha\rangle \equiv \langle \hat{\mathbf{r}}_1\sigma_1 \parallel n_\alpha\ell_\alpha j_\alpha\rangle. \tag{7.50}$$

Since the orbital $a_\alpha^\dagger|0\rangle$ transforms under rotations as $|j_\alpha m_\alpha\rangle$, the orbital creation and annihilation operators behave as follows in view of (3.98).

$$a_\alpha^\dagger \quad \text{transforms as} \quad |j_\alpha m_\alpha\rangle,$$

$$a_\beta \quad \text{transforms as} \quad (-1)^{j_\beta-m_\beta}|j_\beta -m_\beta\rangle. \tag{7.51}$$

We form the tensor product (3.99) of the two operators, using (3.93) to express the Clebsch–Gordan coefficient as a 3-j symbol and using the symmetry properties of the 3-j symbol.

$$T_q^k = \sum_{m_\alpha m_\beta}(-1)^{j_\beta-m_\beta}\langle j_\alpha j_\beta m_\alpha -m_\beta|kq\rangle a_\alpha^\dagger a_\beta$$

$$= \sum_{m_\alpha m_\beta}(-1)^{j_\alpha-m_\alpha}\hat{k}\begin{pmatrix} j_\alpha & k & j_\beta \\ -m_\alpha & q & m_\beta \end{pmatrix}a_\alpha^\dagger a_\beta. \tag{7.52}$$

For inverting relationships involving 3-j symbols it is useful to note the 3-j equivalent of (3.89,3.90).

$$\sum_{m_1 m_2}\begin{pmatrix} j_1 & j_2 & j' \\ m_1 & m_2 & m' \end{pmatrix}\begin{pmatrix} j_1 & j_2 & j \\ m_1 & m_2 & m \end{pmatrix} = \hat{j}^{-2}\delta_{j'j}\delta_{m'm}, \tag{7.53}$$

$$\sum_{jm}\hat{j}^2\begin{pmatrix} j_1 & j_2 & j \\ m_1' & m_2' & m \end{pmatrix}\begin{pmatrix} j_1 & j_2 & j \\ m_1 & m_2 & m \end{pmatrix} = \delta_{m_1'm_1}\delta_{m_2'm_2}. \tag{7.54}$$

We use (7.54) to invert the tensor relation (7.52).

$$a_\alpha^\dagger a_\beta = (-1)^{j_\alpha-m_\alpha}\sum_{kq}\hat{k}\begin{pmatrix} j_\alpha & k & j_\beta \\ -m_\alpha & q & m_\beta \end{pmatrix}T_q^k. \tag{7.55}$$

For the target states we use the notation

$$|i\rangle \equiv |n\ell jm\rangle, \tag{7.56}$$
$$\|i\rangle \equiv \|n\ell j\rangle. \tag{7.57}$$

The reduced density-matrix element is defined in terms of T_q^k by the Wigner—Eckart theorem (3.104).

$$\langle i'|T_q^k|i\rangle = (-1)^{j'-m'} \begin{pmatrix} j' & k & j \\ -m' & q & m \end{pmatrix} \langle i' \parallel \mathbf{T}^k \parallel i \rangle. \tag{7.58}$$

It is

$$\langle i' \parallel \gamma_{\alpha\beta}^k \parallel i \rangle \equiv \hat{k}^{-1} \langle i' \parallel \mathbf{T}^k \parallel i \rangle$$

$$= \sum_{m'm} \sum_{m_\alpha m_\beta} (-1)^{j'+j_\alpha-m-m_\beta} \hat{k}^2 \begin{pmatrix} j' & k & j \\ -m' & q & m \end{pmatrix}$$

$$\times \begin{pmatrix} j_\alpha & k & j_\beta \\ -m_\alpha & q & m_\beta \end{pmatrix} \langle i'|a_\alpha^\dagger a_\beta|i\rangle. \tag{7.59}$$

The reduced density-matrix element for *LS* coupling is found similarly, treating the orbital and spin factors independently.

7.3.1 Direct matrix element

The coordinate representation of the potential in the direct matrix element of (7.35) is

$$N^{-1}V_0(r_0) + v_{01}(|\mathbf{r}_0 - \mathbf{r}_1|) = -\frac{1}{r_0} + \frac{1}{|\mathbf{r}_0 - \mathbf{r}_1|}$$

$$\equiv \sum_{\lambda\eta} v_\lambda^D(r_0, r_1)(-1)^\eta C_{-\eta}^\lambda(\hat{\mathbf{r}}_0)C_\eta^\lambda(\hat{\mathbf{r}}_1), \tag{7.60}$$

where

$$v_\lambda^D(r_0, r_1) = \frac{r_<^\lambda}{r_>^{\lambda+1}} - \frac{1}{r_0}\delta_{\lambda 0}\delta(r_0 - r_1). \tag{7.61}$$

The multipole expansion of the Coulomb potential is given by (3.102, 3.103).

The direct reduced potential matrix element in *jj* coupling is calculated by substituting the full potential matrix elements into (7.37) with the continuum orbitals given by (7.46), the bound orbitals by (7.49), the two-electron potential by (7.60) and the reduced density-matrix element by (7.59).

$$\langle k'L'J'n'\ell'j' \parallel V_{P\pi}^D \parallel j\ell nJLk \rangle = \int d\hat{\mathbf{k}}' \int d\hat{\mathbf{k}}$$

$$\times \sum_{L'J'M'LJM} \sum_{\mu'v'\mu v} \langle L'\mu'|\hat{\mathbf{k}}'\rangle \langle J'M'|L'\tfrac{1}{2}\mu'v'\rangle \langle L\tfrac{1}{2}\mu v|JM\rangle \langle \hat{\mathbf{k}}|L\mu\rangle$$

$$\times \sum_{m'm} \langle PQ|J'j'M'm'\rangle \langle JjMm|PQ\rangle$$

$$\times \sum_{\alpha\beta} \langle i'|a_\alpha^\dagger a_\beta|i\rangle$$

$$\times \int dr_0 r_0^2 \int dr_1 r_1^2 \int d\hat{\mathbf{r}}_0 \int d\hat{\mathbf{r}}_1 \int d^3\sigma_0 \int d^3\sigma_1$$

$$\times \left(\frac{2}{\pi}\right)^{1/2} \frac{1}{k'r_0} \sum_{\bar{L}'\bar{\mu}'\bar{J}'\bar{M}'} i^{-\bar{L}'} U_{\bar{L}'}(k'r_0)\langle\bar{L}'\bar{J}'\bar{M}'|\hat{\mathbf{r}}_0\sigma_0\rangle\langle\bar{L}'\tfrac{1}{2}\bar{\mu}'\nu'|\bar{J}'\bar{M}'\rangle\langle\hat{\mathbf{k}}'|\bar{L}'\bar{\mu}'\rangle$$

$$\times r_1^{-1} u_\alpha(r_1)\langle\ell_\alpha j_\alpha m_\alpha|\hat{\mathbf{r}}_1\sigma_1\rangle$$

$$\times \sum_{\lambda\eta} v_\lambda^D(r_0,r_1)(-1)^\eta C_{-\eta}^\lambda(\hat{\mathbf{r}}_0)C_\eta^\lambda(\hat{\mathbf{r}}_1)$$

$$\times r_1^{-1} u_\beta(r_1)\langle\hat{\mathbf{r}}_1\sigma_1|\ell_\beta j_\beta m_\beta\rangle$$

$$\times \left(\frac{2}{\pi}\right)^{1/2} \frac{1}{kr_0} \sum_{\bar{L}\bar{\mu}\bar{J}\bar{M}} i^{\bar{L}} U_{\bar{L}}(kr_0)\langle\hat{\mathbf{r}}_0\sigma_0|\bar{L}\bar{J}\bar{M}\rangle\langle\bar{J}\bar{M}|\bar{L}\tfrac{1}{2}\bar{\mu}\nu\rangle\langle\bar{L}\bar{\mu}'|\hat{\mathbf{k}}\rangle. \tag{7.62}$$

The spin—angle integrations are performed by (3.104,107). We use the orthogonality of the spherical harmonics (3.71) and the Clebsch—Gordan coefficients (3.89). Expressing the Clebsch—Gordan coefficients as 3-j symbols by (3.93) the direct reduced potential matrix element becomes

$$\langle k'L'J'n'\ell'j' \parallel V_{P\pi}^D \parallel j\ell nJLk\rangle =$$

$$\sum_{M'Mm'm} (-1)^{J'-j'+Q}\hat{P} \begin{pmatrix} J' & j' & P \\ M' & m' & -Q \end{pmatrix} (-1)^{J-j+Q}\hat{P} \begin{pmatrix} J & j & P \\ M & m & -Q \end{pmatrix}$$

$$\times \sum_{\alpha\beta} \sum_{\lambda\eta} (-1)^{J'-M'+\eta} \begin{pmatrix} J' & \lambda & J \\ -M' & -\eta & M \end{pmatrix} \langle L'J' \parallel \mathbf{C}^\lambda \parallel LJ\rangle$$

$$\times \langle\ell_\alpha j_\alpha \parallel \mathbf{C}^\lambda \parallel \ell_\beta j_\beta\rangle R_{L'L\alpha\beta}^{\lambda D}(k',k)$$

$$\times (-1)^{j_\alpha-m_\alpha} \begin{pmatrix} j_\alpha & \lambda & j_\beta \\ -m_\alpha & \eta & m_\beta \end{pmatrix} \langle i'|a_\alpha^\dagger a_\beta|j\rangle. \tag{7.63}$$

The direct radial matrix element is

$$R_{L'L\alpha\beta}^{\lambda D}(k',k) = \frac{2}{\pi k'k} i^{L-L'} \int dr_0 \int dr_1 U_{L'}(k'r_0)u_\alpha(r_1)v_\lambda^D(r_0,r_1)$$
$$\times u_\beta(r_1)U_L(kr_0). \tag{7.64}$$

We use (7.52,7.58,7.59) to obtain the relation between the m-scheme and reduced density-matrix elements

$$(-1)^{j'-m'} \begin{pmatrix} j' & k & j \\ -m' & q & m \end{pmatrix} \langle i' \parallel \gamma_{\alpha\beta}^k \parallel i\rangle = \sum_{m_\alpha m_\beta} (-1)^{j_\alpha-m_\alpha}$$

$$\times \begin{pmatrix} j_\alpha & k & j_\beta \\ -m_\alpha & q & m_\beta \end{pmatrix} \langle i'|a_\alpha^\dagger a_\beta|i\rangle. \tag{7.65}$$

This is substituted into (7.63). The projection quantum number sums over four 3-j symbols reduce to a 6-j symbol by (3.95).

The final form for the direct reduced potential matrix element is

$$\langle k'L'J'n'\ell'_{-}j' \parallel V^D_{P\pi} \parallel j\ell nJLk \rangle =$$

$$\sum_{\alpha\beta\lambda}(-1)^{J+j'+P} \begin{Bmatrix} J' & j' & P \\ j & J & \lambda \end{Bmatrix} \langle L'J' \parallel \mathbf{C}^\lambda \parallel LJ \rangle \langle \ell_\alpha j_\alpha \parallel \mathbf{C}^\lambda \parallel \ell_\beta j_\beta \rangle$$

$$\times R^{\lambda D}_{L'L\alpha\beta}(k',k)\langle i' \parallel \gamma^\lambda_{\alpha\beta} \parallel i \rangle. \tag{7.66}$$

Note that the reduced matrix elements of $\mathbf{C}^\lambda$ (3.107) imply the parity selection rule.

The direct reduced potential matrix element for LS coupling is given by Bray *et al.* (1989). In this case the integrations over the spin coordinates σ_0 and σ_1 result in the factor $\langle v'|v\rangle\langle v_\alpha|v_\beta\rangle$, which prohibits spin flip.

$$\langle k'L'n'\ell's' \parallel V^D_{KS\pi} \parallel s\ell nLk \rangle =$$

$$\sum_{\alpha\beta\lambda}(-1)^{L+\ell'+K} \begin{Bmatrix} L' & \ell' & K \\ \ell & L & \lambda \end{Bmatrix} \langle L' \parallel \mathbf{C}^\lambda \parallel L \rangle \langle \ell_\alpha \parallel \mathbf{C}^\lambda \parallel \ell_\beta \rangle$$

$$\times R^{\lambda D}_{L'L\alpha\beta}(k',k)\langle i' \parallel \gamma^\lambda_{\alpha\beta} \parallel i \rangle \langle s'|s\rangle, \tag{7.67}$$

where the radial matrix element is again given by (7.64) and the reduced density matrix element is defined analogously to the jj-coupling case.

7.3.2 Exchange matrix elements

The reduction of the exchange potential matrix element to the form used for computation parallels that for the direct term. We exchange the coordinates $\mathbf{r}_0$, σ_0 and $\mathbf{r}_1$, σ_1 in the kets of (7.37).

The coordinate representation of the two-electron potential in the exchange matrix elements of (7.35) is

$$-v_{01}(|\mathbf{r}_0 - \mathbf{r}_1|) + \epsilon = -\frac{1}{|\mathbf{r}_0 - \mathbf{r}_1|} + \epsilon$$

$$\equiv \sum_{\lambda\eta} v^E_\lambda(r_0,r_1)(-1)^\eta C^\lambda_{-\eta}(\hat{\mathbf{r}}_0)C^\lambda_\eta(\hat{\mathbf{r}}_1), \tag{7.68}$$

where

$$v^E_\lambda(r_0,r_1) = -\frac{r^\lambda_<}{r^{\lambda+1}_>} + \epsilon\delta_{\lambda 0}\delta(r_0 - r_1), \tag{7.69}$$

$$\epsilon = (1-\theta)E - \tfrac{1}{2}(\epsilon_{i'} + \epsilon_i + \epsilon_\alpha + e_\beta). \tag{7.70}$$

The expression analogous to (7.63) is

$$\langle k'L'J'n'\ell'j' \parallel V_{P\pi}^E \parallel j\ell nJLk\rangle =$$

$$\sum_{M'Mm'm} (-1)^{J'-j'+Q} \widehat{P} \begin{pmatrix} J' & j' & P \\ M' & m' & -Q \end{pmatrix} (-1)^{J-j+Q} \widehat{P} \begin{pmatrix} J & j & P \\ M & m & -Q \end{pmatrix}$$

$$\times \sum_{\alpha\beta} \sum_{\lambda\eta} (-1)^{J'-M'+\eta} \begin{pmatrix} J' & \lambda & j_\beta \\ -M' & -\eta & m_\beta \end{pmatrix} \langle L'J' \parallel \mathbf{C}^\lambda \parallel \ell_\beta j_\beta \rangle$$

$$\times \langle \ell_\alpha j_\alpha \parallel \mathbf{C}^\lambda \parallel LJ \rangle R_{L'L\alpha\beta}^{\lambda E}(k',k)$$

$$\times (-1)^{j_\alpha - m_\alpha} \begin{pmatrix} j_\alpha & \lambda & J \\ -m_\alpha & \eta & M \end{pmatrix} \langle i'|a_\alpha^\dagger a_\beta|i\rangle, \tag{7.71}$$

where the exchange radial matrix element is

$$R_{L'L\alpha\beta}^{\lambda E}(k',k) = \frac{2}{\pi k'k} i^{L-L'} \int dr_0 \int dr_1 U_{L'}(k'r_0) u_\alpha(r_1) v_\lambda^E(r_0, r_1)$$
$$\times u_\beta(r_0) U_L(kr_1). \tag{7.72}$$

We cannot now use the relation (7.65) directly but must use (7.55, 7.58, 7.59) to represent the *m*-scheme density-matrix element. The final form for the exchange reduced potential matrix element is

$$\langle k'L'J'n'\ell'j' \parallel V_{P\pi}^E \parallel j\ell nJLk\rangle =$$

$$\sum_{\alpha\beta\lambda\kappa} (-1)^{-J+j'+\lambda+P+\kappa} \widehat{\kappa}^2 \begin{Bmatrix} J' & j & \kappa \\ j & j' & P \end{Bmatrix} \begin{Bmatrix} J' & J & \kappa \\ j_\alpha & j_\beta & \lambda \end{Bmatrix}$$

$$\times \langle L'J' \parallel \mathbf{C}^\lambda \parallel \ell_\beta j_\beta \rangle \langle \ell_\alpha j_\alpha \parallel \mathbf{C}^\lambda \parallel LJ \rangle$$

$$\times R_{L'L\alpha\beta}^{\lambda E}(k',k)\langle i' \parallel \gamma_{\alpha\beta}^\kappa \parallel i\rangle. \tag{7.73}$$

The corresponding expression for the *LS*-coupling representation is

$$\langle k'L'n'\ell's' \parallel V_{KS\pi}^E \parallel s\ell nLk\rangle =$$

$$\sum_{\alpha\beta\lambda j\sigma} \widehat{\sigma}^2 (-1)^{s'+S+1/2} \begin{Bmatrix} s' & s & \sigma \\ \frac{1}{2} & \frac{1}{2} & S \end{Bmatrix} \widehat{j}^2 (-1)^{-L+\ell'+\lambda+K+j} \begin{Bmatrix} L' & L & j \\ \ell & \ell' & K \end{Bmatrix}$$

$$\times \begin{Bmatrix} L' & L & j \\ \ell_\alpha & \ell_\beta & \lambda \end{Bmatrix} \langle L' \parallel \mathbf{C}^\lambda \parallel \ell_\beta \rangle \langle \ell_\alpha \parallel \mathbf{C}^\lambda \parallel L \rangle$$

$$\times R_{L'L\alpha\beta}^{\lambda E}(k',k)\langle i' \parallel \gamma_{\alpha\beta}^{j\sigma} \parallel i\rangle, \tag{7.74}$$

where the reduced density matrix element is obtained analogously to (7.59), treating orbital and spin degrees of freedom separately. Details are given by Bray *et al.* (1989).

For *LS* coupling it is noteworthy that the direct potential cannot change the spin of the target state while the exchange potential can. The total spin S does not change for LS coupling, whereas it can change in jj coupling because of the spin—orbit potential.

The potential matrix elements for the distorted-wave representation (6.87) of the Lippmann—Schwinger equations are calculated by replacing $U_L(kr)$ in the radial matrix elements (7.63,7.72) by the appropriate solution of the Schrödinger equation for the local, central distorting potential U_0. In the case of scattering by an ion U_0 has charge Z, given by (7.16). This means that its form for large r is Z/r. Scattering solutions are $e^{i\sigma_L}u_L(k,r)$ (4.92). Bound solutions are $u_{NL}(r)$ (4.23), where the quantum numbers N,L are described by (7.39).

7.3.3 *The density matrix in special cases*

In this section we show how to calculate the m-scheme density-matrix element $\langle i'|a_\alpha^\dagger a_\beta|i\rangle$ in cases of particular interest. We also show the simplifications to the potential matrix elements that follow when the density matrix is particularly simple.

One-electron target

For hydrogen or a hydrogenic ion the target states $|i\rangle$ are one-electron orbitals $|\alpha\rangle$. The density matrix is

$$\langle i'|a_\alpha^\dagger a_\beta|i\rangle = \delta_{i'\alpha}\delta_{i\beta}. \tag{7.75}$$

The expression (7.59) for the reduced density matrix has been defined so that in this case it is given by (7.53) as

$$\langle i' \parallel \gamma_{\alpha\beta}^k \parallel i\rangle = \delta_{j'j_\alpha}\delta_{jj_\beta}. \tag{7.76}$$

The LS-coupling case is particularly simple because the target spins s',s are each $\frac{1}{2}$. The potential is spin-independent so the spin coupling is independent of the space coupling. Writing the space-direct and space-exchange amplitudes of the coupled Schrödinger equations (7.24) as D and E respectively we have

$$\langle\tfrac{1}{2}\tfrac{1}{2}v_0'v_1'|SN'\rangle D\langle SN|\tfrac{1}{2}\tfrac{1}{2}v_0v_1\rangle - \langle\tfrac{1}{2}\tfrac{1}{2}v_1'v_0'|SN'\rangle E\langle SN|\tfrac{1}{2}\tfrac{1}{2}v_0v_1\rangle = 0. \tag{7.77}$$

We rearrange the exchange spin-coupling coefficient thus.

$$\langle\tfrac{1}{2}\tfrac{1}{2}v_1'v_0'|SN'\rangle = (-1)^{1+S}\langle\tfrac{1}{2}\tfrac{1}{2}v_0'v_1'|SN'\rangle. \tag{7.78}$$

On factoring out the spin-coupling coefficients we obtain

$$D + (-1)^S E = 0. \tag{7.79}$$

We thus have independent sets of coupled equations for the singlet ($S = 0$) and triplet ($S = 1$) cases with potential matrix elements (7.67,7.74) given by

$$V^S = V^D + (-1)^S V^E. \tag{7.80}$$

Inert target

Configuration-interaction calculations for most atoms are practicable only if they are treated in terms of a few active electrons and a single-determinant closed-shell core, for which $\ell = j = 0$. The core is inert in the sense that the configuration basis omits configurations with unoccupied core orbitals.

Labelling the orbitals $1,...,N$ the density matrix for an inert target is

$$\langle i'|a_\alpha^\dagger a_\beta|i\rangle = \langle 0|a_N..a_1 a_\alpha^\dagger a_\beta a_1^\dagger..a_N^\dagger|0\rangle. \tag{7.81}$$

Each time we move $a_\alpha^\dagger$ μ places to the left and a_β ν places to the right we obtain a term

$$(-1)^{\mu+\nu}\delta_{\alpha\mu}\delta_{\beta\nu}\langle 0|a_N..a_{\mu+1}a_{\mu-1}..a_1 a_1^\dagger..a_{\nu-1}^\dagger a_{\nu+1}^\dagger..a_N^\dagger|0\rangle = \delta_{\alpha\mu}\delta_{\beta\nu}\delta_{\mu\nu}. \tag{7.82}$$

The potential matrix element is

$$\langle \mathbf{k}'i'|V|i\mathbf{k}\rangle = \sum_{\alpha\beta}\langle i'|a_\alpha^\dagger a_\beta|i\rangle\langle \mathbf{k}'\alpha|V|\beta\mathbf{k}\rangle$$

$$= \sum_\mu \langle \mathbf{k}'\mu|V|\mu\mathbf{k}\rangle. \tag{7.83}$$

This is a sum of diagonal terms, one for each orbital. The potential operator V is given by (7.35) and includes direct terms and one- and two-electron exchange terms.

For a closed-shell target the LS-coupling case again simplifies. For the orbital μ the spin integrations give factors $\langle v_0'|v_0\rangle\langle v_\mu'|v_\mu\rangle$ for the space-direct matrix element and $-\langle v_0'|v_\mu\rangle\langle v_\mu'|v_0\rangle$ for the space-exchange matrix element. Since v_μ is equal to v_0' for only half the orbitals we have for space-orbitals α

$$\langle \mathbf{k}'i'|V|i\mathbf{k}\rangle = 2\sum_\alpha\langle \mathbf{k}'\alpha|V^D|\alpha\mathbf{k}\rangle - \sum_\alpha\langle \alpha\mathbf{k}'|V^E|\alpha\mathbf{k}\rangle. \tag{7.84}$$

Closed shell + 1 electron

We represent the inert closed-shell core by $|C\rangle$ and consider the states

$$|i'\rangle = a_\mu^\dagger|C\rangle,$$

$$|i\rangle = a_\nu^\dagger|C\rangle. \tag{7.85}$$

The density matrix is evaluated by applying the fermion commutation rules (3.137–3.139).

$$\langle i'|a_\alpha^\dagger a_\beta|i\rangle = \langle C|a_\mu a_\alpha^\dagger a_\beta a_\nu^\dagger|C\rangle$$
$$= \delta_{\alpha\mu}\delta_{\beta\nu} + \delta_{\mu\nu}\langle C|a_\alpha^\dagger a_\beta|C\rangle. \qquad (7.86)$$

The potential matrix element consists of a hydrogenic term and, if the matrix element is diagonal, a sum of core terms similar to the case of an inert target (7.83).

$$\langle \mathbf{k}'i'|V|i\mathbf{k}\rangle = \langle \mathbf{k}'\mu|V|\nu\mathbf{k}\rangle + \sum_{\rho\in C}\langle \mathbf{k}'\rho|V|\rho\mathbf{k}\rangle. \qquad (7.87)$$

Closed shell + 2 electrons

The state $|i\rangle$ is represented by a linear combination of determinantal configurations $|\rho\rangle$, each of which has an occupied closed-shell core $|C\rangle$ and two active electrons.

$$|i\rangle = \sum_\rho |\rho\rangle\langle\rho|i\rangle. \qquad (7.88)$$

The configuration $|\rho\rangle$ is

$$|\rho\rangle = a_\mu^\dagger a_\nu^\dagger|C\rangle. \qquad (7.89)$$

Primed quantities denote the state $|i'\rangle$.

The density matrix is a linear combination of configuration density matrices.

$$\langle i'|a_\alpha^\dagger a_\beta|i\rangle = \sum_{\rho'\rho}\langle i'|\rho'\rangle\langle\rho'|a_\alpha^\dagger a_\beta|\rho\rangle\langle\rho|i\rangle. \qquad (7.90)$$

Each configuration density matrix is evaluated by repeated application of the fermion commutation rules. It includes normalisation factors $F_{\rho'}, F_\rho$ given by coupling to the symmetry of $|i'\rangle, |i\rangle$.

$$\langle \rho'|a_\alpha^\dagger a_\beta|\rho\rangle = F_{\rho'}F_\rho(\delta_{\alpha\mu'}\delta_{\beta\mu}\delta_{\nu'\nu} - \delta_{\alpha\mu'}\delta_{\beta\nu}\delta_{\nu'\mu} - \delta_{\alpha\nu'}\delta_{\beta\mu}\delta_{\mu'\nu} + \delta_{\alpha\nu'}\delta_{\beta\nu}\delta_{\mu'\mu})$$

$$+ \langle \rho'|\rho\rangle\langle C|a_\alpha^\dagger a_\beta|C\rangle. \qquad (7.91)$$

The expression (7.91) is the density matrix for two active electrons added to the core term (7.82).

We now find the normalisation factor for a two-electron configuration, using the notation $\bar{\mu}$ for the shell quantum numbers $n_\mu \ell_\mu j_\mu$.

$$a_\mu^\dagger|0\rangle = |\bar{\mu}m_\mu\rangle \equiv |n_\mu\ell_\mu j_\mu m_\mu\rangle. \qquad (7.92)$$

The symmetry configuration is denoted by

$$F_\rho |\rho\rangle = |(n_\mu \ell_\mu j_\mu)(n_\nu \ell_\nu j_\nu) jm\rangle$$

$$= F_\rho \sum_{m_\mu m_\nu} \langle j_\mu j_\nu m_\mu m_\nu | jm\rangle a_\mu^\dagger a_\nu^\dagger |0\rangle. \tag{7.93}$$

The normalisation is given by

$$1 = F_\rho^2 \sum_{m'_\mu m'_\nu m_\mu m_\nu} \langle jm | j_\mu j_\nu m'_\mu m'_\nu \rangle \langle j_\mu j_\nu m_\mu m_\nu | jm\rangle \langle 0| a_{\nu'} a_{\mu'} a_\mu^\dagger a_\nu^\dagger |0\rangle. \tag{7.94}$$

Application of the fermion commutation rules gives

$$\langle 0| a_{\nu'} a_{\mu'} a_\mu^\dagger a_\nu^\dagger |0\rangle = \delta_{\mu'\mu} \delta_{\nu'\nu} - \delta_{\mu'\nu} \delta_{\nu'\mu}. \tag{7.95}$$

For the first term of (7.95) the sum rule (3.89) gives 1. The second term is nonzero only for $\bar{\mu} = \bar{\nu}$, in which case we rearrange the left-hand Clebsch–Gordan coefficient and again use the sum rule to obtain $(-1)^J \delta_{\bar\mu\bar\nu}$ if J is even. Odd J is forbidden for electrons occupying orbitals with identical radial parts, i.e. in the same shell. The normalisation factor is

$$F_\rho = (1 + \delta_{\bar\mu\bar\nu})^{-1/2}. \tag{7.96}$$

The potential matrix element is

$$\langle \mathbf{k}' i' | V | i\mathbf{k}\rangle = \sum_{\rho'\rho} \langle i' | \rho'\rangle \left[(1 + \delta_{\bar{\mu}'\bar{\nu}'})(1 + \delta_{\bar\mu\bar\nu}) \right]^{-1/2} \sum_{\alpha\beta} \langle \mathbf{k}'\alpha | V | \beta\mathbf{k}\rangle \langle \rho | i\rangle$$

$$\times (\delta_{\alpha\mu'} \delta_{\beta\mu} \delta_{\nu'\nu} - \delta_{\alpha\mu'} \delta_{\beta\nu} \delta_{\nu'\mu} - \delta_{\alpha\nu'} \delta_{\beta\mu} \delta_{\mu'\nu} + \delta_{\alpha\nu'} \delta_{\beta\nu} \delta_{\mu'\mu})$$

$$+ \langle i' | i\rangle \sum_{\xi \in C} \langle \mathbf{k}'\xi | V | \xi\mathbf{k}\rangle. \tag{7.97}$$

7.3.4 The spin–orbit matrix element

The spin–orbit potential (3.174) is

$$V_{SL}(r_0) \boldsymbol{\sigma} \cdot \mathbf{L} = \frac{1}{4c^2} \frac{1}{r} \frac{dV_0}{dr} \boldsymbol{\sigma} \cdot \mathbf{L}, \tag{7.98}$$

where V_0 is a central potential for the projectile. A reasonable choice of central potential is the electron–nucleus potential plus the direct part of the electron–electron potential given, for example, by (7.87) or (7.95). In practice the latter has a small effect on scattering since its gradient is small.

This form for the spin–orbit potential acts only in diagonal direct potential matrix elements. It may be considered as an additional term in the potential (7.61)

$$V_\lambda^{SL}(r_0, r_1) = L V_{SL}(r_0)\delta_{\lambda 0}\delta(r_0 - r_1), \qquad L = J + \tfrac{1}{2},$$

$$= -(L+1)V_{SL}(r_0)\delta_{\lambda 0}\delta(r_0 - r_1), \qquad L = J - \tfrac{1}{2}. \quad (7.99)$$

The eigenvalues of $\boldsymbol{\sigma} \cdot \mathbf{L}$ are given by (4.171).

7.4 The complete set of target states

Coupled integral equations for a finite set of channels cannot represent the scattering situation exactly since there is a countably infinite number of bound target states and the ionisation continuum. We discuss two ways of circumventing this difficulty. The first is the convergent-close-coupling method. It approximates the complete set of target states by the configuration-interaction expansion of section 5.6. It is in principle convergent with increasing number of basis configurations. The computation becomes very laborious for larger atoms. The second is the coupled-channels-optical method, in which coupled Lippmann–Schwinger equations are solved for a finite set of eigenstates of the target Hamiltonian. The coupling potential is the optical potential, which formally accounts for channels outside the set. The method is feasible for all atoms. Its implementation involves an approximation that has internal verification for discrete channels but whose validity for the target continuum depends on comparison either with experiment or with the first method in cases where it is practicable. It is discussed in section 7.5.

7.4.1 The convergent-close-coupling method

This method simply involves the solution of the Lippmann–Schwinger equations (6.73) or (6.87) with the potential matrix elements (7.35). The states $|i\rangle$ are not eigenstates of the target Hamiltonian. They are configuration-interaction states or pseudostates obtained by diagonalising the target Hamiltonian in a square-integrable basis as described in section 5.6.

So far the method has only been fully tested for one-electron atoms. In the case of hydrogen a complete check is available for a very restricted subset of angular-momentum states, namely LS-coupled collision states with $\ell = K = 0$. This is the Temkin–Poet problem (Temkin, 1962; Poet, 1978, 1981). The three-body potentials are separable in the radial coordinates. This enables a convergent numerical solution to be obtained.

Early numerical investigations of the close-coupling method for the Temkin—Poet problem were made with coordinate-space solutions of coupled equations using pseudostates obtained by diagonalising the hydrogen Hamiltonian in a Slater-function (4.38) basis. Burke and Mitchell (1973) showed that the singlet amplitude was converging with increasing basis size, except at energies near anomalously-varying basis-dependent features called pseudoresonances. Similar observations were made by Oza and Callaway (1983), who used an averaging technique to smooth the amplitudes over the pseudoresonances.

The use of a finite-basis expansion to represent the continuum is reminiscent of the use of quadratures to represent an integration. Heller, Reinhardt and Yamani (1973) showed that use of the Laguerre basis (5.56) is equivalent to a Gaussian-type quadrature rule. The underlying orthogonal polynomials were shown by Yamani and Reinhardt (1975) to be of the Pollaczek (1950) class.

Bray and Stelbovics (1992a) addressed the question of whether the method of using the Sturmians derived by diagonalising the hydrogen Hamiltonian in a Laguerre basis converges in practice, and in particular whether pseudoresonances are a necessary feature of a finite-basis method. They found complete agreement with Poet (1978, 1981) for scattering and ionisation channels at all energies using up to 30 Sturmians. With a sufficiently-large basis there are no pseudoresonances.

There is no difference in principle between the integral equations (6.73) for the Temkin—Poet problem and those for higher angular momenta. The method is thus established as a convergent calculation of electron-atom scattering. The Temkin—Poet solution is restricted to the zero-angular-momentum case, whereas the convergent-close-coupling method is perfectly general.

7.5 The optical potential

The Lippmann—Schwinger equations (6.73) are written formally in terms of a discrete notation $|i\rangle$ for the complete set of target states, which includes the ionisation continuum. For a numerical solution it is necessary to have a finite set of coupled integral equations. We formulate the coupled-channels-optical equations that describe reactions in a channel subspace, called P space. This is projected from the channel space by an operator P that includes only a finite set of target states. The entrance channel $|0\mathbf{k}_0\rangle$ is included in P space. The method was first discussed by Feshbach (1962). Its application to the momentum-space formulation of electron—atom scattering was introduced by McCarthy and Stelbovics

(1983*b*). Here we generalise the symmetric formulation of Bray, Konovalov and McCarthy (1991*c*), which applied to one-electron targets.

The coupled-channels-optical equations are formally analogous to the Lippmann–Schwinger equivalent of (7.29) in which the coupling potential includes the potential V (7.40) and a polarisation potential $\tilde{V}^{(Q)}$ that describes the real (on-shell) and virtual (off-shell) excitation of the complementary channel space, called Q space. The total coupling potential is the optical potential

$$V^{(Q)} = V + \tilde{V}^{(Q)}, \tag{7.100}$$

so called because early applications to the elastic scattering of nucleons by nuclei reminded physicists of the scattering of light by a cloudy (absorptive) crystal (refractive) ball.

The polarisation potential is complex and nonlocal. The imaginary part is due to on-shell amplitudes for the excitation of Q space from P space. At long range the potential is real. We will show its relationship for large r, where it is due to virtual dipole excitations, to the classical dipole potential $-\alpha/2r^4$ where α is the polarisability.

7.5.1 *The formal polarisation potential*

The channels to be included in P space are the entrance channel, those for which we want to describe experimental observations and others that are so strongly coupled that numerical investigation shows their inclusion to be necessary.

The channel projection operator P and its complementary operator Q have the following properties

$$P + Q = 1, \quad P^2 = P, \quad Q^2 = Q, \quad PQ = QP = 0. \tag{7.101}$$

They are defined in terms of the projection operators $\bar{P}$ and $\bar{Q}$ for the N-electron target space as follows.

$$P = I\bar{P}, \quad Q = I\bar{Q}, \tag{7.102}$$

where I is the identity for the one-electron orbital space.

$$\bar{P} = \sum_{i \in \bar{P}} |i\rangle\langle i| \quad , \quad \bar{Q} = \sum_{i \in \bar{Q}} |i\rangle\langle i|. \tag{7.103}$$

They obey relations analogous to (7.101).

The collision Schrödinger equation (7.1) is written

$$(E^{(+)} - H)|\Psi_0^{(+)}(\mathbf{k})\rangle = 0. \tag{7.104}$$

The coordinate–spin representation of H (7.12) is symmetric in the coordinates x_r of the $N + 1$ electrons and $\langle x|\Psi_0^{(+)}(\mathbf{k})\rangle$ is antisymmetric. The boundary condition is

$$\lim_{r_0 \to \infty} \langle x | \Psi_0^{(+)}(\mathbf{k}) \rangle = \langle x_0 | v\mathbf{k} \rangle \langle \bar{x}_0 | 0 \rangle, \qquad (7.105)$$

where $|0\rangle$ indicates the entrance-channel target state, usually the ground state, and the projectile spin projection v has been restored to the projectile-state notation. The channel projection operators (7.101) satisfy

$$\lim_{r_0 \to \infty} \langle x | P | \Psi_0^{(+)}(\mathbf{k}) \rangle = \langle x_0 | v\mathbf{k} \rangle \langle \bar{x}_0 | 0 \rangle, \qquad (7.106)$$

$$\lim_{r_0 \to \infty} \langle x | Q | \Psi_0^{(+)}(\mathbf{k}) \rangle = 0. \qquad (7.107)$$

We insert $P + Q$ in (7.104) and premultiply by P and Q respectively to obtain

$$P(E^{(+)} - H)P|\Psi_0^{(+)}(\mathbf{k}) \rangle = PHQ|\Psi_0^{(+)}(\mathbf{k}) \rangle, \qquad (7.108)$$

$$Q(E^{(+)} - H)Q|\Psi_0^{(+)}(\mathbf{k}) \rangle = QHP|\Psi_0^{(+)}(\mathbf{k}) \rangle. \qquad (7.109)$$

Using the boundary condition (7.107) we write (7.109) in the form

$$Q|\Psi_0^{(+)}(\mathbf{k}) \rangle = QG_Q(E^{(+)})QHP|\Psi_0^{(+)}(\mathbf{k}) \rangle, \qquad (7.110)$$

where the Q-projected Green's function $G_Q(E^{(+)})$ satisfies

$$Q(E^{(+)} - H)QG_Q(E^{(+)}) = Q. \qquad (7.111)$$

Since P and Q spaces are orthogonal we have

$$PHQ = PWQ,$$
$$QHP = QWP, \qquad (7.112)$$

where W is the symmetric two-electron operator

$$W = \sum_{r<s} v_{rs}. \qquad (7.113)$$

The P-projected Schrödinger equation is written by substituting (7.110) in (7.108).

$$P(E^{(+)} - H - \tilde{V}^{(Q)})P|\Psi_0^{(+)}(\mathbf{k}) \rangle = 0. \qquad (7.114)$$

The polarisation potential $\tilde{V}^{(Q)}$ is given by

$$\tilde{V}^{(Q)} = WQG_Q(E^{(+)})QW. \qquad (7.115)$$

Since H is symmetric $\tilde{V}^{(Q)}$ is also a symmetric two-electron operator

$$\tilde{V}^{(Q)} = \sum_{r<s} \tilde{v}_{rs}^{(Q)}. \qquad (7.116)$$

We may therefore use the arguments of section 7.1, replacing (7.1) by (7.114) to obtain the explicitly-antisymmetrised set of P-projected coupled Schrödinger equations analogous to (7.24,7.35).

$$\sum_j \langle \mathbf{k}'i| P(E^{(+)} - \epsilon_j - K_0 - V^{(Q)}) P | j u_j^{(+)}(\mathbf{k}) \rangle = 0, \qquad (7.117)$$

where the optical potential $V^{(Q)}$ is given by

$$\langle \mathbf{k}'i| P V^{(Q)} P | j u_j^{(+)}(\mathbf{k}) \rangle = \sum_{\alpha\beta} \langle i| \bar{P} a_\alpha^\dagger a_\beta \bar{P} | j \rangle$$

$$\times \Big[\langle \mathbf{k}'\alpha| N^{-1} V_0 + v_{01}^{(Q)} - \theta E I_0 | \beta u_j^{(+)}(\mathbf{k}) \rangle$$

$$+ \langle \alpha \mathbf{k}' | (1 - \theta) E - \tfrac{1}{2}(\epsilon_i + \epsilon_j + \epsilon_\alpha + \epsilon_\beta) - v_{01}^{(Q)} | \beta u_j^{(+)}(\mathbf{k}) \rangle \Big]. \qquad (7.118)$$

The two-electron optical potential is given by (7.116) as

$$v_{01}^{(Q)} = v_{01} + \tilde{v}_{01}^{(Q)}. \qquad (7.119)$$

7.5.2 *The coupled-channels-optical equations*

The integral equations corresponding to (7.117) are obtained by using the unsymmetrised multichannel expansion (7.27), considering (7.117) as a projection onto the channel state $\langle \mathbf{k}'i|$ of

$$\bar{P}(E^{(+)} - K)\bar{P} |\psi_0^{(+)}(\mathbf{k})\rangle = \bar{P} V^{(Q)} \bar{P} |\psi_0^{(+)}(\mathbf{k})\rangle. \qquad (7.120)$$

Here we have used (7.102,7.103) to replace the channel projection operators by the target projection operators. Since (7.120) contains the same information as (7.117) this projection onto the channel space has not destroyed the symmetry.

Using the boundary condition (7.105) and the notation (6.7) we write the integral equation corresponding to (7.120) for a particular entrance-channel momentum $\mathbf{k}_0$.

$$|\psi_0^{(+)}(\mathbf{k}_0)\rangle = |0\mathbf{k}_0\rangle + \frac{1}{\bar{P}(E^{(+)} - K)\bar{P}} \bar{P} V^{(Q)} \bar{P} |\psi_0^{(+)}(\mathbf{k}_0)\rangle. \qquad (7.121)$$

The P-projected T-matrix element is defined by

$$\langle \mathbf{k}i| T |0\mathbf{k}_0\rangle = \langle \mathbf{k}_i| \bar{P} V^{(Q)} \bar{P} |\psi_0^{(+)}(\mathbf{k}_0)\rangle, \quad i \in \bar{P}. \qquad (7.122)$$

Substituting (7.121) into (7.122) we obtain the P-projected Lippmann–Schwinger equations

$$\langle \mathbf{k}i| T |0\mathbf{k}_0\rangle = \langle \mathbf{k}i| V^{(Q)} |0\mathbf{k}_0\rangle + \sum_{j \in \bar{P}} \int d^3k' \langle \mathbf{k}i| V^{(Q)} | j\mathbf{k} \rangle$$

$$+ \frac{1}{E^{(+)} - \epsilon_j - \tfrac{1}{2}k'^2} \langle \mathbf{k}'j| T |0\mathbf{k}_0\rangle, \quad i \in \bar{P}. \qquad (7.123)$$

These are the coupled-channels-optical equations, which are formally identical to (6.73) except that the channels are restricted to P space and the potential V is replaced by the optical potential $V^{(Q)}$ (7.118). The extension of (7.123) to the distorted-wave representation is analogous to the extension of (6.73) to (6.87).

The formal coupled integral equations (6.73) and their explicitly-antisymmetric form (7.35) require a discrete notation for the target continuum. In (7.123) discrete notation is used only for discrete states and the continuum states in the expansion of $V^{(Q)}$ may be treated by integration.

7.5.3 *The Q-space weak-coupling approximation*

The optical potential, defined by (7.111,7.115) can only be calculated exactly if we can solve the whole collision problem to find the spectral representation of H. We must approximate it as closely as possible with the rationale that, since strongly-coupled channels are treated explicitly in P space, a reasonable approximation for the remaining channels should not cause significant errors in the amplitudes for the excitation of P-space channels from the entrance channel.

Our first task in constructing the reduced matrix elements of the polarisation potential is to find the reduced matrix elements of the Q-projected Green's function (7.111). Using the notation (7.48) we have

$$\langle L'J'i'P \parallel Q(E^{(+)} - H)QG_Q(E^{(+)}) \parallel PiJL\rangle = \langle L'J'i'P \parallel Q \parallel PiJL\rangle.$$

$$(7.124)$$

The reduced form of the projection operator Q is found by considering its coordinate—spin representation.

$$\langle \mathbf{r}_0'\sigma_0'\bar{x}_0'|Q|\bar{x}_0\sigma_0\mathbf{r}_0\rangle = \int d^3k \sum_v \sum_{i\in\bar{Q}} \langle \mathbf{r}_0'\sigma_0'\bar{x}_0'|iv\mathbf{k}\rangle\langle \mathbf{k}vi|\bar{x}_0\sigma_0\mathbf{r}_0\rangle$$

$$= \sum_{PLJ} \int dk \frac{2}{\pi} r_0'^{-1} U_L(kr_0')U_L(kr_0)r_0^{-1} \sum_{i\in\bar{Q}} \langle \hat{\mathbf{r}}_0'\sigma_0'\bar{x}_0' \parallel PiJL\rangle$$

$$\times \langle LJiP \parallel \bar{x}_0\sigma_0\hat{\mathbf{r}}_0\rangle,$$

$$(7.125)$$

where we have substituted (7.47) into the first line of (7.125), performed the integration over $\hat{\mathbf{k}}$ and summed over projection quantum numbers. The reduced form of Q is written

$$Q = \sum_{PLJ} \sum_{i\in\bar{Q}} I_{0L} \parallel PiJL\rangle\langle LJiP \parallel,$$

$$(7.126)$$

where the radial identity in the space of the continuum electron is defined by (7.125) and written

$$I_{0L} = \frac{2}{\pi} \int dk |U_L(k)\rangle\langle U_L(k)|. \tag{7.127}$$

In choosing the partition $x_0, \bar{x}_0$ of the set x of coordinate–spin variables we have broken the symmetry of the problem. It will be restored by explicit symmetrisation of the expression for the optical potential.

Equn. (7.124) defining the reduced form of the Q-projected Green's function $G_Q(E^{(+)})$ is written using (7.127) and equating terms in the P sum.

$$\sum_{i'' \in \bar{Q}} \sum_{L''J''} \langle L'J'i'P \parallel I_{0L'}(E^{(+)} - H)I_{0L''} \parallel Pi''J''L'' \rangle$$

$$\times \langle L''J''i''P \parallel G_Q(E^{(+)}) \parallel PiJL \rangle$$

$$= \delta_{i'i}\delta_{J'J}\delta_{L'L}I_{0L}. \tag{7.128}$$

We partition the Hamiltonian using the second form of (7.12) and defining the interaction of the projectile with the target electrons by

$$v_0 = \sum_r v_{0r}. \tag{7.129}$$

It is convenient to eliminate the factor r_0^{-1} from the coordinate representation of the projectile state and use the radial Hamiltonian (4.10).

$$K_{0L} = -\frac{1}{2}\frac{d^2}{dr_0^2} + \frac{L(L+1)}{2r_0^2}. \tag{7.130}$$

The first factor in (7.128) becomes

$$\langle L'J'i'P \parallel I_{0L'}(E^{(+)} - H)I_{0L''} \parallel Pi''J''L'' \rangle$$

$$= I_{0L'}(E^{(+)} - \epsilon_{i'} - K_{0L'} - V_0)I_{0L'}\delta_{i'i''}\delta_{J'J''}\delta_{L'L''}$$
$$- I_{0L'}\langle L'J'i'P \parallel v_0 \parallel Pi''J''L'' \rangle I_{0L''}. \tag{7.131}$$

We now make the approximation of weak coupling in Q space for the reduced matrix elements of v_0 in (7.131).

$$\langle L'J'i'P \parallel v_0 \parallel Pi''J''L'' \rangle = V_{0Pi'J'L'}(r_0)\delta_{i'i''}\delta_{J'J''}\delta_{L'L''}. \tag{7.132}$$

This approximation neglects the coupling of channels within Q space. The integral equations (7.123) of course couple channels within P space and P-space channels to Q-space channels. The reduced matrix element (7.132) involves a state-dependent local, central potential, whose definition is a further part of the approximation.

Substituting (7.131,7.132) in (7.128) we obtain the expression for the Green's function

$$\langle L'J'i'P \parallel G_Q(E^{(+)}) \parallel PiJL \rangle = \delta_{i'i}\delta_{J'J}\delta_{L'L}I_{0L}G_{0PiJL}(E_i^{(+)})I_{0L}, \quad (7.133)$$

where

$$E_i^{(+)} = E_0 - \epsilon_i = \tfrac{1}{2}k_i^2. \quad (7.134)$$

and the coordinate representation of G_{0PiJL} is given by

$$(E_i^{(+)} - K_{0L} - V_0 - V_{0PiJL})G_{0PiJL}(E_i^{(+)}; r_0', r_0) = \delta(r_0' - r_0). \quad (7.135)$$

This is the definition of the Green's function of a local, central potential. The solution of (7.135) is

$$G_{0PiJL}(E_i^{(+)}; r_0', r_0) = -k_i^{-1}f_{PiJL}(k_ir_<)\Big[g_{PiJL}(k_ir_>) + if_{PiJL}(k_ir_>)\Big], \quad (7.136)$$

where f and g denote respectively the regular and irregular solutions, discussed in section 4.4.2, of

$$(E_i^{(+)} - K_{0L} - V_{0PiJL})u_{PiJL}(k_ir) = 0. \quad (7.137)$$

The approximation to the Q-projected Green's function is based on the unsymmetrised expression (7.133). We explicitly symmetrise it to obtain

$$G_Q(E^{(+)}) = \tfrac{1}{2}\sum_{PLJ}\sum_{i \in \bar{Q}} \parallel PiJL \rangle \big(G_{0PiJL} + G_{1PiJL}\big)\langle LJiP \parallel. \quad (7.138)$$

The reduced matrix element of the two-electron polarisation potential (7.115,7.116) is

$$\langle k'L'J'i \parallel \tilde{v}_{01P\pi}^{(Q)}(1-X) \parallel iJLk \rangle = \tfrac{1}{2}\sum_{PL''J''}\sum_{i'' \in \bar{Q}}$$

$$\times \Big\{\int_0^\infty dr_0' \int_0^\infty dr_0 \langle k'L'J'i' \parallel v_{01}(r_0') \parallel Pi''J''L'' \rangle G_{0Pi''J''L''}(E_{i''}^{(+)}; r_0', r_0)$$

$$\times \langle L''J''i''P \parallel v_{01}(r_0)(1-X) \parallel iJLk \rangle$$

$$+ \int_0^\infty dr_1' \int_0^\infty dr_1 \langle k'L'J'i' \parallel v_{01}(r_1') \parallel Pi''J''L'' \rangle G_{1Pi''J''L''}(E_{i''}^{(+)}; r_1', r_1)$$

$$\times \langle L''J''i''P \parallel v_{01}(r_1)(1-X) \parallel iJLk \rangle\Big\}, \quad (7.139)$$

where the operator X exchanges the coordinates of the projectile and target orbital in the sense of (7.118).

In order to implement the approximation to the optical potential we must choose a form for the potential V_{OPiJL} (7.132) in which the Green's function is calculated (7.135). We choose different types of potential for the discrete and continuum channels of Q space, projected respectively by Q^- and Q^+. For Q^- space we choose the average potential for the target state $|i\rangle$. Its coordinate representation is obtained from (7.48,7.63,7.66)

$$V_{OPiJL}(r_0) = \sum_{\alpha\beta\lambda}(-1)^{J+j+P}\begin{Bmatrix} J & j & P \\ j & J & \lambda \end{Bmatrix}\langle LJ \parallel \mathbf{C}^\lambda \parallel LJ\rangle\langle \ell_\alpha j_\alpha \parallel \mathbf{C}^\lambda \parallel \ell_\beta j_\beta\rangle$$

$$\times \langle i \parallel \gamma_{\alpha\beta}^\lambda \parallel i\rangle \int_0^\infty dr_1 u_\alpha(r_1)v_\lambda^D(r_0,r_1)u_\beta(r_1). \tag{7.140}$$

In (7.140) we have added the electron–nucleus potential to the definition (7.132), using (7.61).

The weak-coupling approximation (7.132,7.140) can be verified within the context of the coupled-channels-optical method. Equns. (7.123) may be solved with a particular channel, defined by the target state $|i\rangle$, included in either P space or Q^- space. If it is in P space the channel i is fully coupled. The approximation is verified if the two solutions agree. In practice the lowest dipole-excited channels should be included in P space with the experimentally-observed channels, but the approximation is closely verified for higher channels in Q^- space. However, computation of (7.123) is not difficult and it is common to include all discrete channels in P space that are necessary for convergence.

For continuum channels the radial orbitals u_α in (7.140) are not bounded and the integral is divergent. The choice of radial functions in (7.136) must be based on intuition obtained from a study of ionisation, which is treated in chapter 10. A necessary condition for a reasonable distorted wave in the distorted-wave Born approximation for ionisation is that it should be orthogonal to the initial state in the ionisation amplitude. For computational simplicity we set V_{OPiJL} equal to zero and orthogonalise the resulting Ricatti–Bessel functions to all the states of P space using (5.83).

7.5.4 *The long-range dipole polarisation potential*

Important effects in electron–atom elastic scattering are due to the long-range polarisation potential that results from virtual excitation of dipole channels. For illustration it is sufficient to follow Allen, Bray and McCarthy (1988) in considering the large-angular-momentum contribution to the elastic T-matrix element from the excitation of a single $\ell = 1$ orbital with radial part $u_1(r)$ from a single $\ell = 0$ orbital with radial part $u_0(r)$.

The elastic reduced T-matrix element T_L for LS coupling is a solution of (7.44).

The case under consideration requires special values of the angular-momentum quantum numbers in the reduced potential matrix elements (7.67), namely $\ell = 0$, $\ell' = \lambda = 1$, $L = K$, $L' = L \pm 1$. The corresponding direct radial matrix element (7.63) may be written

$$R^{1D}_{L'L10}(k, q) = \frac{2}{\pi k q} \int_0^\infty dr_0 U_{L'}(kr_0) U_L(qr_0)$$

$$\times \left[r_0^{-2} \int_0^{r_0} dr_1 r_1 u_1(r_1) u_0(r_1) \right.$$

$$\left. + r_0 \int_{r_0}^\infty dr_1 r_1^{-2} u_1(r_1) u_0(r_1) \right]. \tag{7.141}$$

For notational convenience we have replaced k_0 and k' in (7.44) by q and k respectively. Since the radial orbitals are negligible beyond a value of r_1 that we call r_∞ the second term of (7.141) is negligible for $r_0 > r_\infty$. We consider only the case $L \gg qr_\infty$, for which

$$R^{1D}_{L'L10}(k, q) = \frac{2}{\pi k q} \beta \int_0^\infty dr_0 r_0^{-2} U_{L'}(kr_0) U_L(qr_0), \tag{7.142}$$

$$\beta = \int_0^\infty dr_1 r_1 u_1(r_1) u_0(r_1), \tag{7.143}$$

since $U_L(qr_0)$ is negligible for $r_0 < r_\infty$.

For $L \gg qr_\infty$ we make the following approximations.

 i) The direct reduced potential matrix elements are so small that the second Born approximation is valid.
 ii) The exchange radial matrix elements (7.72) are negligible since each radial integration is cut off by the bound orbitals.
 iii) The on-shell potential matrix elements, which give the imaginary parts and on-shell subtractions in the multichannel analogue of (4.124), are negligible. This is justified below equn. (7.150).
 iv) The entrance-channel or static potential matrix element is small since the average potential in the entrance channel is of short range.

With these approximations the elastic reduced T-matrix element of (7.44) is

$$T_L = \int_0^\infty dk \frac{2k^2}{x^2 - k^2} \left\{ [V_L^+(k, q)]^2 + [V_L^-(k, q)]^2 \right\}, \tag{7.144}$$

where $V_L^\pm$ are the contributions to the integrand of (7.44) in the second Born approximation $T_L = V_L$, with $L' = L \pm 1$ respectively. The on-shell momentum x is defined by

$$\tfrac{1}{2} x^2 = E_0 + \epsilon_0 - \epsilon_1, \tag{7.145}$$

where ϵ_0 and ϵ_1 are the energies of the ground and dipole-excited states respectively and E_0 is the incident kinetic energy.

Using the special values of the angular-momentum quantum numbers in (7.67) we find

$$V_L^+(k,q) = \frac{2}{\pi k q}\left[\frac{L+1}{3(2L+1)}\right]^{1/2} R_L^+(k,q),$$

$$V_L^-(k,q) = \frac{2}{\pi k q}\left[\frac{L}{3(2L+1)}\right]^{1/2} R_L^-(k,q), \quad L \gg q r_\infty, \qquad (7.146)$$

where the radial matrix elements (7.142) corresponding to $V_L^\pm$ are

$$R_L^\pm(k,q) = \beta(kq)^{1/2}(\pi/2)\int_0^\infty dr\, r^{-1} J_{L\pm1+1/2}(kr) J_{L+1/2}(qr). \qquad (7.147)$$

Here we have converted the Ricatti–Bessel functions of (7.142) to cylindrical Bessel functions in order to make use of analytic identities (Luke, 1973).

$$U_L(\rho) = (\pi\rho/2)^{1/2} J_{L+1/2}(\rho). \qquad (7.148)$$

Substituting (7.146) in (7.144) we obtain

$$T_L = \frac{2}{3}\left[\frac{2}{\pi q}\right]^2 \int_0^\infty dk\, \frac{1}{x^2 - k^2}\frac{1}{2L+1}\left[(L+1)R_L^{+2} + LR_L^{-2}\right]. \qquad (7.149)$$

The radial matrix elements $R_L^\pm$ may be evaluated analytically using the identity

$$\int_0^\infty dr\, r^{-1} J_\nu(\alpha r) J_\mu(\beta r) = \frac{1}{2}\left[\frac{\alpha}{\beta}\right]^\nu \frac{\Gamma(\nu+\mu)/2}{\Gamma(1-(\nu-\mu)/2)\Gamma(\nu+1)}$$
$$\times F\left[\frac{\nu+\mu}{2},\frac{\nu-\mu}{2},\nu+1;\frac{\alpha}{\beta}\right], \qquad (7.150)$$

where $\alpha < \beta$ and F is the hypergeometric function (Oberhettinger, 1973). The k integration in (7.149) may then be evaluated numerically. The expression (7.150) enables us to see how to approximate (7.149) accurately. The function $\left[(L+1)R_L^{+2} + LR_L^{-2}\right]$ is sharply peaked at $k = q$ and peaks more sharply as L increases. The reason for this is the factor $(\alpha/\beta)^\nu$, which becomes $(k/q)^\nu$ for $k < q$ and $(q/k)^\nu$ for $k > q$. It is this peaking that makes the on-shell radial matrix elements relatively small, since the on-shell value $k = x$ is substantially different from q.

The approximation for the elastic T-matrix element is

$$T_L^{(1)} = \frac{2}{3}\left[\frac{2}{\pi q}\right]^2 \frac{1}{x^2 - q^2}\int_0^\infty dk\, \frac{1}{2L+1}\left[(L+1)R_L^{+2} + LR_L^{-2}\right], \qquad (7.151)$$

which may be evaluated using the orthonormality identity

$$\int_0^\infty dk k J_{n+1/2}(kr') J_{n+1/2}(kr) = \delta(r' - r)/r. \tag{7.152}$$

We substitute (7.147) into (7.151) to obtain

$$T_L^{(1)} = \frac{-\alpha q}{(2L+3)(2L+1)(2L-1)}, \tag{7.153}$$

where the polarisability α is given according to (7.143) by

$$\alpha = 2\beta^2/3(\epsilon_1 - \epsilon_0). \tag{7.154}$$

The polarisability is additive for all orbital pairs and dipole channels.

The result (7.153) was first derived by O'Malley, Spruch and Rosenberg (1961). It is identical to the first Born approximation T-matrix element for scattering from the local polarisation potential

$$V_\alpha(r) = -\alpha/2r^4. \tag{7.155}$$

This is

$$T_L^{(1)} = \int d\hat{\mathbf{q}}' \int d\hat{\mathbf{q}} \sum_M \langle LM|\hat{\mathbf{q}}'\rangle \langle \mathbf{q}'|V_\alpha|\mathbf{q}\rangle \langle \hat{\mathbf{q}}|LM\rangle$$

$$= \frac{-\alpha}{2q} \int dr r^{-3} J_{L+1/2}(qr) J_{L+1/2}(qr), \tag{7.156}$$

which may be evaluated to yield (7.153). It is valid for large angular momenta.

We may consider long-range elastic scattering to be due to a polarisation potential that is strictly nonlocal, but is equivalent to (7.155) in the sense that it gives the same T-matrix element or phase shift.

This result is very useful in solving the coupled integral equations (7.38) or (7.44). It enables us to cut the calculation off at angular momenta where the Born approximation is valid for dipole channels, where the T-matrix elements for nondipole excitations are negligible, and where elastic T-matrix elements may be validly extrapolated by (7.153). The Born approximation for dipole channels may be evaluated analytically enabling their T-matrix elements to be summed over all angular momenta.

The range of validity of the approximation (7.153) can be understood from a comparison (Allen *et al.*, 1988) of T_L, calculated by solving (7.44) for two states, with $T_L^{(1)}$ at different energies. This is illustrated in fig. 7.1 for a small-polarisability case, the $1s - 2p$ excitation of hydrogen ($\alpha = 2.9596$, $\epsilon_1 - \epsilon_0 = 10.2$ eV) and a large-polarisability case, the $3s - 3p$ excitation of sodium ($\alpha = 176.63$, $\epsilon_1 - \epsilon_0 = 2.1$ eV). At the excitation threshold $x^2 = 0$, $T_L = T_L^{(1)}$ for all L where the second Born

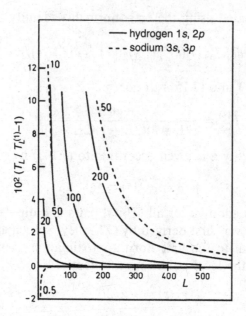

Fig. 7.1.　Comparison of T_L with the approximation $T_L^{(1)}$ (7.153) for the $2p$ excitation of hydrogen and the $3p$ excitation of sodium. Curves are drawn for the energies (eV) indicated (Allen *et al.*, 1988).

approximation is valid, roughly $L > 20$. As the energy increases the angular momentum above which T_L is within a few percent of $T_L^{(1)}$ increases to some hundreds. The distance at which (7.153) is valid is of order 100 a.u.

7.6 Alternative methods for restricted energy ranges

The convergent-close-coupling and coupled-channels-optical methods for electron—atom scattering apply to all atom or ion targets at all energies. Their computational practicability has been demonstrated for hydrogen and alkali-metal targets. A simpler coupled-channels-optical calculation involving equivalent-local approximations in the optical potential has been successfully applied to these and two-electron targets, giving a strong indication of the probable validity of the more-detailed method for larger atoms. This model is designed to apply well above the ionisation threshold, but is quite successful at all energies.

The distorted-wave Born series, obtained by iterating the distorted-wave Lippmann—Schwinger equations (6.87), has been discussed in section 6.10. Its validity is expected to improve with increasing energy. Approximations in the spirit of the convergent-close-coupling method involve choosing pseudostate bases to satisfy theoretical criteria that depend on the energy range. Results are basis-dependent but a judicious choice of basis in particular circumstances is capable of an excellent description of experimental data.

At energies below the first excitation threshold the variational principles discussed for bound states in chapter 5 can be extended to scattering (Callaway, 1978). We do not discuss this because of its restricted validity. However, there is another extension of bound-state methods into the positive-energy range that applies at least up to the ionisation threshold and somewhat beyond. This is the *R*-matrix method. Its possible extension to higher energies is discussed.

Examples of the application of all these methods are given in chapters 8 and 9. In this section we outline the methods.

7.6.1 *Distorted-wave Born and related approximations*

The driving terms of the distorted-wave Lippmann–Schwinger equations (6.87) are the distorted-wave Born approximation for the excitation of each target state $|i\rangle$ from the ground state. This is a weak-coupling approximation, which ignores the coupling of all but the entrance and exit channels. The first iteration of (6.87) produces the distorted-wave second Born approximation, whose implementation involves all channels. A related approximation that again involves all channels is the unitarised distorted-wave Born approximation, so called because the *S*-matrix (4.73), generalised to many channels, is unitary in the approximation, thus conserving probability flux. Here the real parts of all the Green's functions in (6.87) are set to zero, reducing the integral equations to a set of coupled algebraic equations in which the on-shell imaginary parts account for the influence on one channel of real excitations into other channels. In referring to the Lippmann–Schwinger equations (6.87) it is useful to consider them in the sense of the convergent-close-coupling method as a finite set of equations that can be expanded to convergence for the excitation of low-lying states.

It is possible to estimate the validity of the various distorted-wave approximations by investigating a two-channel model. Fig. 7.2 does this for the 3*s* and 3*p* channels of sodium at an incident energy of 54.42 eV. The approximations for the differential cross sections (6.55) are compared with the solution of the two-channel integral equations. The distorted-wave Born approximation is semiquantitative for the 3*p* channel but not for the 3*s* channel since it ignores the strong effect of dipole coupling (section 7.5.4). The distorted-wave second Born approximation involves coupling and therefore represents the 3*s* channel better, although it is not

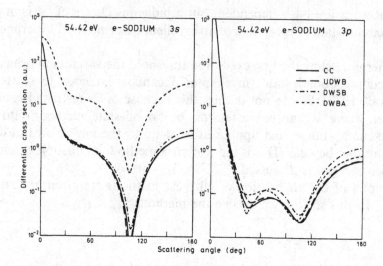

Fig. 7.2. Distorted-wave approximations to a two-channel (CC) calculation of electron–sodium scattering. UDWB, unitarised distorted-wave Born; DWSB, distorted-wave second Born; DWBA, distorted-wave Born (Bray *et al.*, 1989).

noticeably better for the 3*p* channel. The unitarised distorted-wave Born approximation is very good in both channels.

In implementing distorted-wave Born approximations choices of the distorting potentials other than the ground-state average (6.91) may be made in an attempt to improve realism. Choosing the final-state average potential for the final-state distorted wave restores time-reversal invariance. More coupling may be included by calculating the distorted waves in various approximations to the optical potential. Choices of distorting potentials that describe experimental data better in particular cases are discussed·by Madison, Bray and McCarthy (1991). The approximation is not sufficiently detailed to take them seriously.

In applying the distorted-wave second Born approximation we have the same difficulty as in calculating the optical potential. We must calculate the spectrum of the Green's function of (6.87). The first iteration of (6.87) is written as

$$\langle \mathbf{k}^{(-)}i|T|0\mathbf{k}_0^{(+)}\rangle = \langle \mathbf{k}^{(-)}i|V - U + (V - U)\frac{1}{E^{(+)} - K - U}(V - U)|0\mathbf{k}_0^{(+)}\rangle.$$

$$(7.157)$$

The Q-projected Green's function (7.111) in the polarisation potential (7.115) contains the collision Hamiltonian H, but practical implementation of it involves the weak-coupling approximation (7.132). A second-Born calculation is equivalent to a calculation of the polarisation potential with $Q=1$ and V_{0PiJL} being the reduced matrix element of U. This has been done for hydrogen by Madison *et al.* (1991) with an unsymmetrised Green's function but explicitly-antisymmetrised channel states. The use of different potentials in the three factors of the second-order term of (7.157) was found to affect cross sections appreciably.

If we introduce the complete set $|j\rangle$ of target states into (7.157) the second-order term becomes

$$\sum_j \langle \mathbf{k}^{(-)}i|V - U|j\rangle \frac{1}{E^{(+)} - \epsilon_j - K_0 - U} \langle j|V - U|0\mathbf{k}_0^{(+)}\rangle. \qquad (7.158)$$

The closure approximation consists in replacing ϵ_j by an average energy ϵ_C, thereby eliminating the target states by the closure or representation theorem (3.17). This approximation has been calculated for example by Kingston and Walters (1980) for hydrogen. Tests of its validity are described by Madison, Winters and Downing (1989). The closure approximation is compared with explicit calculations for the $n=2$ states of hydrogen that omitted exchange amplitudes. They found errors of 50–100% in the second-order amplitude for the $2s$ excitation at larger scattering angles. Much of this error is due to the closure treatment of nearby states. The approximation improves if excitation amplitudes for low-lying states are calculated explicitly in some approximation and subtracted from the closure sum.

The unitarised eikonal–Born series (Byron, Joachain and Potvliege, 1982) is a nonperturbative scattering method valid at energies higher than several times the ionisation threshold. It is based on the many-body generalisation of the Wallace (1973) amplitude, which has the advantage that the corresponding S matrix is unitary. Its disadvantage is that it does not take proper account of the long-range dipole polarisation potentials. This is remedied by replacing the second-order Wallace amplitude by the plane-wave second Born amplitude in the direct partitions of the electron–hydrogen problem. Exchange amplitudes are calculated from the direct amplitudes by suitably partitioning the Lippmann–Schwinger equations.

Stelbovics (1990) has shown that there is no energy for which the Born or distorted-wave Born series converges. The reason is that the one-electron exchange potential in (7.35) depends linearly on the energy. This potential acts only in reduced Lippmann–Schwinger equations where the total angular momentum is close to that of the basis orbitals describing the target states. The Born series converges at higher angular momenta. The divergence shows up numerically in the iteration of the reduced Lippmann–Schwinger equations for low angular momenta. The correct solution is of course obtained by the matrix methods (4.125). The iteration of the scattering equations involves half-off-shell T-matrix elements and its divergence does not imply that on-shell amplitudes diverge. There are strong indications from comparison with experiment for hydrogen (Madison *et al.*, 1991) that the distorted-wave second Born approximation is valid at energies higher than several times the ionisation threshold.

7.6.2 *The equivalent-local coupled-channels-optical method*

The weak-coupling approximation to the polarisation potential (7.115, 7.116,7.119) may be summarised as

$$
\langle \mathbf{k}'i|\tilde{v}^{(Q)}|j\mathbf{k}\rangle = \sum_{\ell \in Q} \int d^3q \langle \mathbf{k}'i|v|\ell\mathbf{q}^{(-)}\rangle \frac{1}{E^{(+)} - \epsilon_\ell - \frac{1}{2}q^2} \langle \mathbf{q}^{(-)}\ell|v|j\mathbf{k}\rangle
$$
$$
+ \int d^3q' \int d^3q \langle \mathbf{k}'i|v|\Psi^{(-)}(\mathbf{q}',\mathbf{q})\rangle \frac{1}{E^{(+)} - \frac{1}{2}(q'^2 + q^2)}
$$
$$
\times \langle \Psi^{(-)}(\mathbf{q}',\mathbf{q})|v|j\mathbf{k}\rangle. \tag{7.159}
$$

Here we have summed the partial-wave series whose terms are represented by (7.139). The first term of (7.159) is the polarisation potential due to the discrete part Q^- of Q space. The second term is due to the ionisation space Q^+, and $|\Psi^{(-)}(\mathbf{q}',\mathbf{q})\rangle$ is the exact solution of the Schrödinger equation for an ionised channel.

The amplitudes in (7.159) can be calculated analytically in certain approximations (McCarthy and Stelbovics, 1983*b*). This allows a sufficiently-fast computation for performing the integrations by a multidimensional method such as the diophantine method (Conroy, 1967). For Q^- space $|\mathbf{q}^{(-)}\rangle$ is approximated by a plane wave. For Q^+ the slower electron is represented by a Coulomb function (4.85) orthogonalised by (5.83) to the bound orbital in the same amplitude, and the faster electron by a plane wave. The direct amplitudes factorise into an electron–electron amplitude and a structure transform involving the target orbitals. It is necessary to make the same factorisation to approximate the exchange amplitudes. This is an equivalent-local approximation for the exchange part of the two-electron potential.

Computation is too laborious for the necessary range of $\mathbf{k}',\mathbf{k}$, so a further approximation is necessary. The polarisation potential matrix element is calculated only at about 10 points in the variable K, where

$$
K = |\mathbf{k} - \mathbf{k}'|, \quad \tfrac{1}{2}k^2 = E - \epsilon_0. \tag{7.160}
$$

This is achieved by an angular-momentum projection, whose *LS*-coupling form is

$$
\langle \mathbf{k}'i|\tilde{v}^{(Q)}|j\mathbf{k}\rangle = \sum_{\ell''m''} i^{\ell''} \langle \ell'\ell''m'm''|\ell m\rangle U_{\ell''\ell'\ell}(K)\langle \hat{\mathbf{K}}|\ell''m''\rangle. \tag{7.161}
$$

The calculation is done for the one-dimensional functions

$$U_{\ell''\ell'\ell}(K) = \sum_{m''m'} \langle \ell'\ell''m'm''|\ell m\rangle \int d\widehat{\mathbf{K}} \langle \mathbf{k}'i|\tilde{v}^{(Q)}|j\mathbf{k}\rangle i^{-\ell''} \langle \ell''m''|\widehat{\mathbf{K}}\rangle. \quad (7.162)$$

Again this is an equivalent-local approximation since the momentum representation (3.40) of a local potential depends only on K.

This approximation has been shown to have at least semiquantitative validity over the whole energy range. It is unacceptable only for excitations involving a change of target spin s in the LS-coupling representation. Here exchange amplitudes make the only contribution and the factorisation approximation is too severe.

7.6.3 Pseudostate approximations to the target spectrum

For some years the method in widespread use for the electron—hydrogen problem was to solve the coupled integrodifferential equations of the co-ordinate representation with the appropriate boundary and orthogonality conditions (Seaton, 1973). This was so laborious that the use of Sturmians as a discrete approximation to the target spectrum could not be implemented to convergence.

The diagonalisation of the hydrogen Hamiltonian in a Slater-function (4.38) basis has been reviewed by Callaway (1978) in the context of variational solutions of the integrodifferential equations. This basis has useful features. The inclusion of all the Slater functions necessary for the radial eigenstate $u_{n\ell}(r)$ produces exact eigenstates for principal quantum numbers up to n in the ℓ manifold. The remainder are pseudostates, which represent the higher discrete states and the continuum. Since Slater functions are not orthonormal there are linear-dependence difficulties that severely limit the size of the basis for which the diagonalisation is numerically feasible.

With such pseudostate expansions the T-matrix elements are basis-dependent. Choosing a basis is an art, which can be assisted by fitting the results of approximate calculations that are relevant to the problem under consideration without actually treating the basis parameters as phenomenological fitting parameters for the experiment to be described.

For energies up to just above the ionisation threshold Callaway (1978) determined the basis by variational considerations, including the fact that for elastic scattering below the first inelastic threshold each phase shift can be approximated to arbitrary accuracy since the best basis yields the largest phase shift. It proved feasible to find a basis that gave an excellent description of elastic and inelastic scattering in the energy region where the cross sections fluctuate rapidly due to resonances

(Williams, 1988). At higher energies where larger bases were needed the calculation was simplified by calculating the optical potential for higher pseudostate excitations. Callaway and Unnikrishnan (1989) obtained excellent agreement with elastic scattering experiments from 30 eV to 400 eV in this way.

For higher-energy reactions involving the $n=1$ and 2 states of hydrogen, van Wyngaarden and Walters (1986) determined the pseudostate basis to agree with plane-wave second-Born calculations in the closure approximation. Very good agreement was obtained for a wide range of experimental data.

Madison and Callaway (1987) compared the results of pseudostate calculations with those of explicit distorted-wave second-Born calculations, omitting exchange amplitudes. They concluded that it is possible to find basis sets of a managable size whose results are quite close to the second-Born results at the energy of detailed investigation and which give close results also at different energies.

Pseudostate calculations have the advantage over Born and optical-potential methods that they constitute a numerically-exact solution of a problem. The problem is not identical to a scattering problem but can be made quite realistic for useful classes of scattering phenomena by an appropriate basis choice. The state vectors, or equivalently the set of half-off-shell T-matrix elements, for such a calculation contain quite realistic information about the ionisation space.

7.6.4 The R-matrix method

The R-matrix method is a multichannel generalisation of the calculation of potential scattering described in section 4.4.3. It was introduced by Wigner and Eisenbud (1947) to describe neutron-nucleus reactions at low energy. Its application to electron–atom scattering has been described by Burke and Robb (1975).

The $(N + 1)$-electron collision problem is solved in a sphere of radius a, which is chosen to be larger than the distance beyond which the radial orbitals of a chosen set of bound states become negligibly small. The solution is essentially a configuration-interaction method where the basis configurations consist of determinants of bound orbitals representing the N target electrons and continuum orbitals for the continuum electron. The radial continuum orbitals are solutions of a potential problem

$$\left(\frac{d^2}{dr^2} - \frac{L(L+1)}{r^2} + 2V(r) + k_{NL}^2\right)u_{NL}(r) = \sum_{N'} \lambda_{NN'L} \, u_{N'L}(r), \quad (7.163)$$

with the boundary conditions

$$u_{NL}(0) = 0, \quad du_{NL}(a)/dr = 0. \tag{7.164}$$

The quantities $\lambda_{NN'L}$ are Lagrange multipliers ensuring that the continuum orbitals are orthogonal to bound orbitals of the same symmetry. The basis is given further flexibility by including short-range $(N+1)$-electron functions to allow for possible resonant states of the $(N+1)$-electron system.

The internal solutions are matched at $r = a$ to solutions of the scattering problem in the external region. Here it is a simple coupled-channels problem in which exchange and target-correlation terms are negligible. The matching matrix is the R matrix.

The method describes reactions quite well at energies below the threshold of the lowest target state omitted from the basis. Apart from the short-range $(N+1)$-electron functions it is an alternative numerical method for solving the coupled-channels problem. However, it omits virtual excitation of the continuum, which has a significant effect, even at very low energies. It has the computational feature that the time-consuming internal solution is independent of the incident energy. To vary the energy in small steps it is necessary only to solve the external and matching problems, both of which are relatively fast.

The R-matrix method may be extended to higher energies by a method introduced by Burke, Noble and Scott (1987). Here the basis for solving the internal $(N+1)$-electron problem includes two continuum orbitals, one representing a target electron in an ionised state. In contrast to the convergent-close-coupling method the introduction of a large number of continuum states to the target basis does not result in smoothly-varying T-matrix elements. The T matrix for an increasingly-large basis has increasingly-many sharp fluctuations that are again called pseudoresonances. Fig. 7.3 shows the real part of the $^3P^o$ elastic-scattering T-matrix element in a calculation for hydrogen with 2438 basis configurations involving orbital angular momenta for each orbital up to 3. The sphere radius is 25 a.u.

The intermediate-energy R-matrix method averages the T-matrix elements over energy using a Lorentzian averaging function of width I.

$$\rho(E - E') = \frac{I}{\pi} \frac{1}{(E - E')^2 + I^2}. \tag{7.165}$$

The averaged T-matrix element $\langle T(E) \rangle$ is given by

$$\langle T(E) \rangle = \int_{-\infty}^{\infty} dE' \rho(E - E') T(E') = T(E + iI). \tag{7.166}$$

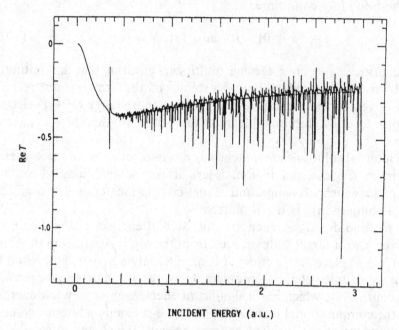

Fig. 7.3. The real part of the $^3P^o$ elastic-scattering T-matrix element in the intermediate-energy R-matrix calculation of Scholz, Scott and Burke (1988).

As I tends to zero the averaging procedure corresponds to evaluating the T-matrix element just above the branch cut on the real energy axis. It is numerically possible to choose such a large basis that the width I can be reduced to a value for which each averaged T-matrix element is independent of I. Such a calculation, however, involves enormous computational labour. The method has not yet been implemented up to the angular-momentum values required for a realistic calculation.

8

Spin-independent scattering observables

To understand an electron–atom collision means to be able to calculate correctly the T-matrix elements for excitations from a completely-specified entrance channel to a completely-specified exit channel. Quantities that can be observed experimentally depend on bilinear combinations of T-matrix elements. For example the differential cross section (6.55) is given by the absolute squares of T-matrix elements summed and averaged over magnetic quantum numbers that are not observed in the final and initial states respectively. This chapter is concerned with differential and total cross sections and with quantities related to selected magnetic substates of the atom.

In the study of electron–atom collisions there has been a constant emphasis on increasing the state selectivity of the particles in both the initial and final states. Thus while total cross section measurements define the initial kinetic energies, measurements of differential cross sections as a function of angle give additional information on the final momentum states of the separating particles. Added state selectivity is obtained through the use of spin-polarised electrons, or spin-polarised atoms, and with spin analysis of scattered particles (see e.g. Kessler (1985)). The great progress that has been made with the use of spin-selected particles will be discussed in the next chapter. Much of the progress that has been made has been for the case of elastic scattering.

8.1 Collisional alignment and orientation

The status for inelastic collisions is a little less satisfactory than for elastic collisions. Collisional excitation of atoms involves excited states with several magnetic substates. Standard total or differential cross section measurements sum over the magnetic substates, and thus give no information on the shape of the excited state and direction of the angular momentum transferred to the excited atom. These can be determined in collisional alignment and orientation studies, which avoid these sums and therefore access detailed information not available by other techniques. The term atomic *alignment* is used to describe the shape of the excited-state charge cloud and its direction of alignment in space, whereas the term *orientation* refers to the direction and magnitude of the angular momentum transferred to the atom during the collision.

The study of alignment and orientation of atoms by electron impact has yielded a wealth of detailed information on the mechanism and dynamics of electron–atom collisions. Information on atomic alignment and orientation after a collisional excitation process can be obtained by electron–photon coincidence experiments or alternatively by scattering from laser-excited atoms. Fano and Macek (1973) summarised the methods for extracting the dynamically-relevant observables from photon–electron coincidence experiments, allowing these observables to be disentangled from the geometry of each experiment and from the quantum numbers that are not relevant in the collision process. The scattering from optically-pumped targets, which represents the time-inverse experiment, may be similarly exploited by using a suitable modification of their theory (Macek and Hertel, 1974).

Complete information requires the detection of the polarisation of the emitted decay photon for any given scattering angle of the electron, and the relative momentum of the particles before and after the collision. In favourable cases a complete set of excitation amplitudes for the excited atomic states, including their relative phases, can be determined. This allows a much closer comparison to be made with scattering calculations, thus providing very stringent and constructive tests for the further development of scattering theories. The progress that has been achieved in these alignment and orientation studies has been the subject of many reviews and progress reports, of which the most comprehensive is the review by Andersen, Gallagher and Hertel (1988). These excitation studies have recently been supplemented by studies of ejected electrons, which in some circumstances can yield similar information. They are discussed in chapter 10.

8.1.1 Basic experimental schemes

There are two equivalent experimental arrangements for determining the alignment parameters of an excited atom A_i in the process

$$A_0 + e_0(\mathbf{k}_0, v_0) \rightarrow A_i + e_i(\mathbf{k}_i, v_i), \tag{8.1}$$

where $|0\rangle$ and $|i\rangle$ describe the states of the atom before and after the collision respectively, and v_0 and v_i are the initial and final spin projections of the electron.

The first arrangement uses photon—electron coincidence techniques.

$$A_0 + e_0 \;\rightarrow\; \left.\begin{array}{c} A_i \;+\; e_i \\ \downarrow \\ A_0 \;+\; \gamma, \end{array}\right\} \text{ coincidence.} \qquad (8.2)$$

The second, time-reversed, process uses scattering of electrons by atoms optically pumped by a laser, i.e.

$$\gamma + A_0 \rightarrow A_i \text{ (optical pumping)}$$
$$e_i + A_i \rightarrow A_0 + e_0. \qquad (8.3)$$

Both methods can give identical information on A_i excited in process (8.1). The schemes for measuring directional correlations between two particles resulting from a collision process by coincidence techniques are discussed in some detail in chapter 2. Similarly details of optical pumping techniques can also be found in chapter 2.

8.1.2 General concepts

We now introduce the general concepts and definitions behind studies of alignment and orientation in electron—atom collisions. Let us consider the scattering process (8.1) in some detail. We must first understand the concept of coherence. We illustrate it by considering the initial state of the atom

$$|0\rangle = |n\ell jm\rangle. \qquad (8.4)$$

This state depends on the magnetic quantum number m, which is defined by choosing a particular orientation of the coordinate frame. A randomly-oriented gas target has equal probability of an atom being found in any magnetic substate. The target contains an incoherent sum of magnetic substates. If we now choose a different coordinate frame the magnetic substates are linear combinations of those for the new frame.

$$|n\ell jm\rangle = \sum_{m'} a_{n\ell jm'} |n\ell jm'\rangle. \qquad (8.5)$$

The original magnetic substates are a coherent superposition of the new set, which we may consider as a basis for representing the magnetic substates.

The target part of the entrance-channel state $|0v_0\mathbf{k}_0\rangle$ is a coherent superposition of magnetic substates defined by an arbitrary choice of coordinate frame. It is transferred into another fully-coherent superposition of states in the exit channel $|iv_i\mathbf{k}_i\rangle$. The scattering amplitude is defined as a generalisation of (4.46), so that its absolute square is the corresponding differential cross section.   Equn. (6.55) gives the scattering amplitude in terms of the T-matrix element.

$$f_{i0} = (2\pi)^2 (k_i/k_0)^{1/2} \langle \mathbf{k}_i v_i | T | 0 v_0 \mathbf{k}_0 \rangle. \tag{8.6}$$

The scattering amplitude may be considered as a matrix in the space of magnetic substates of the target and projectile: the scattering matrix.   The excitation process is coherent since the collision time is much shorter than any characteristic time associated with the excited state.

In the terminology of Bederson (1969, 1970), a 'perfect scattering experiment' is one which completely determines the scattering matrix.   In this case both $\mathbf{k}_0$ and $\mathbf{k}_i$ and all the quantum numbers of the atom and electron before and after the collision are measured.   This is only achieved in exceptional cases.   Normally at least some amplitudes are averaged or summed over and much less detailed information is available.   In this chapter we limit ourselves to the case where the spin states of the electrons are not determined and must therefore be averaged over.   In such cases the particles after the scattering process cannot be described by pure states, but only by mixed states, i.e. by a superposition of basis states that is only partially coherent and may even be fully incoherent. One must, however, be careful in describing the degree of coherence or incoherence in a collision process, since statements about the degree of coherence depend on the choice of coordinate frame and basis states used to describe the scattering process.

The least ambiguous and most appropriate description of the atom after the collision is in terms of the density matrix (Blum, 1981), whose elements are bilinear combinations of scattering amplitudes for different magnetic substates.   For the sake of simplicity we restrict ourselves to the most common case, in which the target is initially in an S state and the excitation involves the transfer of one electron from an s orbital to a p orbital in the independent-particle approximation.   In atoms with one active electron the transition is $^2S - {}^2P$.   If there are two active electrons it is $^1S - {}^1P$.   We use the LS-coupling scheme.

The scattering amplitudes or density matrix elements required for the description of the excited states are given with respect to a suitable reference frame. The optimum solution is to use a frame-independent parametrisation of the dynamical information, closely related to the symmetry process.

The atomic charge cloud after excitation and the collision kinematics are illustrated in fig. 8.1. The scattering plane is defined by $\mathbf{k}_0$ and $\mathbf{k}_i$. From parity conservation it can be seen that the scattering plane is a plane of symmetry. Parity conservation also requires that the orbital

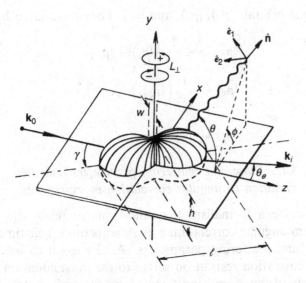

Fig. 8.1. Schematic representation of the collisionally-induced charge cloud in a P-state atom. The collision plane is determined by the ingoing and outgoing electron momenta $\mathbf{k}_0$ and $\mathbf{k}_i$. The excited atom is characterised by the alignment angle γ, its inherent angular momentum $\langle \ell_y \rangle = \langle L_\perp \rangle$, and the shape of the charge cloud P_ℓ. The direction of emission of the photon and its polarisation vectors are also shown.

angular momentum component of the atom in the excited state satisfy $\delta \ell_i \cdot \mathbf{k}_i = 0$. The final angular momentum (or orientation), for colliding particles with no initial orientation, must be perpendicular to the collision plane. It is conventionally denoted by $L_\perp$. If the excited state is in a nonisotropic distribution of magnetic substates (as shown in fig. 8.1), it also has an alignment. The standard frame of reference used in most calculations is called the collision frame with the z_c axis parallel to $\mathbf{k}_0$, the x_c axis in the scattering plane and perpendicular to z_c, and the y_c axis perpendicular to the scattering plane. This is the frame shown in fig. 8.1. The so-called natural frame (Hermann and Hertel, 1982) has $x_n = z_c$ along the incident direction, $z_n = y_c$ perpendicular to the scattering plane and parallel to the only symmetry axis, namely the momentum transfer. A third frame, called the atomic frame, is sometimes used. Here the x_a axis is chosen parallel to the symmetry axis of the charge cloud and $z_a = z_n = y_c$.

In addition to a choice of coordinate frames we have a choice of basis sets. For atomic p-orbitals we could choose the 'atomic' basis set defined by the magnetic quantum numbers m, i.e. $|p_0\rangle$, $|p_1\rangle$ and $|p_{-1}\rangle$, or by the collision-frame orbitals $|p_x\rangle$, $|p_y\rangle$, and $|p_z\rangle$. They are related by

$$|p_z\rangle = -\frac{1}{\sqrt{2}}\Big(|p_1\rangle - |p_{-1}\rangle\Big),$$

$$|p_x\rangle = \frac{i}{\sqrt{2}}\Big(|p_1\rangle + |p_{-1}\rangle\Big),$$

$$|p_y\rangle = |p_0\rangle. \tag{8.7}$$

8.1.3 State and scattering parameters and their relation to angular-correlation measurements

We now describe a formalism that allows us to relate the observables extracted from angular correlation experiments (and their time inverse) to atomic state and scattering parameters. As discussed earlier we will for the sake of illustration restrict ourselves to the most common case, where the initial state of the atom is an S state, and the excited state is a P state. We will also assume for the moment that explicit spin-dependent forces can be neglected during the collision so that the scattering is adequately described in the LS-coupling scheme and changes in spin are caused by electron exchange only. Since collision times are of the order of 10^{-14}s at intermediate energies, and decay times for atoms in excited states are $\sim 10^{-9}$s, the excitation and decay can be treated as independent processes.

The transition (8.4) is completely characterised by the scattering amplitudes (8.6) normalised according to

$$|f_{i0}|^2 = \sigma_{i0}, \tag{8.8}$$

where σ_{i0} is the differential cross section for the indicated transition. For brevity we characterise the scattering amplitudes in the following discussion by the orbital angular momentum quantum numbers of the excited state only, omitting those of the ground state, which is usually an S state.

The amplitude for excitation of a state with orbital angular momentum ℓ can be described in general as a coherent superposition of amplitudes for degenerate magnetic substates.

$$f_i = \Sigma_m f_{\ell m}. \tag{8.9}$$

The $f_{\ell m}$ are complex amplitudes for the excitation of the state $|n\ell mv\rangle$. Observables for the state $|i\rangle$ are related to $|f_i|^2$. We can generalise this to observables for different excited states $|i\rangle$ and $|i'\rangle$, which are expressed in terms of all bilinear combinations $f_{i'}f_i^*$ of the scattering amplitudes. Because in this chapter we restrict ourselves to cases where no spin analysis is performed on the electrons, and we are interested in the orbital states of the excited atoms A_i immediately after the interaction, we must sum and average over the final and initial unobserved spin projections.

$$\langle f_{\ell'm'} f^*_{\ell m} \rangle = \frac{1}{2j_0 + 1} \frac{1}{2} \sum_{m_0 v_0 v_i} f_{i'} f^*_i. \tag{8.10}$$

Neglecting the nuclear spin, the spin-unresolved excitation of a P state is fully described by the density matrix.

$$\rho = \begin{pmatrix} \langle f_{11} f^*_{11} \rangle & \langle f_{11} f^*_{10} \rangle & \langle f_{11} f^*_{1-1} \rangle \\ \langle f_{10} f^*_{11} \rangle & \langle f_{10} f^*_{10} \rangle & \langle f_{10} f^*_{1-1} \rangle \\ \langle f_{1-1} f^*_{11} \rangle & \langle f_{1-1} f^*_{10} \rangle & \langle f_{1-1} f^*_{1-1} \rangle \end{pmatrix}. \tag{8.11}$$

The density matrix ρ contains all the information on the state of the excited subensemble of atoms A_i immediately after the excitation. The diagonal elements of ρ are the differential cross sections

$$\rho_{mm} = \langle |f_{\ell m}|^2 \rangle = \sigma_m, \tag{8.12}$$

where σ_m denotes the cross section for excitation of the state $|\ell m\rangle$, and the trace of ρ gives the total differential cross section

$$\mathrm{tr}\rho = \sum_m \langle |f_{\ell m}|^2 \rangle = \sigma, \tag{8.13}$$

which is the cross section normally determined from measurement of the scattered electron current alone.

Not all of the off-diagonal elements of ρ are independent, since hermiticity requires that

$$\rho_{mm'} = \rho^*_{m'm}. \tag{8.14}$$

This, together with the normalisation (8.13), means that the excited state is described by eight real parameters.

Reflection invariance provides further constraints, the total wave function of the two colliding particles retaining reflection symmetry through the collision plane. Since we have already assumed that electron spin plays no role in the collision, this means that the reflection symmetry of the excited P state is the same as that for the original S state, i.e. symmetric. Thus the coefficient of the antisymmetric $|p_y\rangle$ orbital in the expansion of the excited state must vanish. From (8.6) we see that this requires $f_{11} = -f_{1-1}$. In general reflection symmetry requires that the elements of the density matrix satisfy the condition (Blum and Kleinpoppen, 1979)

$$\rho_{mm'} = (-1)^{m'} \rho_{m-m'} = (-1)^{m'+m} \rho_{-m-m'}. \tag{8.15}$$

ρ is then specified by five real parameters, for example $\rho_{11} = \sigma_1$, $\rho_{00} = \sigma_0$, $\rho_{10} = \langle f_{11} f^*_{10} \rangle$ (which is complex) and $\rho_{1-1} = \langle f_{11} f^*_{1-1} \rangle$ (which is real from conditions 8.14 and 8.15). It follows that five independent measurements must be made in order to completely determine the density matrix. Only two (σ_0 and σ) can be determined in experiments with axial symmetry. The determination of the remaining three requires coincident measurement of the electrons and photons.

The five independent parameters can also be chosen so that two of them give the shape of the charge cloud of the excited state shown schematically in fig. 8.1 (e.g. the height h or width w and length ℓ), one gives the alignment (i.e. the angle γ), one its orientation or angular momentum expectation value $\langle L_\perp \rangle$, and the final one is the differential cross section (8.13).

A convenient parametrisation due to Hermann and Hertel (1980) is

$$\lambda = \sigma_0/\sigma, \qquad \cos\chi = \frac{\mathrm{Re}\rho_{10}}{\sqrt{\sigma_1\sigma_0}},$$

$$\sin\phi = \frac{\mathrm{Im}\rho_{10}}{\sqrt{\sigma_1\sigma_0}}, \qquad \cos\theta = \frac{\rho_{1-1}}{\sigma_1}. \qquad (8.16)$$

Together with the total differential cross section (8.13) these parameters completely specify the density matrix.

With reflection invariance $f_{1-1} = -f_{11}$ and $\rho_{1-1} = \langle f_{11}f_{1-1}^* \rangle = -\sigma_1$ and $\cos\theta = -1$ in (8.16). In the case of $^2S - {}^2P$ transitions in an atom with one active electron (as in atomic hydrogen), the corresponding four independent magnetic substate parameters are generally taken to be those defined by Morgan and McDowell (1977), i.e.

$$\sigma = \sigma_0 + 2\sigma_1, \quad \lambda = \sigma_0/\sigma, \quad R = \mathrm{Re}\langle f_{10}f_{11}^* \rangle/\sigma, \quad I = \mathrm{Im}\langle f_{10}f_{11}^* \rangle/\sigma. \quad (8.17)$$

The off-diagonal elements of ρ give a measure of the interference that exists between the various magnetic substates following excitation, since they contain information on the relative phases of the excitation amplitudes. The normal convention is to say that coherent excitation has occurred if the corresponding off-diagonal elements of ρ are not zero. Incoherent excitation is said to have taken place if the density matrix is diagonal in m.

If the photons and scattered electrons are detected independently of each other, the incident-electron-beam direction is an axis of symmetry, and the excited ensemble must also have this axial symmetry. Then all the off-diagonal terms of ρ are zero (Blum, 1981). Coherently-excited magnetic substates can only be produced by excitation processes, such as (8.2), which are not axially symmetric. The determination of these off-diagonal elements of ρ has been the main goal of electron–photon coincidence experiments.

8.1.4 Completely-coherent excitation of P states

In certain cases, as in singlet–singlet $^1S - {}^1P$ transitions in an atom with two active electrons, where $s_0 = s_i = 0$, the electron spin remains unchanged during the collision ($v_0 = v_i$) and the sum over electron spin in equn. (8.9) is superfluous. Thus

$$\langle f_{\ell'm'} f^*_{\ell m} \rangle = f_{\ell'm'} f^*_{\ell m}. \tag{8.18}$$

In this case we have from equns. (8.15, 8.16)

$$\cos\theta = -1 \tag{8.19}$$

and

$$\cos^2\chi + \sin^2\phi = 1 \tag{8.20}$$

so that $\chi = \phi$ is then the relative phase between f_{10} and f_{11}. ρ is thus specified in terms of three parameters, which can be chosen in the manner of Eminyan *et al.* (1974), to be σ, $\lambda = \gamma$ and χ (8.16), where

$$\sigma = \sigma_0 + 2\sigma_1, \quad \lambda = \gamma = \sigma_0/\sigma, \quad \cos\chi = \frac{\mathrm{Re} f_{10} f^*_{11}}{\sigma \left[\frac{1}{2}\lambda(1-\lambda)\right]^{\frac{1}{2}}}. \tag{8.21}$$

If ^3P states have been excited from a ^1S ground state, again three parameters are sufficient for a complete specification of the excited system, since only one spin channel with total spin $1/2$ is allowed and equn. (8.10) again reduces to a single term (8.18).

8.1.5 State multipole description

The above description of the excited states in terms of excitation amplitudes is frame and basis set dependent. A more convenient description is in terms of state multipoles. It can be generalised to excited states of different orbital angular momentum and provides more physical insight into the dynamics of the excitation process and the subsequent nature of the excited ensemble. The angular distribution and polarisation of the emitted photons are closely related to the multipole parameters (Blum, 1981). The representation in terms of state multipoles exploits the inherent symmetry of the excited state, leads to simple transformations under coordinate rotations, and allows for easy separation of the dynamical and geometric factors associated with the radiation decay.

State multipoles are components of a spherical tensor, which is the following tensor product (3.99), averaged over unobserved magnetic quantum numbers.

$$\langle T^k_q(\ell'\ell) \rangle = \sum_{m'm} (-1)^{\ell'-m'} \langle \ell'\ell m' - m | kq \rangle \langle f_{\ell'm'} f^*_{\ell m} \rangle, \tag{8.22}$$

where ℓ and ℓ' are the orbital angular momenta of final states $|i\rangle$ and $|i'\rangle$. It is assumed that the orbital angular momentum of the initial state $|0\rangle$ is 0. The Clebsch–Gordan coefficients impose conditions on the tensor operator $T_q^k(\ell'\ell)$ of rank k and component q through the restrictions

$$k \le \ell' + \ell, \quad -k \le q \le k. \tag{8.23}$$

The monopole $(k = 0)$ is proportional to the differential cross section

$$T_0^0(\ell\ell) = \sigma/\sqrt{2\ell + 1}, \tag{8.24}$$

the three parameters of rank 1, $\langle T_q^1(\ell'\ell)\rangle$ with $q = \pm 1, 0$, are the components of the orientation vector, and the five parameters of rank 2 are the components of the alignment tensor. For $\ell' = \ell = 1$ these are the only components that contribute.

The physical meaning of the first-rank tensor can be seen when these are related to the components of the mean angular momentum vector $\langle \ell \rangle$. With the help of the Wigner–Eckart theorem one can show that

$$\langle T_q^1(\ell\ell)\rangle = \frac{\sqrt{3}\sigma}{\sqrt{\ell(\ell + 1)(2\ell + 1)}}\langle \ell_q \rangle, \tag{8.25}$$

where the 'spherical' vector components of $\langle \ell \rangle$ are defined by (see 8.7)

$$\langle \ell_{\pm 1}\rangle = \mp \frac{1}{\sqrt{2}}(\langle \ell_x \rangle \pm i\langle \ell_y \rangle), \quad \langle \ell_0 \rangle = \langle \ell_z \rangle. \tag{8.26}$$

If the system is oriented we mean that $\langle \ell \rangle \ne 0$. For axially symmetric systems only, $\langle \ell_0 \rangle = \langle \ell_z \rangle$ and hence $\langle T_0^1(\ell\ell)\rangle$ is nonvanishing. For the case shown in fig. 8.1 where no axial symmetry exists and only $\langle \ell_y \rangle \ne 0$,

$$\langle T_1^1(\ell\ell)\rangle = \langle T_{-1}^1(\ell\ell)\rangle = \frac{-i\sigma\sqrt{3}\langle \ell_y \rangle}{\sqrt{2\ell(\ell + 1)(2\ell + 1)}} \tag{8.27}$$

is the only nonvanishing element of rank 1.

Hermiticity imposes further constraints. For $\ell = 1$ there are only five independent multipoles, e.g. $\langle T_0^0\rangle$, $\langle T_1^1\rangle$, $\langle T_2^2\rangle$, $\langle T_1^2\rangle$, $\langle T_0^2\rangle$, with $\langle T_1^1\rangle$ imaginary and the components of the alignment tensor real (Blum, 1981). When reflection invariance holds in the collision plane the state multipoles can be related to the orientation O and alignment parameters A, first introduced by Happer (1972) and Fano and Macek (1973), through

$$\langle T_1^1(\ell\ell)\rangle = -i\sigma\sqrt{\frac{3\ell(\ell + 1)}{2(2\ell + 1)}}O_1^- = -i\sigma\sqrt{\frac{3}{2\ell(2\ell + 1)(\ell + 1)}}\langle \ell_y \rangle,$$

$$\langle T_0^2(\ell\ell)\rangle = \frac{N\sigma}{\sqrt{6\ell(\ell + 1)}}\langle 3\ell_z^2 - \ell^2 \rangle = N\sigma\frac{\sqrt{\ell(\ell + 1)}}{\sqrt{6}}A_0,$$

$$\langle T_1^2(\ell\ell)\rangle = \frac{N\sigma}{\sqrt{2\ell(\ell+1)}}\langle \ell_x\ell_z + \ell_z\ell_x\rangle = N\sigma\frac{\sqrt{\ell(\ell+1)}}{\sqrt{2}}A_1,$$

$$\langle T_2^2(\ell\ell)\rangle = \frac{N\sigma}{\sqrt{2\ell(\ell+1)}}\langle \ell_x^2 - \ell_y^2\rangle = N\sigma\frac{\sqrt{\ell(\ell+1)}}{\sqrt{2}}A_2, \tag{8.28}$$

where

$$N = \sqrt{\frac{30}{(2\ell+3)(2\ell+1)(2\ell-1)}}.$$

For completely coherent excitation and $\ell' = \ell = 1$ one obtains the following relation between the state multipoles and the parameters σ, λ, χ (8.21).

$$\langle T_0^0\rangle = \frac{\sigma}{\sqrt{3}}, \quad \langle T_1^1\rangle = +i\sigma\langle \ell_y\rangle/2 = -i\sigma\sqrt{\lambda(1-\lambda)}\sin\chi, \quad \langle T_2^2\rangle = \frac{\sigma}{2}(\lambda-1)$$

$$\langle T_1^2\rangle = -\sigma\sqrt{\lambda(1-\lambda)}\cos\chi, \quad \langle T_0^2\rangle = \frac{\sigma}{\sqrt{6}}(1-3\lambda). \tag{8.29}$$

8.1.6 *Characterisation of emitted radiation*

We now consider the radiative decay of the excited ensemble of atoms. The angular distribution and polarisation of the emitted photons can be conveniently described in terms of the Stokes parameters I, η_1, η_2, and η_3 (Born and Wolf, 1970). The emitted photons can be observed in the direction $\hat{n}$ making polar angles θ and azimuthal angles ϕ with respect to the collision frame (fig. 8.1). It is convenient to choose the coordinate system in which the direction of observation $\hat{n}$ of the radiation is chosen as the z axis. The polarisation vector of the photons is restricted to the plane perpendicular to $\hat{n}$ by the two unit vectors $\hat{\epsilon}_1 = (\theta + 90°, \phi)$ and $\hat{\epsilon}_2 = (\theta, \phi + 90°)$. The direction $\hat{\epsilon}_1$ is conveniently chosen to lie in the plane spanned by $\hat{n}$. If I is the intensity of light emitted in the direction $\hat{n}$ and $I(\gamma)$ the intensity transmitted by a linear polariser oriented at an angle γ with respect to the $\hat{\epsilon}_1$-axis, then the Stokes parameters are defined by

$$I = I(0°) + I(90°), \tag{8.30a}$$

$$I\eta_3 = I(0°) - I(90°), \tag{8.30b}$$

where η_3 is the degree of linear polarisation with respect to the $\hat{\epsilon}_1$ and $\hat{\epsilon}_2$-axes,

$$I\eta_1 = I(45°) - I(135°), \tag{8.30c}$$

where η_1 is the degree of linear polarisation with respect to two orthogonal axes oriented at 45° to the ϵ_1-axis, and

$$I\eta_2 = I_+ - I_-, \tag{8.30d}$$

where η_2 is the degree of circular polarisation, $I_+(I_-)$ denoting the intensity of light with positive (negative) helicity, or right (left) circular polarisation.

In order to specify the polarisation of the emitted radiation, Born and Wolf (1970) introduced the degree of polarisation

$$P = \sqrt{\eta_1^2 + \eta_2^2 + \eta_3^2}, \tag{8.31}$$

and the coherence-correlation factor

$$\mu = (\eta_1 + i\eta_2)/\sqrt{1 - \eta_3^2} = |\mu|e^{i\beta}, \tag{8.32}$$

where $|\mu|$ is called the degree of coherence and β the effective phase difference. In general these parameters are restricted by

$$P \leq 1, \quad |\mu| \leq 1. \tag{8.33}$$

The equality sign holds if the radiation results from a transition between two pure atomic states, such as a $^1P \rightarrow {}^1S$ transition. Then the radiation is said to be completely coherent. An experimental measurement of P or $|\mu|$ allows us to obtain some information on the coherence properties of the excitation process.

The Stokes parameters are directly related to the state multipoles characterising the ensemble of excited-state atoms. Explicit equations are given by Blum and Kleinpoppen (1979), who also give equations where the effects of fine and hyperfine structure are taken into account. As discussed in subsection 8.1.3 five parameters are in general required for determining the excited state. These can be the cross section σ and the four Stokes parameters, which are all independent of each other.

For $^1P - {}^1S$ transitions, however, the excited atoms are completely described by three parameters as discussed in subsection 8.1.4. Thus in this case the four Stokes parameters are not independent quantities and only three measurements are required for a complete determination of the excited state, e.g. σ, I and $I\eta_2$. The intensity of light radiated in the transition $|i\rangle \rightarrow |f\rangle$ and emitted into a solid angle $d\Omega$ at (θ, ϕ) measured by an ideal detector sensitive only to polarisation $\hat{\epsilon}$ in the time interval $t, t + dt$, is given by

$$I(\hat{\epsilon}, \theta, \phi) \sim |\langle f|\hat{\epsilon} \cdot \mathbf{r}|i\rangle|^2 e^{-\gamma t}, \tag{8.34}$$

where γ is the decay constant.

Expressing the Stokes parameters in terms of the parameters σ, λ and χ (8.21) one obtains for a $^1P - {}^1S$ transition (e.g. Slevin, 1984)

$$I = C\{\frac{1}{2}(1 - \lambda)(1 + \cos^2\theta - \sin^2\theta\cos 2\phi) + \lambda\sin^2\theta$$
$$+ \sqrt{\lambda(1 - \lambda)}\cos\chi\sin 2\theta\cos\phi\} \tag{8.35a}$$
$$I\eta_1 = C\{-(1 - \lambda)\cos\theta\sin 2\phi - 2\sqrt{\lambda(1 - \lambda)}\cos\chi\sin\theta\sin\phi\} \tag{8.35b}$$

$$I\eta_2 = C\{2\sqrt{\lambda(1-\lambda)}\sin\chi\sin\theta\sin\phi\} \tag{8.35c}$$

$$I\eta_3 = -C\{\frac{1}{2}(1-\lambda)[\sin^2\theta - (1+\cos^2\theta)\cos2\phi] - \lambda\sin^2\theta$$
$$- \sqrt{\lambda(1-\lambda)}\cos\chi\sin2\theta\cos\phi\}, \tag{8.35d}$$

where

$$C = \frac{\sigma}{3}\frac{\omega^4 d\Omega}{2\pi c^3}|\langle 0 \parallel r \parallel \ell\rangle|^2(-1)e^{-\gamma t}. \tag{8.36}$$

Equations similar to (8.32) can be derived in terms of state multipoles or other parameters. They completely specify the radiation field and show how the properties of this field are related to the dynamical observables of the collision. In practice the time evolution of the excited state is not measured, since it is of no interest in determining λ and χ, and the total integrated intensity is usually determined experimentally. This means that the factor $e^{-\gamma t}$ in C is replaced by $1/\gamma$.

Two different kinds of measurements can be made in order to determine λ and χ.

i) Angular correlation measurements where the angular correlation between the scattered electron and emitted photon is measured without regard to the polarisation of the light (e.g. (8.35a)).

ii) Polarisation correlation measurements in which the polarisation of the emitted photons is determined. The degree of circular polarisation is given by dividing expression (8.35c) by (8.35a) and for $\hat{n}$ in the y direction ($\theta = \phi = 90°$)

$$\eta_2 = +2\sqrt{\lambda(1-\lambda)}\sin\chi$$
$$= -\langle L_\perp\rangle, \tag{8.37}$$

where we have used (8.29) in the second line. The relation $\eta_2 = -\langle L_\perp\rangle$ follows from conservation of angular momentum. Since the final $^1S_0$ state is spherically symmetric, the photons must carry away the angular momentum of the excited state.

A measurement of circular polarisation also gives the sign of χ and $\langle L_\perp\rangle$, whereas the measurement of the angular correlation alone (expression (8.35a)) does not. Similarly at $\theta = \phi = 90°$ the Stokes parameters for linear polarisation are given by

$$\eta_3 = 2\lambda - 1, \tag{8.38}$$

and

$$\eta_1 = -2\sqrt{\lambda(1-\lambda)}\cos\chi. \tag{8.39}$$

For transitions in the vacuum ultraviolet it is usually easier to measure angular rather than polarisation correlations, due to the difficulty in measuring polarisation in this wavelength region.

Instead of λ and χ it is informative to choose the electron impact coherence parameters, which describe the excited-state charge cloud shown schematically in fig. 8.1. In the absence of spin-analysis they are $\langle P_\ell \rangle$ describing the relative difference between the length ℓ and width w, $\langle \gamma \rangle$ describing the alignment and $\langle L_\perp \rangle$ describing the orientation. In terms of the Stokes parameters the electron impact coherence parameters are

$$\langle P_\ell \rangle = (\eta_1^2 + \eta_3^2)^{1/2},$$
$$\langle \gamma \rangle = \frac{1}{2}\tan^{-1}\left(\frac{\eta_1}{\eta_3}\right),$$
$$\langle L_\perp \rangle = -\eta_2. \tag{8.40}$$

8.2 Hydrogen

The electron–hydrogen problem is the prototype for electron–atom scattering calculations. It has both characteristic difficulties: identical electrons and the target continuum. Its simplification is that no approximations need to be made to obtain an orbital description of the target states. However, this does not over-simplify the scattering problem, since equn. (7.35) shows that all electron–atom scattering problems may be formulated in terms of potential matrix elements that are linear combinations of single-orbital problems with coefficients given by a calculation of the target structure.

8.2.1 Differential cross sections

For a hydrogen target scattering is purely elastic up to the $n=2$ excitation threshold at 10.2 eV. The ionisation threshold is 13.6 eV. Fig. 8.2 compares experimental differential cross sections for elastic scattering at low energies with the coupled-channels-optical calculation, which describes the data excellently. Since the excitation and ionisation thresholds are atypically high for hydrogen it is tempting to consider it as a case where low-energy scattering may be easy to describe by a coupled-channels or R-matrix calculation that accounts for virtual excitation to discrete channels but not to the continuum. That this is not the case is shown at the lowest energy 0.582 eV in fig. 8.2 by comparing the full calculation with one that omits the continuum and another that omits all nonelastic channels. The continuum must be included, even at very low energy, to achieve satisfactory agreement with experiment.

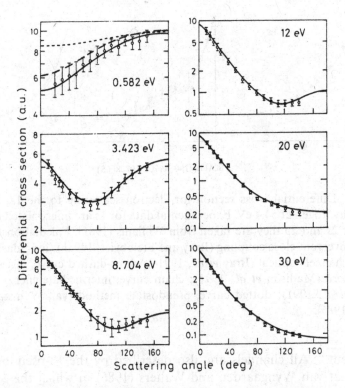

Fig. 8.2. Differential cross section for the elastic scattering of electrons on hydrogen. Circles, Williams (1975); solid curve, coupled-channels-optical calculation; long-dashed curve, one channel with discrete polarisation potential only; short-dashed curve, one channel without polarisation potential. Adapted from Bray *et al.* (1991*b*).

The energy at which the most comprehensive range of experimental quantities has been observed is 54.4 eV. This energy tests scattering theories quite severely because it is well above the ionisation threshold but not so high that one might naively expect Born terms to dominate the amplitudes. Fig. 8.3 compares the experimental differential cross sections for the 1*s*, 2*s* and 2*p* channels with various calculations. The experimental data for the 1*s* channel have been obtained by cubic spline interpolation in the data of Williams (1975). We note first that the convergent-close-coupling calculation achieves close, but not perfect, agreement.

It is perhaps as interesting to compare the approximate calculations with the convergent-close-coupling calculation as with experiment. The one that takes all channels into account most completely is the coupled-channels-optical calculation (Bray, Konovalov and McCarthy, 1991*c*) in which *P* space consists of the *n*=1, 2 and 3 channels. It agrees closely, but not completely, with the convergent calculation and similarly with

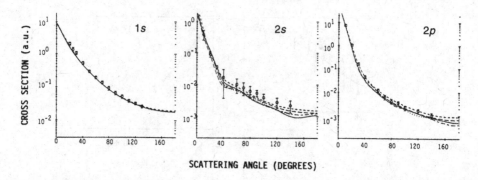

Fig. 8.3.   Differential cross section for electron scattering to the 1s, 2s and 2p states of hydrogen at 54.4 eV. Experimental data for 1s are interpolated (Williams, 1975), for 2s and 2p they are taken from Williams (1981). Calculations are: solid curve, convergent close coupling (Bray and Stelbovics, 1992b); long-dashed curve, coupled channels optical (Bray *et al.*, 1991c); short-dashed curve, distorted-wave second Born (Madison *et al.*, 1991); chain curve, intermediate-energy R matrix (Scholz *et al.*, 1991); dotted curve, pseudostate method (van Wyngaarden and Walters, 1986).

experiment.   All channels are also described by the 18-pseudostate calculation of van Wyngaarden and Walters (1986) in which the 1s, 2s and 2p states are exact target states.   Similar agreement is achieved.   The intermediate-energy R-matrix calculation of Scholz *et al.* (1991) only uses the full method up to total angular momentum $K=4$ and simpler approximations beyond.   The need for a full solution in this case is assessed by comparing the more-complete approximations with the distorted-wave second Born calculation of Madison, Bray and McCarthy (1991). Differences are not large.   In fact the spread of disagreement of all these calculations is no larger than the experimental error.

The coupled-channels-optical method is a completely *ab initio* method that has been shown to be computationally feasible at all energies.   It is compared with experimental differential cross sections for the $n=1$ and 2 channels at a range of energies in figs. 8.4 and 8.5.   Here P space consists of the $n=1$, 2 and 3 channels.   The contribution of the polarisation potential may be assessed in these figures by comparing the curves calculated by coupling the six lowest-energy channels with and without it.

Significant discrepancies exist for absolute differential cross sections between all calculations on one hand and all experiments on the other. The implications for using the hydrogen target as a test of scattering theory are serious.   An attempt to resolve the difficulty was made by Lower, McCarthy and Weigold (1987), who reported very accurate direct

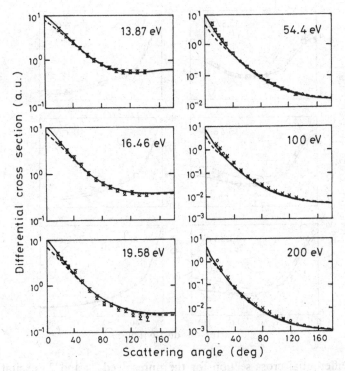

Fig. 8.4. Differential cross section for the elastic scattering of electrons on hydrogen. Experimental data: Callaway and Williams (1976) and Williams (1976a, 1981). Solid curve, coupled channels optical; broken curve, six-state coupled channels (Bray *et al.*, 1991c).

measurements of the ratio of $n=1$ to $n=2$ differential cross sections at three angles. They are compared with earlier measurements and with calculations in table 8.1. Once again the calculations are in excellent agreement with each other and in reasonable agreement with the new experiment but the ratios are not so different from those of the earlier experiments that they suggest a trend.

8.2.2 Integrated and total cross sections

Calculations that take all channels into account to describe scattering give values for the total cross section in addition to differential and integrated cross sections for particular low-lying channels. The total cross section is calculated from the entrance-channel T-matrix element by the optical theorem (6.47). This provides an important check on the validity of the description of higher channels without which a calculation of a limited subset of cross-section data cannot be taken seriously.

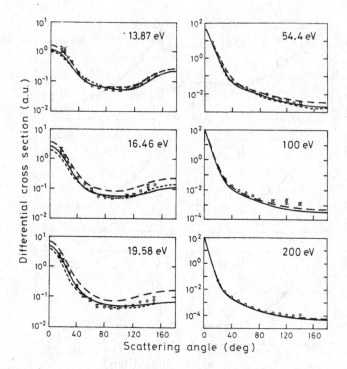

Fig. 8.5. Differential cross section for the unresolved 2s and 2p excitations of hydrogen. Experimental data: $E < 50$ eV, Williams (1976a); $E > 50$ eV, Williams and Willis (1975); solid curve, coupled channels optical; long-dashed curve, six-state coupled channels (Bray et al., 1991c); short-dashed curve, intermediate-energy R matrix (Scholz et al., 1991). From Bray et al. (1991c).

The difficulties associated with making an atomic-hydrogen target have precluded direct measurements of total cross sections for hydrogen. Estimates may be made by adding the best available estimates for the integrated cross sections of particular channels and the total ionisation cross section.

There are only isolated measurements of integrated cross sections, but there are absolute measurements of differential cross sections. We adopt the procedure of using the coupled-channels-optical calculation of Bray et al. (1991c) to interpolate and extrapolate these measurements, since it agrees quite well with differential cross sections in figs. 8.4 and 8.5.

For the 1s channel we estimate the integrated cross section from the differential cross sections of Williams (1975). These estimates are compared in table 8.2 with the results of the three calculations considered in this section. The calculations are in close agreement but the experimental estimates are significantly higher at 54.4 eV and 100 eV.

Table 8.1. *The ratio of n=1 to n=2 differential cross sections for hydrogen.*
The experimental cases are LMW, Lower, McCarthy and Weigold (1987);
W/WW, Williams (1975) for n=1 and Williams and Willis (1975) for n=2;
vW/WW, van Wingerden et al. (1977) for n=1 and Williams and Willis (1975)
for n=2. Calculations are CCO, coupled-channels-optical (Bray et al. 1991c);
CCC, convergent-close-coupling (Bray and Stelbovics 1992b); vWW, pseudostate
method (van Wyngaarden and Walters, 1986)

case	100 eV			200 eV		
	30°	45°	60°	30°	45°	60°
LMW	11.4±0.7	14.5±0.8	14.8±0.8	25.4±1.4	28.8±1.5	35.8±2.5
W/WW	9.8±1.2	15.6±2.7	11.7±1.6	25.6±3.4	26.4±6.6	25.7±3.6
vW/WW	10.1±1.0	14.4±2.1	12.0±1.7	29.7±3.5	29.9±5.8	32.6±4.5
CCO	12.3	18.6	17.5	29.0	33.9	32.6
CCC	10.5	17.2	17.5	—	—	—
vWW	13.0	19.3	18.2	28.4	34.4	31.0

For the summed $n=2$ channels there are absolute differential cross section measurements by Williams (1976a) and Williams and Willis (1975). Estimates of the $2s$ and $2p$ integrated cross sections are obtained here by dividing the $n=2$ estimate in the ratio given by the coupled-channels-optical calculation. They are given in table 8.2. Once again the calculations agree quite closely but the experimental estimates are high at 54.4 eV and 100 eV. There is an independent measurement of the $2p$ integrated cross section at 54.4 eV by Williams (1981). The result is $0.89\pm0.08\ \pi a_0^2$, which compares with the estimate $1.01\pm0.09\ \pi a_0^2$ of table 8.2.

To estimate the total cross section for hydrogen we use the $n=1$ and 2 estimates above. For $n=3$ we interpolate in the direct measurements of integrated cross sections by Mahan, Gallagher and Smith (1976). For higher discrete channels we use the roughly-valid rule that integrated cross sections for principal quantum number n are proportional to n^{-3}. Very accurate measurements of the total ionisation cross section have been made by Shah *et al.* (1987). These total cross section estimates are shown in table 8.3 in comparison with the three calculations we are considering.

The comparison of theory and experiment in table 8.3 is somewhat unsatisfactory. The coupled-channels-optical and pseudostate calculations agree with each other and with the convergent-close-coupling calculation within a few percent, yet there are noticeable discrepancies with the experimental estimates. The convergent-close-coupling method calculates total ionisation cross sections in complete agreement with the measurements

Table 8.2. *Integrated cross sections (πa_0^2) for the 1s, 2s and 2p states of hydrogen. Experimental values (EXP) are obtained as described in the text. Calculations are: CCO, coupled channels optical (Bray et al., 1991c); CCC, convergent close coupling (Bray and Stelbovics, 1992b); vWW, pseudostate method (van Wyngaarden and Walters, 1986)*

E_0(eV)	EXP	CCO	CCC	vWW
1s				
19.6	3.25±0.32	3.45	–	
54.4	1.27±0.13	1.00	0.96	–
100	0.61±0.06	0.46	0.45	0.480
200	0.21±0.02	0.20	–	0.197
2s				
19.6	0.112±0.016	0.109	–	–
54.4	0.076±0.007	0.0545	0.0675	0.0651
100	0.046±0.004	0.0414	0.0446	0.0404
200	0.026±0.003	0.0251	–	0.0250
2p				
19.6	0.64±0.09	0.617	–	–
54.4	1.01±0.09	0.729	0.719	0.739
100	0.73±0.06	0.630	0.624	0.638
200	0.47±0.05	0.443	–	0.446

of Shah *et al.* (see chapter 10). The discrepancies are therefore in the integrated cross sections for scattering.

We summarise the situation for integrated and total cross sections by remarking that direct absolute measurements of these quantities are very important for testing scattering theories. There is room for doubt that present estimates are sufficiently accurate.

8.2.3 Magnetic substate parameters

Detailed measurements of the magnetic substate parameters λ, R and I (8.17) have been made for the 2p channel. Our example considers these parameters at 54.4 eV. They are more sensitive to the calculation

Table 8.3. *Total cross sections* (πa_0^2) *for hydrogen. The column headings are as in table 8.2*

E_0(eV)	EXP	CCO	CCC	vWW
19.6	4.55±0.45	4.81	–	–
54.4	3.34±0.33	2.85	2.71	–
100	2.21±0.22	2.06	1.96	2.13
200	1.40±0.14	1.28	–	1.31

method than the differential cross sections in the sense that the spread of disagreement is larger.

Fig. 8.6 compares experimental data with the same calculations as were illustrated in fig. 8.3.   The disagreement between different calculations and between calculations and experiment is larger at larger angles, but the experiment is very difficult in this region.   No calculation reproduces the low experimental values of λ and R beyond 80°.   The disagreement between the two integral-equation methods, convergent-close-coupling and coupled-channels-optical, is again smaller than the corresponding experimental error.

8.2.4 State multipoles

The measurement of the complete set of state multipoles for the $n=2$ states of hydrogen at 350 eV has been reported by Williams and Heck (1988). The electron scattering angle was 3°.   The scattered electrons with 10.2

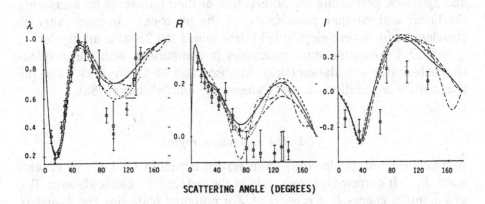

Fig. 8.6.   The magnetic substate parameters λ, R and I (8.17) for 54.4 eV electron scattering to the 2p state of hydrogen. Squares, Weigold, Frost and Nygaard (1979) and Hood, Weigold and Dixon (1979); circles, Williams (1981,1986). The theoretical curves are as for fig. 8.3.

Table 8.4. *State multipoles at $E_0=350$ eV, $\theta=3°$ for the $n=2$ excitation of hydrogen using positive, negative and zero electric fields. The experimental data are due to Williams and Heck (1988). Errors in the final significant figures are given in parentheses. Calculations are: CCO, method of Bray, Madison and McCarthy (1990); and vWW, van Wyngaarden and Walters (1986)*

Multipole	+250 Vcm^{-1}	-250 Vcm^{-1}	zero field	CCO	vWW
$\mathrm{Re}\langle T_0^1(10)\rangle$	0.17(71)	0.77(88)	–	0.421	0.39
$\mathrm{Im}\langle T_0^1(10)\rangle$	0.82(62)	0.29(74)	–	1.20	1.15
$\mathrm{Re}\langle T_1^1(10)\rangle$	0.87(33)	0.36(31)	–	1.19	1.02
$\mathrm{Im}\langle T_1^1(10)\rangle$	1.9(3)	1.90(47)	–	2.69	2.6
$\mathrm{Im}\langle T_1^1(11)\rangle$	0.36(9)	0.22(8)	0.34(1)	0.508	0.31
$\langle T_0^2(11)\rangle$	–	–	6.5(8)	7.03	6.99
$\langle T_1^2(11)\rangle$	–	–	−7.3(8)	−6.42	−6.46
$\langle T_2^2(11)\rangle$	–	–	−10.7(8)	−10.57	−10.54
λ	–	–	0.11(3)	0.085	–
R	–	–	0.21(1)	0.197	–
I	–	–	−0.010(3)	−0.016	–
σ_{2s}/σ_{2p}	0.087(14)	0.055(17)	0.030(1)	0.035	0.032
σ_{2p}(a.u.)	–	–	24.1(11)	23.11	23.05
σ_{2s}(a.u.)	–	–	0.73(5)	0.82	0.74

eV energy loss ($2s$ and $2p$ states) were measured in coincidence with the 10.2 eV photons either from the fast decay of the $2p$ state or the delayed Stark-quenched $2s$ state.   An external electrostatic field mixed the $2s$ and $2p$ states, permitting the observation of their coherence by measuring the linear and circular polarisation of the photons.   In-plane angular correlations for a zero electric field determined the $2p$-state multipoles.

Table 8.4 shows the state multipoles in comparison with the coupled-channels-optical calculation (Bray, Madison and McCarthy, 1990) and the pseudostate calculation of van Wyngaarden and Walters (1986).

8.2.5 The resonance region

A resonance r in an electron–atom system occurs at total energy ϵ_r with width Γ_r.   It corresponds to a pole in the reduced T-matrix element $T_{P\pi}$ at a complex energy $E = \epsilon_r - \frac{1}{2}i\Gamma_r$.   For potential scattering the T-matrix element is given by (4.142).   The resonance index r corresponds to the symmetry $P\pi$, where P is the total angular momentum and π is the parity, and a principal quantum number identifying the particular resonance in the symmetry manifold.   In the case of a hydrogen target it is convenient to use *LS* coupling, for which the symmetry is characterised by the total orbital angular momentum *K* and electron spin *S* using the spectroscopic notation of section 5.8.

A resonance has much in common with a bound state of the electron–atom compound system, which is represented by a pole on the negative energy axis. For hydrogen the compound system is the negative ion H⁻. The relationship of the low-energy resonances for hydrogen to states of the compound system is illustrated in the independent-electron approximation by fig. 8.7. The illustration considers only *s* orbitals and represents a state of H⁻ by the orbital pair *msns*. The notation *ms∞s* means that the second electron's energy is at the limit $n = \infty$ of the Rydberg series and is just unbound. In such a state the one-electron state *ms* is an exact state of the hydrogen atom. Note that the ground state of the H⁻ ion is bound by 0.754 eV.

The resonance illustrated in fig. 8.7 has the Hartree–Fock configuration 2s2s and symmetry ¹S. The compound system in this state can decay by an electron dropping into the ground state of H (1s∞s). The system gains 9.55 eV, which appears as kinetic energy of the second electron. The process is called autoionisation. The resonance is at a projectile energy $E_0 = 9.55$ eV or a total energy $\epsilon = -4.06$ eV.

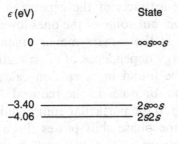

ϵ (eV)	State
0	∞s∞s
−3.40	2s∞s
−4.06	2s2s
−13.61	1s∞s
−14.36	1s1s

Fig. 8.7. Energy level diagram of the H⁻ ion showing states where Hartree–Fock configurations have two *s* orbitals.

Between the $2s2s$ and $2s\infty s$ states there is a sequence of resonances with Hartree—Fock configurations $2sns$, $n = 3, \infty$. They occur just below the $n=2$ threshold at 10.20 eV in E_0 and condense to this energy. A similar sequence of ^1S resonances occurs just below each inelastic threshold. Similar sequences occur in the other symmetry manifolds with Hartree—Fock configurations consisting of different orbitals.

Since the compound system for a resonant state has an electron that is unbound initially and finally, resonances may be investigated by a scattering calculation. Methods of calculation that are appropriate for rapid energy dependence are the R-matrix method (Burke and Robb, 1975), the variational pseudostate method (Callaway, 1982) and the coupled-channels-optical method (McCarthy and Shang, 1992). Of these only the last two methods treat the target continuum, which must be taken into account at all energies to obtain correct absolute cross sections. The coupled-channels-optical method uses the same optical potential and quadrature points for an energy range over which the optical potential changes imperceptibly. Only on-shell and half-on-shell potential matrix elements need to be recalculated when the energy is changed in this range.

The T-matrix element for a particular reaction may be represented perfectly generally by a linear combination of resonance terms (Bloch, 1957), most of which overlap considerably. The methods we have considered in chapter 7 may be put into this form. Most resonances in the expansion are artifacts of the expansion and have no individual physical manifestation, but some of the ones lowest in energy are isolated, at least from others in the same symmetry manifold. They appear as anomalies in the energy dependence of cross sections.

Resonances may be found in a reaction calculation by searching the complex energy plane for poles in the reduced T-matrix elements or by using a generalisation to the particular compound system of the potential-scattering rule that the phase shift passes through $\pi/2$ as the energy is varied over the resonance. The ones that have physical manifestation appear as visible anomalies in the energy dependence of the reduced T-matrix elements for each symmetry. The parameters of such a resonance can be found by fitting a partial cross section of the form suggested by (4.145) over a small energy region near resonance.

$$\sigma_r = |aE + b + \frac{c + id}{E - \epsilon_r + i\Gamma_r/2}|^2. \qquad (8.41)$$

Tables 8.5 and 8.6 show resonances that have been identified experimentally and theoretically for hydrogen below the $n=2$ and 3 thresholds respectively. The orbital angular momentum of a resonance may be identified by the fact that the resonant reduced T-matrix element for elastic

Table 8.5. *Resonance energies and widths for electron–hydrogen elastic scattering below the n=2 threshold. Column headings for experimental data are: Williams, Williams (1976b); Warner, Warner et al. (1986). Errors in the last significant figures are shown in parentheses. Column headings for calculations are: PS, pseudostate method (Seiler, Oberoi and Callaway, 1971); R-matrix, Pathak, Kingston and Berrington (1980); CCO, McCarthy and Shang (1992)*

Symmetry	Williams	Warner	PS	R-matrix	CCO
Energies (eV)					
^1S	9.557(10)	9.549(13)	9.574	9.557	9.553
	–	–	–	–	9.875
	–	–	10.178	10.177	10.172
^3S	–	–	10.151	10.147	–
^1P	–	–	–	–	9.531
	–	–	10.185	10.176	10.177
^3P	9.735(10)	9.736(13)	9.768	9.741	9.743
^1D	–	10.115(13)	10.160	10.126	10.144
Widths(10^{-3}eV)					
^1S	45.0(5)	63.0(8)	54.0	52.0	48.0
	–	–	–	–	8.9
	–	–	2.3	2.6	2.4
^3S	–	–	0.02	–	–
^1P	–	–	–	–	9.1
	–	–	0.02	–	0.04
^3P	6.0(5)	5.0(2)	8.0	7.1	4.5
^1D	–	6.0(2)	7.7	8.8	6.9

scattering (4.145) does not contribute to the differential cross section at an angle for which the corresponding Legendre polynomial is zero. An example is shown in fig. 8.8 which compares differential cross sections at 30° and 90° over the first ^1S and ^3P resonances below the n=2 threshold. The ^3P resonance has no contribution at 90°.

Fig. 8.8 shows that the coupled-channels-optical method with the equivalent-local polarisation potential (McCarthy and Shang, 1992) gives a good semiquantitative description of the experimental data of Williams (1976b) for elastic differential cross sections below the n=2 threshold. At energies just below the n=3 threshold the resonances affect the n=2 excitations. Fig. 8.9 shows the energy dependence of the integrated cross sections for the 2s and 2p channels. Since a resonance is a property of the compound system, not the channel, the resonances observed in

Table 8.6. *Resonance energies $E_0(eV)$ and widths $\Gamma(10^{-3}eV)$ for the electron–hydrogen 2s and 2p channels below the $n=3$ threshold. The experimental data are due to Williams (1988). Errors in the last significant figures are shown in parentheses. Column headings for calculations are: PS, pseudostate calculation by Callaway (1982); R-matrix, Pathak, Kingston and Berrington (1980); CCO, McCarthy and Shang (1992)*

Symmetry	EXP 2p	EXP 2s	PS	R-matrix	CCO 2p	CCO 2s
E_0						
^1S	11.722(9)	11.724(12)	11.7218	11.7218	11.726	11.726
^1D	11.807(9)	11.803(9)	11.8048	11.8049	11.805	11.803
^1P	11.902(6)	–	11.8929	11.8930	11.892	11.891
^3F	11.925(2)	11.926(2)	11.9255	11.9258	11.927	11.927
^3D	11.997(5)	12.00(5)	11.9949	11.9952	11.996	11.996
^1S	12.029(5)	12.024(5)	12.0272	12.0273	12.028	12.030
^3P	12.040(6)	12.036(4)	12.0369	12.0370	12.037	12.038
^1D	12.049(4)	12.048(4)	12.0532	12.0534	12.047	12.046
Γ						
^1S	45(9)	37(8)	38.89	40.93	39.0	37.0
^1D	45(8)	37(8)	44.46	43.52	49.4	47.8
^1P	33(10)	–	32.5	34.13	29.7	30.9
^3F	4(2)	4(2)	2.96	3.10	3.7	3.5
^3D	10(3)	15(10)	10.2	10.5	10.0	12.0
^1S	9(3)	9(3)	8.31	7.89	9.5	9.8
^3P	10(4)	10(4)	8.31	8.25	12.7	13.0
^1D	7(3)	7(3)	6.57	5.82	7.9	7.9

both channels should be identical. Table 8.6 shows that they are very similar experimentally and in the coupled-channels-optical calculation. Fig. 8.9 shows also that a pseudostate basis can be found that gives an excellent description of the 2s and 2p integrated cross sections below the $n=3$ threshold. The coupled-channels-optical calculation included the channels 1,2,3,4s; 2,3,4p and 3,4d in P space and only the continuum in Q space. Its semiquantitative success shows that the equivalent-local approximation for the polarisation potential, which is designed for higher energies, has sufficient validity at low energy to reduce the absolute cross section from the over-large values obtained by omitting the continuum to values near the experiment. Similar over-large values are obtained by the 15-state *R*-matrix calculation of Fon, Aggarwal and Ratnavelu (1992), which also omits the continuum.

8.3 Sodium

The calculation of electron–sodium scattering gives a good example of the treatment of a many-electron target by a practical method. The potential matrix elements (7.35) may be treated, according to (7.87), as matrix elements for scattering by two potentials, an electron–electron potential and an electron–core potential. In practice the one-electron exchange terms are negligible in the electron-core potential. We describe the core by the same model as we use for the electron-core bound-state problem in section 5.8. This is a frozen-core Hartree–Fock potential with the addition of a polarisation potential (5.82). The parameters of the polarisation potential are determined phenomenologically from bound-state data. This reflects an incomplete treatment of the structure problem, but not of the scattering problem. In fact it is interesting to

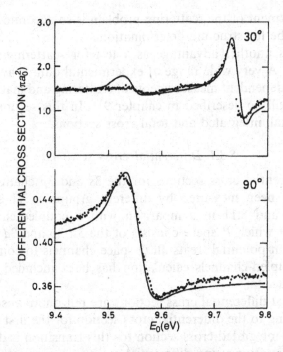

Fig. 8.8. Differential cross section at 30° and 90° for electron–hydrogen elastic scattering below the $n = 2$ threshold. Experiment, Williams (1976*b*); solid curve, coupled channels optical (equivalent local) (McCarthy and Shang, 1992). From McCarthy and Shang (1992).

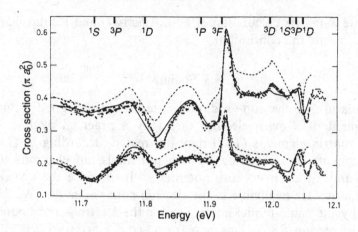

Fig. 8.9. Integrated cross section for electron scattering to the 2s (above) and 2p (below) states of hydrogen below the $n=3$ threshold. The positions and quantum numbers of resonances are shown on the upper scale. Experiment, Williams (1988); solid curve, coupled channels optical (equivalent local) (McCarthy and Shang, 1992); long-dashed curve, pseudostate method (Callaway, 1982); short-dashed curve, 9-state coupled channels.

see if the treatment of the scattering problem is so accurate that it can be used as a probe for structure determination.

Sodium has another advantage as a test for scattering and structure calculations. A very wide range of experimental data is available. This includes spin-dependent measurements of scattering and magnetic substate data, which will be described in chapter 9. In this section we consider only differential, integrated and total cross sections.

8.3.1 *Differential cross sections*

Relative differential cross sections for the 3s and 3p channels at several energies have been measured by different groups. These are shown in figs. 8.10 and 8.11 in comparison with a coupled-channels-optical calculation for which P space consists of the 3s, 3p and 3d channels and the polarisation potential treats all Q space channels to convergence. A 3s, 3p, 3d coupled-channels calculation has been included to assess the effect of Q space.

Experimental differential cross sections are put on an absolute scale by first normalising to the differential cross section for the first dipole transition (3p). The integrated cross section for this transition is determined by numerical integration using differential cross sections measured as close to $\theta = 0$ as possible, supplemented by shape extrapolation based on a calculation. Integrated cross sections are determined in ways that ultimately depend on measurements of the optical oscillator strength (5.84). They

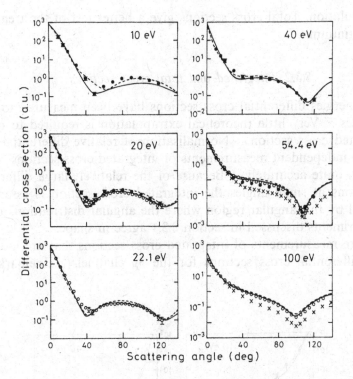

Fig. 8.10. Differential cross section for elastic electron scattering on sodium. Open circles, Lorentz and Miller (1991); closed circles, Srivastava and Vušković (1980); crosses, 54.4 eV, Allen *et al.* (1987), and 100 eV, Teubner, Buckman and Noble (1978); solid curve, coupled channels optical; broken curve, 3-state coupled channels (Bray *et al.*, 1991*d*). From Bray *et al.* (1991*d*).

are discussed in section 8.3.2. For the data shown in fig. 8.11 those of Srivastava and Vušković (1980) and Teubner *et al.* (1986) are normalised to the integrated cross sections of Enemark and Gallagher (1972), those of Buckman and Teubner (1979) to the optical oscillator strength and those of Lorentz and Miller (1991) to the coupled-channels-optical calculation.

All the experimental differential cross sections agree with each other and the calculation over at least three orders of magnitude at angles less than about 20°. However, there are large discrepancies between experiments at larger angles. The calculation agrees well with the data of Srivastava and Vušković at lower energies and with Lorentz and Miller at higher energies.

In general Q space does not have a very large effect on cross sections. This is true also of later calculations (illustrated in chapter 9) where P space includes discrete channels up to convergence. Differential cross sections do not critically test the need to include the ionisation continuum in the calculation. Total cross sections give a better test of the treatment of the continuum.

8.3.2 Integrated and total cross sections

At most energies differential cross sections have been measured to quite small angles.   Very little theoretical extrapolation is required to obtain an integrated cross section.   Normalisation of relative differential cross sections to independent measurements of integrated cross sections can be carried out quite accurately.   Because of the relatively-large differential cross sections at small angles the integrated cross section is essentially determined by this angular region where the angular distributions of the three experiments discussed in section 8.3.1 agree in shape.

There are measurements of integrated cross sections that are independent of differential cross sections for the $3p$ channel.   Enemark and

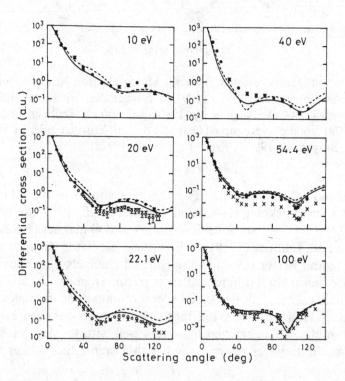

Fig. 8.11.   Differential cross section for electron scattering to the $3p$ state of sodium. Open circles, Lorentz and Miller (1991); closed circles:,Srivastava and Vušković (1980); crosses, 22.1 eV; Teubner et al. (1986), other energies Buckman and Teubner (1979); calculations are as for fig. 8.10. From Bray et al. (1991d).

Table 8.7. *The integrated cross section σ_{3p} for the 3p channel of electron—sodium scattering and the total cross section σ_T ($10^{-16} cm^2$). EXP (σ_{3p}), interpolation in the data of Enemark and Gallagher (1972); EXP (σ_T), Kwan et al. (1991); CCO, coupled-channels-optical calculation (Bray et al., 1991d)*

E(eV)	3p channel		total	
	EXP	CCO	EXP	CCO
4.1	30.8±1.3	30.0	67.1±14.1	89.5
5.9	32.1±1.7	30.6	66.5±14.0	78.6
10.8	32.0±1.4	30.8	55.9±11.7	61.1
20.7	28.7±1.0	27.6	43.3±9.1	49.1
30.7	25.2±0.9	23.6	32.6±6.8	40.5
40.8	22.1±0.6	21.0	30.0±6.3	34.6
50.8	19.8±0.5	19.3	26.2±5.5	31.6
60.9	18.0±0.4	17.7	22.9±4.8	28.8
76.1	15.7±0.2	15.1	22.0±4.6	24.6
100	13.2±0.2	11.3	—	—

Gallagher (1972) measured the relative optical excitation function at energies from 2.5 eV to 1003 eV. After subtracting cascade contributions estimated from experiment, the $3p$ integrated cross section was normalised to the Born approximation (calculated from the known optical oscillator strength) at 1003 eV. Since this dipole transition dominates the reaction cross section at high energy and the Born term dominates the T-matrix element, this normalisation is reasonable. Further confirmation is given by the fact that the generalised oscillator strength is independent of incident energy up to quite large values of K^2 (Buckman and Teubner, 1979), where K is the momentum transfer. This is the prediction of the Born approximation. The data are compared with the coupled-channels-optical calculation in table 8.7.

Absolute measurements of total cross sections have been made by beam-transmission techniques. The results of Kwan *et al.* (1991) are compared with the coupled-channels-optical calculation in table 8.7. In most cases the coupled-channels-optical cross section is within one standard deviation of the experimental result.

8.4 Two-electron atoms

The calculation of electron scattering on atoms whose structures can be represented by two electrons and an inert closed-shell core is an example of the general case where a configuration-interaction calculation of the target states is required. The prototype is helium, a pure two-electron target.

8.4.1 Helium

Since helium occurs naturally as an atomic gas it has been used for many years as a target for electron-collision experiments. However detailed calculations involving approximations for the complete set of target states have been performed only by the equivalent-local coupled-channels-optical method, described in section 7.6.2.

Fig. 8.12 shows the example of 50 eV electron scattering to the three lowest singlet states (McCarthy, Ratnavelu and Zhou, 1991). In the calculation P space consisted of ten channels: $1,2,3^1S$; $2,3^3S$; $2,3^1P$; $2,3^3P$; 3^1D. Polarisation potentials for ionisation were included for all couplings in the $n=1$ and 2 subspace. The basis used in the configuration-

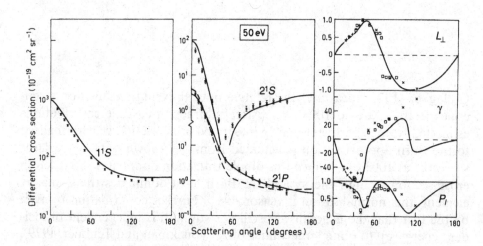

Fig. 8.12. Differential cross section for the 1^1S, 2^1S and 2^1P states of helium and electron impact coherence parameters (8.40) for the 2^1P state at $E_0=50$ eV. Experimental data for differential cross sections are: 1^1S, Register, Trajmar and Srivastava (1980); $2^1S,2^1P$, Cartwright *et al.* (1992). Experimental data for electron impact coherence parameters are: crosses, McAdams *et al.* (1980); squares, Beijers *et al.* (1987); plus signs, Eminyan *et al.* (1974). Solid curves, coupled channels optical (equivalent local) (McCarthy *et al.*, 1991); broken curve, distorted-wave Born (Cartwright *et al.*, 1992). From McCarthy *et al.* (1991).

interaction representation of the *P*-space target states consisted of all allowed excitations in the space defined by the 1,2,3,4*s*; 2,3,4*p* and 3*d* Hartree—Fock orbitals, with higher excitations allowed for by $\bar{s}, \bar{p}$ and $\bar{d}$ pseudo-orbitals.

The calculation describes the differential cross sections and 2^1P electron impact coherence parameters quite well. For the 2^1P differential cross section it is contrasted with a variant of the distorted-wave Born approximation, first-order many-body theory, where the distorted waves are both calculated in the initial-state Hartree—Fock potential.

The equivalent-local form of the coupled-channels-optical method does not give a satisfactory description of the excitation of triplet states (Brunger *et al.*, 1990). Here only the exchange part of the polarisation potential contributes. The equivalent-local approximation to this is not sufficiently accurate. It is necessary to check the overall validity of the treatment of the complete target space by comparing calculated total cross sections with experiment. This is done in table 8.8. The experiments of Nickel *et al.* (1985) were done by a beam-transmission technique (section 2.1.3). The calculation overestimates total cross sections by about 20%, due to an overestimate of the total ionisation cross section. However, an error of this magnitude in the (second-order) polarisation potential does not invalidate the coupled-channels-optical calculation for low-lying discrete channels.

8.4.2 Magnesium

The example of magnesium at $E_0 = 40$ eV illustrates the application of the coupled-channels-optical method to a two-electron atom with a core. It

Table 8.8. *Total cross sections for electron-helium scattering. CCO, coupled-channels-optical (equivalent local) method (McCarthy et al.,1991); experiment, Nickel et al. (1985). Units are $10^{-16}cm^2$*

E_0(eV)	CCO	Experiment
30	2.69	2.391 ± 0.072
40	2.23	2.001 ± 0.060
50	2.06	1.715 ± 0.051
80	1.46	1.269 ± 0.038
100	1.38	1.120 ± 0.034
200	0.76	0.734 ± 0.022

is necessary first to construct the configuration-interaction states describing the target. The basis used (Mitroy and McCarthy, 1989) consisted of $1s$, $2s$ and $2p$ orbitals calculated in the Hartree–Fock approximation for the $(3s^2)^1S^e$ ground state, $3s, 3p, 3d, 4s$ and $4p$ orbitals defined by Hartree–Fock calculations of the $(3sn\ell)^1L$ states and $\overline{3p}, \overline{3d}$ and $\overline{4s}$ orbitals obtained by performing a natural orbital transformation on a large-basis configuration-interaction wave function for the ground state. All these orbitals were orthogonalised to each other. The size of the basis is restricted by computational constraints, since it is necessary to calculate potential matrix elements for every orbital pair in the scattering calculation. Nevertheless the structure calculation is a considerable improvement on Hartree–Fock, as shown in the energy-level table 8.9. Energy levels obtained by a configuration-interaction calculation with the above basis are compared with experiment and a large multiconfiguration Hartree–Fock calculation. It is interesting to contrast the configuration-interaction results in table 8.9 with those of a much larger calculation in table 5.5, which achieves very close energy agreement with experiment and other large structure calculations. The optical oscillator strength (5.84) for the 3^1P state of table 8.9 is within 2% of the multiconfiguration Hartree–Fock value.

The coupled-channels-optical calculation of Zhou (1992) had a P space consisting of the following 10 channels: $3,4,5^1S^e$; 4^3S^e; $3,4,5^1P^o$; $3,4^3P^o$; 3^1D^e. The equivalent-local polarisation potential for the continuum was included in the following couplings: $3^1S^e - 3^1S^e$, 3^1P^o, 3^3P^o, 4^1S^e; $3^1P^o - 3^1P^o$, 3^1D^e, 4^1S^e; $4^1S^e - 4^1S^e$.

Magnesium is much more like sodium than helium, since the reaction cross section is dominated by the first dipole excitation. The respective integrated cross sections for the 2^1P, 3^2P and 3^1P states of helium, sodium

Table 8.9. *Energies (in a.u.) of the lowest-lying singlet states of magnesium. HF, Hartree–Fock; MCHF, multiconfiguration Hartree–Fock (Froese-Fischer, 1975); CI, configuration interaction (see text); EXP, Moore (1949)*

State	HF	MCHF	CI	EXP
$(3s^2)^1S^e$	−0.2427	−0.276 60	−0.274 05	−0.280 99
$(3s3p)^1P^o$	−0.0994	−0.119 49	−0.114 73	−0.121 28
$(3s4s)^1S^e$	−0.0755	−0.082 13	−0.079 57	−0.082 78
$(3s3d)^1D^e$	−0.0548	−0.071 13	−0.065 76	−0.069 56
$(3s4p)^1P^o$		−0.055 66	−0.052 16	−0.055 55

and magnesium are about 0.09, 22.1 and 11.3 πa_0^2. The corresponding excitation energies are 21.2, 2.1 and 4.3 eV. An experiment with very small statistical errors by Brunger *et al.* (1988) observed the relative differential cross section for the 3^1P^o state of magnesium. As was the case for the corresponding experiment at 54.4 eV for sodium by Buckman and Teubner (1979) (fig. 8.11) there is close shape agreement with the coupled-channels-optical calculation over four orders of magnitude at small angles, but agreement worsens at larger angles. The comparison is made in fig. 8.13. In this connection the close agreement of the coupled-channels-optical calculation with spin-dependent data for sodium at all angles (chapter 9) is perhaps significant.

The experimental data of Brunger *et al.* (1988) and those of an earlier experiment by Williams and Trajmar (1978) for the 3^1P^o state were normalised by equating the integrated cross section (obtained by integrating under the extrapolated differential cross section) with the result of the optical excitation experiment by Leep and Gallagher (1976). This measurement included cascades, which contribute about 10 per cent according to the coupled-channels-optical calculation. It was normalised originally at 1400 eV to the generalised oscillator strength from a Born-approximation calculation by Robb (1974), which used configuration interaction. The coupled-channels-optical value of 11.3 πa_0^2 for the 3^1P^o integrated cross section (12.4 πa_0^2 with cascades) is to be compared with the optical-excitation estimate of 15.70 $\pm$ 0.16 πa_0^2.

Also shown in fig. 8.13 are comparisons of the differential cross sections of the coupled-channels-optical method with the experimental values of

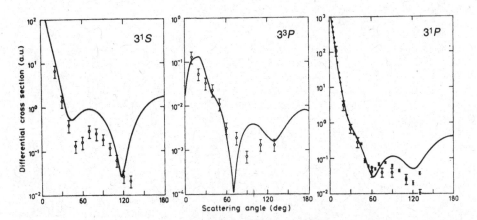

Fig. 8.13. Differential cross section for electron—magnesium scattering at $E_0 =$ 40 eV. Open circles, Williams and Trajmar (1978); closed circles, Brunger *et al.* (1988); full curves, coupled channels optical (Zhou, 1992).

Williams and Trajmar (1978) for the 3^1S and 3^3P channels. The figure gives a good idea of the present state of theory and experiment for all but the most-detailed investigations. There is a need for more experimental data and for a full coupled-channels-optical calculation.

In the absence of independent measurements of the total cross section the total ionisation cross section gives an estimate of the validity of the equivalent-local polarisation potential used for the coupled-channels-optical calculation of fig. 8.13. The calculated value at 40 eV is 5.2 πa_0^2, compared with 4.66±0.47 πa_0^2 measured by Karstensen and Schneider (1975).

9

Spin-dependent scattering observables

In the last chapter we discussed how our understanding of electron impact excitation of atoms has substantially improved in recent years. Sophisticated experimental techniques are available for revealing sensitive details of the collision process, in addition to providing accurate and reliable differential and total cross section data. These details include the shape and inherent angular momentum of the excited atoms after the scattering process, measured as a function of the scattering angle and incident energy. These studies have provided stringent tests of current scattering theories, particularly at intermediate energies and backward angles.

In conventional collision experiments the strong Coulomb interaction generally masks the much weaker relativistic spin-dependent interactions. The role of the spin-dependent interactions, such as the exchange and spin—orbit interactions, has also been clarified by sophisticated measurements with spin-polarised electrons and/or spin-polarised targets, sometimes employing spin analysis after the collision process (Kessler, 1985, 1991; Hanne, 1983).

Such measurements were first applied with considerable success to elastic scattering. Indeed one was able to discuss experiments which would determine all the theoretically calculable amplitudes (Bederson, 1970). For inelastic processes, such measurements necessitate the simultaneous application of spin selection techniques and the alignment and orientation measurements discussed in the previous chapter. The experiments have become feasible with the advancement of experimental techniques. The first successful differential electron impact excitation study with spin-polarised electrons and alignment and orientation measurements was performed by Goeke et al. (1983) for the e—Hg case. McClelland, Kelley and Celotta (1985, 1986) carried out a systematic study for superelastic scattering of polarised electrons from polarised laser-excited Na (3^2P) atoms. This system is essentially a two-electron collision system in which spin exchange is the dominant spin-dependent interaction. It thus allows one to obtain information on the alignment and orientation parameters for both the singlet and triplet contributions.

A polarised electron beam is one with a preferred orientation of the electron spin direction. If there are $N_\uparrow$ electrons with spin components parallel to a given direction and $N_\downarrow$ with spins antiparallel to that direction then

$$P = \frac{N_\uparrow - N_\downarrow}{N_\uparrow + N_\downarrow} \tag{9.1}$$

is the component of the electron polarisation vector $\mathbf{P}$ in that direction. $|\mathbf{P}|$ is the degree of polarisation.

The use of polarised beams in collision studies has enabled experimentalists to perform very detailed tests of theoretical models, particularly with regard to the role of electron exchange and the spin—orbit interaction in spin-dependent scattering. We will now briefly discuss the role of these interactions before using the general density matrix method to describe the more general case where more than one mechanism may contribute to the spin-dependent effects.

9.1 Origin of spin-dependent effects

9.1.1 Spin—orbit interaction

The role of the spin—orbit interaction (i.e. relativistic effects) can be most clearly observed when an unpolarised beam of electrons is scattered by spinless heavy atoms and the polarisation of the electrons is observed after the collision. We can consider an unpolarised beam as a mixture of two equal fractions of opposing spin directions. We can choose this arbitrary spin direction to be perpendicular to the scattering plane. In this direction the polarisation remains unchanged in the scattering process. This can easily be seen if we consider the electric field which electrons experience in their rest frame. The positively-charged scattering centre (nucleus) moves with a velocity $-\mathbf{v}$ with respect to the electron and there is an electric field $\mathbf{E}$ between them. Thus the current that is represented by the moving charge produces a magnetic field $\mathbf{B} = \mathbf{E} \times \mathbf{v}/c$ which acts on the magnetic moment of the electron. This field is perpendicular to the scattering plane since $\mathbf{E}$ and $\mathbf{v}$ lie in the scattering plane. If the polarisation $\mathbf{P}$ of the incident electron is not parallel or antiparallel to $\mathbf{B}$, the magnetic moment associated with $\mathbf{P}$ will experience a torque causing $\mathbf{P}$ to precess. Thus only if the polarisation is perpendicular to the scattering plane does it retain its direction.

The cross sections for scattering of the two beams with opposite polarisation differ from each other because of the spin—orbit part of the scattering potential, which is proportional to the scalar product $\mathbf{L} \cdot \mathbf{S}$ of the electron's orbital and spin angular momenta. The scattering potential will therefore be either higher or lower for spin-up electrons (e↑) or spin-down electrons (e↓) depending on which side of the atom they pass, since this changes the sign of **L** (see section 2.5.2 and fig. 2.13). The differential cross sections will therefore be slightly different for the two spin directions. The scattered beam will thus in general be polarised, with the polarisation given by

$$\mathbf{P}' = S_P(\theta)\hat{\mathbf{n}}, \tag{9.2a}$$

where

$$S_P(\theta) = P' = \frac{N_\uparrow - N_\downarrow}{N_\uparrow + N_\downarrow} = \frac{\sigma_\uparrow - \sigma_\downarrow}{\sigma_\uparrow + \sigma_\downarrow} \tag{9.2b}$$

is the polarisation function and $\hat{\mathbf{n}}$ is the unit vector normal to the scattering plane. For the present purpose we denote the differential cross section by σ.

The polarisation will be particularly high where one of the two cross sections has a deep and sharp minimum, so that its value is small compared to the other cross section at the same angle. The positions of these minima due to diffraction (see e.g. figs. 8.10, 8.13) are determined by the effective radius of the atom, which in turn depends on the effective potential. This produces a small shift in the position of the minimum depending on the spin direction. Thus near the minimum of the complete differential cross section $\sigma(\theta) = \sigma_\uparrow(\theta) + \sigma_\downarrow(\theta)$ there will be a small angular region where either $\sigma_\uparrow$ or $\sigma_\downarrow$ dominate in turn, leading to large P' which changes its sign between the two minima.

The above comments for an unpolarised incident beam are obviously still valid for any partially polarised beam ($|\mathbf{P}| < 1$). Since the cross sections for e↑ and e↓ scattering are different, the relative proportion, i.e. the polarisation, will change with scattering. An existing polarisation **P** can be analysed through a left–right asymmetry in the differential cross section since the contribution of the spin–orbit term to the scattering potential differs in sign for the two directions.

$$A = \frac{\sigma_\ell(\theta) - \sigma_r(\theta)}{\sigma_\ell(\theta) + \sigma_r(\theta)} = S_A(\theta)\mathbf{P} \cdot \hat{\mathbf{n}}. \tag{9.3}$$

The polarisation function $S_P(\theta)$ and the asymmetry parameter $S_A(\theta)$ (also known as the analysing power or Sherman function) are identical and denoted by S for elastic scattering, due to time-reversal invariance of the projectile–target interaction (Kessler, 1985). S is a complicated function of the electron energy E, atomic number Z of the target and the scattering angle. The Mott detector, which uses scattering by a high Z material, is often used to determine the polarisation of a beam of electrons (see section 2.5.2). The precise measurement of S is, however, difficult.

With careful measurements and analysis, Fletcher *et al.* (1986) were able to obtain a best value for the uncertainty in S of $\pm 5\%$.

9.1.2 Electron exchange

The role of exchange scattering can be observed most clearly when polarised electrons are scattered from a target of polarised light atoms. If one observes a spin-flip process like

$$e \uparrow + A \downarrow \longrightarrow e \downarrow + A \uparrow, \tag{9.4a}$$

then it is most likely due to exchange between the incident electron and the atomic electron, since other spin-dependent interactions are negligible in light atoms. The amplitude for this process is usually defined to be g. Electrons can, of course, also be scattered by the direct process

$$e \uparrow + A \downarrow \longrightarrow e \uparrow + A \downarrow, \tag{9.4b}$$

where no change in spin direction occurs. This process is described by the direct scattering amplitude f. In any scattering process f and g must of course be added coherently, and only the relative phase γ between them can be observed.

If explicitly-spin-dependent forces, such as the spin–orbit interaction, are negligible, the cross section for scattering of electrons of polarisation $\mathbf{P}_e$ by a one-electron target of polarisation $\mathbf{P}_A$ is given in terms of the cross section σ_u for an unpolarised beam by (McClelland, Kelley and Celotta, 1987),

$$\sigma(\theta) = \sigma_u(\theta)\left[1 - A_{\text{ex}}(\theta)\mathbf{P}_e \cdot \mathbf{P}_A\right], \tag{9.5}$$

where

$$\sigma_u = \tfrac{1}{2}|f|^2 + \tfrac{1}{2}|g|^2 + \tfrac{1}{2}|f-g|^2 = \tfrac{1}{4}|f+g|^2 + \tfrac{3}{4}|f-g|^2. \tag{9.6}$$

Here $f + g$ and $f - g$ are the singlet and triplet amplitudes respectively. The exchange asymmetry is given by (Kessler, 1985)

$$A_{\text{ex}}(\theta) = \frac{fg^* + f^*g}{2\sigma_u} = \frac{|f| \, |g| \cos\gamma}{\sigma_u} = \frac{|f+g|^2 - |f-g|^2}{|f+g|^2 + 3|f-g|^2}. \tag{9.7}$$

The polarisation of the scattered electrons is given by

$$\mathbf{P}'_e = \frac{(1 - |f|^2/\sigma_u)\mathbf{P}_A + (1 - |g|^2/\sigma_u)\mathbf{P}_e + i(fg^* - f^*g)\mathbf{P}_e \times \mathbf{P}_A}{1 - A_{\text{ex}}\mathbf{P}_e \cdot \mathbf{P}_A}. \tag{9.8a}$$

For an unpolarised target ($\mathbf{P}_A = 0$)

$$\mathbf{P}'_e = (1 - |g|^2/\sigma_u)\mathbf{P}_e \tag{9.8b}$$

and the incident electron beam is partially depolarised by exchange with the unpolarised target electrons. If the target is polarised and the initial beam is unpolarised, the electron polarisation after scattering is

$$\mathbf{P}'_e = (1 - |f|^2/\sigma_u)\mathbf{P}_A. \qquad (9.8c)$$

For a complete experiment yielding $|f|, |g|$ and γ, one has to measure σ_u and three observables. The measurement of the asymmetry, which from equn. (9.5) is given by

$$A(\theta) = \frac{\sigma_{\uparrow\downarrow} - \sigma_{\uparrow\uparrow}}{\sigma_{\uparrow\downarrow} + \sigma_{\uparrow\uparrow}}, \qquad (9.9)$$

(where the subscripts denote cross sections for antiparallel and parallel polarisation vectors) gives information on $\cos\gamma$ (equn. (9.7)). Measuring the polarisation component normal to $\mathbf{P}_e$ and $\mathbf{P}_A$ gives information on $\sin\gamma$ through the term in $fg^* - f^*g$ in (9.8a). Measurements with one of the colliding beams unpolarised gives information on $|f|^2$ and $|g|^2$ (equns. (9.8b) and (9.8c)). Measurements of this kind have been performed for elastic scattering (e.g. McClelland *et al.*, 1987, 1990), inelastic scattering (e.g. Baum, Raith and Schröder, 1988) and ionisation (e.g. Crowe *et al.*, 1990). Some of these measurements will be discussed later.

It is possible to observe the polarisation $\mathbf{P}'_A$ of the scattered atoms rather than performing the measurements on the electrons. $\mathbf{P}'_A$ is given by (9.8a) on interchanging $\mathbf{P}_e$ and $\mathbf{P}_A$. Such measurements have been carried out in a series of pioneering experiments by Bederson and co-workers (Bederson, 1973).

9.1.3 The fine-structure effect

Hanne (1976, 1983) showed that electron scattering from individual fine-structure states of a multiplet can lead to significant polarisation effects, even for unpolarised very light targets for which the spin–orbit interaction is negligible. The target states of a fine-structure multiplet for spin s and orbital angular momentum ℓ are distinguished by their total angular momentum j. In the absence of explicit spin-dependent interactions these effects depend on (a) nonvanishing orbital angular momentum orientation of the target (i.e. $\langle L_\perp \rangle \neq 0$), (b) electron exchange, and (c) resolution of fine-structure levels in the final and/or the initial state. The sum of the contributions over the entire multiplet, i.e. the 'average' polarisation and asymmetry, vanishes.

The physical mechanism underlying the fine-structure effect can be seen if we consider say the excitation of the $2^3P_{0,1,2}$ fine-structure states of helium from the singlet ground state 1^1S_0. This can only occur by exchange processes, since in the ^3P_j states the atoms have their electron spins aligned, while in the singlet state they are antiparallel. As discussed in chapter 8, excitation of a P state by electrons scattered through a certain angle will in general leave the atoms with an orbital angular momentum orientation $\langle L_\perp \rangle$ normal to the scattering plane (see fig. 8.1).

Let us assume that our detection system can select electrons from a single fine-structure state, say the $^3P_0$ state. Atoms in this state have their spins and orbital angular momenta antiparallel to each other, so that $\langle S_\perp \rangle = -\langle L_\perp \rangle$. Since the ^3P_j states can only be excited from the $^1S_0$ ground state by exchange, the process must take place by the capture of an electron with spin orientation opposite to that of the ejected electron and parallel to that of the other bound electron. Thus when the spin orientation of the incident electrons corresponds to that of the excited state, the excitation probability is higher than in the opposite case. If, as in fig. 8.1, the scattered electron leaves $\langle L_\perp \rangle$ oriented 'up' from the plane of scattering, then $\langle S_\perp \rangle$ is oriented 'down'. Thus the 'down' component of the incident beam will be preferentially scattered through the given angle. Indeed for total atomic orientation, $\langle S_\perp \rangle = 1$, the cross section $\sigma_\uparrow$ for spin-up electrons to excite the $^3P_0$ fine structure level vanishes. The resulting asymmetry, which depends on the scattering angle and energy, can be shown (Hanne, 1983) to be in this case given by

$$A = \frac{\sigma_\uparrow - \sigma_\downarrow}{\sigma_\uparrow + \sigma_\downarrow} = -\langle L_\perp \rangle = \langle S_\perp \rangle. \tag{9.10}$$

The fine-structure effect also results in a polarisation of the scattered electron. Since the spin-down electron is preferentially captured, the atomic spin-up electron is preferentially released, and the electron arriving at the detector has an average spin-up component, i.e. positive polarisation, and indeed

$$P' = -A = \langle L_\perp \rangle. \tag{9.11}$$

We see that a measurement of the asymmetry of a scattered polarised beam or the polarisation after scattering of an initially-polarised beam yields directly the orientation $\langle L_\perp \rangle$ of the excited atomic state. This is usually obtained by coincidence experiments (chapter 8). The above case is, however, a special case and in more complicated situations the two techniques yield complementary information.

The existence of the fine-structure effect has been demonstrated for sodium (Hanne, Szmytkowski and van der Wiel, 1982; McClelland et al., 1985; Nickich et al., 1990) using the time-reversed arrangement. A polarised electron beam is superelastically scattered from sodium atoms excited to $3^2P_{1/2}$ or $3^2P_{3/2}$ states by a single-frequency laser. McClelland et al. (1985) measured the spin asymmetry of polarised electrons that de-excite unpolarised atoms from the $3^2P_{3/2}$ fine-structure state over the angular range $-35° \leq \theta \leq 35°$. As expected from reflection symmetry, the asymmetry for scattering to the left (positive angles) differed only in sign from that for scattering to the right.

9.2 Combined effects of several polarisation mechanisms

9.2.1 Generalised formalism for the scattering of polarised electrons by unpolarised targets

For heavy targets one usually has to take into account exchange and spin–orbit interactions as well as internal spin–orbit coupling in the target. We will now follow the approach of Bartschat and Madison (1988), who applied the formalism of reduced density matrices to describe the scattering (either elastic or inelastic) of polarised electrons by unpolarised targets. We treat the case where only the scattered electrons are observed and the target electrons may have spin and orbital angular momentum in the initial and/or final state.

We will again work in the collision system where the z-axis of quantisation is parallel to k_0 and the scattering plane is the zx-plane (see fig. 8.1). We can write the density matrix (Blum, 1981) for the final state as

$$\rho_{v_i' v_i}^{m_i' m_i}(\mathbf{k}_i) = \sum_{v_0' v_0 m_0' m_0} f(m_i' v_i', m_0' v_0') f^*(m_i v_i, m_0 v_0) \rho_{v_0' v_0} \rho_{m_0' m_0}, \qquad (9.12)$$

where $\rho_{v_0' v_0} \rho_{m_0' m_0}$ describes the preparation of the initial state, i.e. of the projectile electrons and the target atom. The density-matrix elements contain the total information that can be obtained from the scattering process for a given set of elements $\rho_{v_0' v_0} \rho_{m_0' m_0}$ describing the initial state. If only the scattered electrons are observed, the corresponding reduced density-matrix elements for the outgoing electrons are obtained by summing over the atomic quantum numbers.

$$\rho_{v_i' v_i}(\mathbf{k}_i) = \sum_{m_i' m_i} \rho_{v_i' v_i}^{m_i' m_i}(\mathbf{k}_i) = \sum_{v_0' v_0} \langle v_i' v_0'; v_i v_0 \rangle \rho_{v_0' v_0}, \qquad (9.13)$$

where we have defined

$$\langle v_i' v_0'; v_i v_0 \rangle = (2j_0 + 1)^{-1} \sum_{m_i' m_i m_0' m_0} f(m_i v_i', m_0' v_0') f^*(m_i v_i, m_0 v_0). \qquad (9.14)$$

For spin 1/2 particles there are 16 possible combinations of $(v_i' v_0' v_i v_0)$ and therefore 32 real (or 16 complex) parameters. From (9.14) it can be seen that hermiticity requires that

$$\langle v_i' v_0'; v_i v_0 \rangle = \langle v_i v_0; v_i' v_0' \rangle^*. \qquad (9.15)$$

Reflection invariance with respect to the scattering plane (i.e. parity conservation) yields the further restriction that

$$f(m_i \nu_i, m_0 \nu_0) = \Pi_i \Pi_0 (-1)^{j_i - m_i + 1/2 - \nu_i + j_0 - m_0 + 1/2 - \nu_0}$$
$$\times f(-m_i - \nu_i; -m_0 - \nu_0). \tag{9.16}$$

where Π_i and Π_0 are the parities of the final and initial atomic states. It therefore follows from (9.14) that

$$\langle \nu_i' \nu_0'; \nu_i \nu_0 \rangle = (-1)^{\nu_i' - \nu_i + \nu_0' - \nu_0} \langle -\nu_i' - \nu_0'; -\nu_i - \nu_0 \rangle. \tag{9.17}$$

Equns. (9.15) and (9.17), which correspond to (8.14) and (8.15), restrict the number of parameters so that only eight independent parameters are required to characterise in general the reduced density matrix of the scattered electrons. In special situations the number of independent parameters can be reduced even further as discussed earlier.

The differential cross section for scattering of unpolarised incident particles is

$$\sigma_u(\theta) = \frac{1}{2} \sum_{\nu_i \nu_0} \langle \nu_i \nu_0; \nu_i \nu_0 \rangle = \frac{1}{2(2j_0 + 1)} \sum_{m_i m_0 \nu_i \nu_0} |f(m_i \nu_i, m_0 \nu_0)|^2. \tag{9.18}$$

The differential cross section for the scattering of polarised incident particles is given by

$$\sigma(\theta) = \text{tr}\rho(\mathbf{k}_i) = \sum_{\nu_i} \sum_{\nu_0' \nu_0} \langle \nu_i \nu_0'; \nu_i \nu_0 \rangle \rho_{\nu_0' \nu_0}. \tag{9.19}$$

The spin density matrix of the projectiles is given in terms of the polarisation vector $\mathbf{P}_e = (P_x, P_y, P_z)$ by (e.g. Kessler, 1985).

$$\rho_{\nu_0' \nu_0} = \frac{1}{2} \begin{pmatrix} 1 + P_z & P_x - iP_y \\ P_x + iP_y & 1 - P_z \end{pmatrix}. \tag{9.20}$$

Using this and the relationships (9.15) and (9.17) one obtains

$$\sigma(\theta) = \frac{1}{2} \sum_{\nu_i} \left[\langle \nu_i \tfrac{1}{2}; \nu_i \tfrac{1}{2} \rangle + \langle \nu_i -\tfrac{1}{2}; \nu_i -\tfrac{1}{2} \rangle - iP_y (\langle \nu_i \tfrac{1}{2}; \nu_i -\tfrac{1}{2} \rangle - \langle \nu_i -\tfrac{1}{2}; \nu_i \tfrac{1}{2} \rangle) \right]$$
$$= \frac{1}{2} \sum_{\nu_i \nu_0} \langle \nu_i \nu_0; \nu_i \nu_0 \rangle - 2P_y \text{Im} \langle \tfrac{1}{2} -\tfrac{1}{2}; \tfrac{1}{2} \tfrac{1}{2} \rangle = \sigma_u (1 + P_y S_A), \tag{9.21}$$

where the asymmetry function has been defined to be

$$S_A \equiv \frac{-2}{\sigma_u} \text{Im} \langle \tfrac{1}{2} -\tfrac{1}{2}; \tfrac{1}{2} \tfrac{1}{2} \rangle$$
$$= \frac{-2}{\sigma_u (2j_0 + 1)} \text{Im} \sum_{m_i m_0} f(m_i \tfrac{1}{2}, m_0 -\tfrac{1}{2}) f^*(m_i \tfrac{1}{2}, m_0 \tfrac{1}{2}). \tag{9.22}$$

The expectation value of an observable O referring to the scattered electrons alone is given by $\text{tr}[\rho O]/\text{tr}\rho$. Thus the polarisation component P'_y normal to the scattering plane after the collision is given by

$$iP'_y\sigma(\theta) = \langle-\tfrac{1}{2}\tfrac{1}{2};\tfrac{1}{2}\tfrac{1}{2}\rangle + \langle-\tfrac{1}{2}-\tfrac{1}{2};\tfrac{1}{2}-\tfrac{1}{2}\rangle + iP_y\left[\langle-\tfrac{1}{2}-\tfrac{1}{2};\tfrac{1}{2}\tfrac{1}{2}\rangle - \langle-\tfrac{1}{2}\tfrac{1}{2};\tfrac{1}{2}-\tfrac{1}{2}\rangle\right]. \tag{9.23}$$

For the case of no polarisation component perpendicular to the scattering plane before the collision ($P_y = 0$) we get

$$P'_y \equiv S_P = \frac{-2}{\sigma_u(2j_0+1)}\text{Im} \sum_{m_im_o} f(m_i\tfrac{1}{2},m_0\tfrac{1}{2})f^*(m_i-\tfrac{1}{2},m_0\tfrac{1}{2}), \tag{9.24}$$

which defines the polarisation function S_P. In contrast to the discussion in section 9.1.1, the asymmetry (9.22) and polarisation function (9.24) are no longer described by the same Sherman function S.

We can rewrite (9.23) in the form

$$P'_y = (S_P + T_yP_y)/(1+S_AP_y), \tag{9.25}$$

with T_y being a real parameter given by

$$T_y = \sigma_u^{-1}\left[\langle-\tfrac{1}{2}-\tfrac{1}{2};\tfrac{1}{2}\tfrac{1}{2}\rangle - \langle-\tfrac{1}{2}\tfrac{1}{2};\tfrac{1}{2}-\tfrac{1}{2}\rangle\right]. \tag{9.26}$$

The observable T_y ($-1 \leq T_y \leq 1$) describes the contraction or even inversion of the original component of polarisation normal to the plane. Not only σ_u and S_P but also T_y and S_A can therefore be measured with projectiles that have no polarisation component in the scattering plane.

Performing the corresponding analysis for the polarisation components in the scattering plane one obtains (Bartschat, 1989)

$$P'_x = (T_xP_x + U_{xz}P_z)/(1+S_AP_y) \tag{9.27}$$

and

$$P'_z = (T_zP_z + U_{zx}P_x)/(1+S_AP_y), \tag{9.28}$$

where the signs have been chosen so that for the special case of elastic scattering from targets with zero angular momentum the results are obtained in terms of the normal STU parameters, defined by equations (9.34).

Thus, in addition to the three observables σ_u, S_A, and S_P, one needs the following five observables to describe in general the change in polarisation caused by scattering of spin 1/2 particles by an unpolarised target:

$$T_{x,y} = \frac{1}{\sigma_u(2j_0+1)} \sum_{m_im_0}\left[f(m_i-\tfrac{1}{2},m_0\tfrac{1}{2})f^*(m_i\tfrac{1}{2},m_0\tfrac{1}{2})\right.$$
$$\left. \pm f(m_i-\tfrac{1}{2},m_0-\tfrac{1}{2})f^*(m_i\tfrac{1}{2},m_0-\tfrac{1}{2})\right], \tag{9.29a}$$

$$T_z = \frac{1}{\sigma_u(2j_0+1)} \sum_{m_i m_0} \Big[f(m_i \tfrac{1}{2}, m_0 \tfrac{1}{2}) f^*(m_i \tfrac{1}{2}, m_0 \tfrac{1}{2})$$

$$- f(m_i \tfrac{1}{2}, m_0 -\tfrac{1}{2}) f^*(m_i \tfrac{1}{2}, m_0 -\tfrac{1}{2}) \Big], \tag{9.29b}$$

$$U_{xz} = \frac{2}{\sigma_u} \mathrm{Re} \langle \tfrac{1}{2}\tfrac{1}{2}; -\tfrac{1}{2}\tfrac{1}{2} \rangle$$

$$= \frac{1}{\sigma_u(2j_0+1)}$$

$$\times \sum_{m_i m_0} \Big[f(m_i -\tfrac{1}{2}, m_0 \tfrac{1}{2}) f^*(m_i \tfrac{1}{2}, m_0 \tfrac{1}{2})$$

$$- f(m_i -\tfrac{1}{2}, m_0 -\tfrac{1}{2}) f^*(m_i \tfrac{1}{2}, m_0 -\tfrac{1}{2}) \Big]. \tag{9.29c}$$

$$U_{zx} = \frac{2}{\sigma_u} \mathrm{Re} \langle \tfrac{1}{2} -\tfrac{1}{2}, \tfrac{1}{2}\tfrac{1}{2} \rangle$$

$$= \frac{1}{\sigma_u(2j_0+1)} \sum_{m_i m_0} \Big[f(m_i \tfrac{1}{2}, m_0 -\tfrac{1}{2}) f^*(m_i \tfrac{1}{2}, m_0 \tfrac{1}{2})$$

$$+ f(m_i \tfrac{1}{2}, m_0 \tfrac{1}{2}) f^*(m_i \tfrac{1}{2}, m_0 -\tfrac{1}{2}) \Big]. \tag{9.29d}$$

The polarisation $\mathbf{P}'$ after scattering of an electron beam with initial polarisation $\mathbf{P} = (P_x, P_y, P_z)$ is then given by

$$\mathbf{P}' = \Big[(S_P + T_y P_y) \hat{\mathbf{y}} + (T_x P_x + U_{xz} P_z) \hat{\mathbf{x}} + (T_z P_z - U_{zx} P_x) \hat{\mathbf{z}} \Big] / (1 + S_A P_y). \tag{9.30}$$

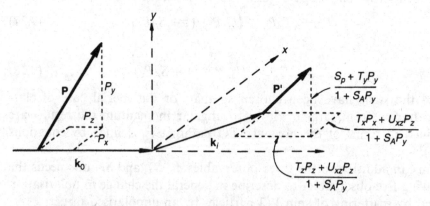

Fig. 9.1. Physical meaning of the generalised STU parameters and the polarisation function S_P and asymmetry function S_A for scattering of a beam with initial polarisation $\mathbf{P}$, the final polarisation being $\mathbf{P}'$. The contraction parameters T_x, T_y and T_z describe the change of initial polarisation along the three axes, while the U parameters describe the rotation in the scattering plane.

Here T_x, T_y, T_z describe the change in length of the polarisation components, while U_{xz} and U_{zx} describe the rotation of the polarisation in the scattering plane. This is shown schematically in fig. 9.1.

9.2.2 Elastic scattering from a spinless target

For elastic scattering from a target with $j_0 = j_i = 0$ we get from (9.18), after dropping the arguments $m_i = m_0 = 0$,

$$\sigma_u(\theta) = |f(\tfrac{1}{2}, \tfrac{1}{2})|^2 + |f(-\tfrac{1}{2}, \tfrac{1}{2})|^2, \tag{9.31}$$

where we have used the relation

$$f(v_i, v_0) = (-1)^{1-v_i-v_0} f(-v_i, -v_0), \tag{9.32}$$

which follows from the more general relation (9.16). With the notation for the spin-flip amplitude $f(-\tfrac{1}{2}, \tfrac{1}{2}) = g$ and for the non-spin-flip amplitude $f(\tfrac{1}{2}, \tfrac{1}{2}) = f$, one obtains

$$\sigma_u(\theta) = |f|^2 + |g|^2. \tag{9.33}$$

It similarly follows from relations (9.22) and (9.24) that

$$S(\theta) = S_A = S_P = \frac{1}{\sigma_u} \mathrm{Im}\left[f(\tfrac{1}{2}, \tfrac{1}{2})f^*(\tfrac{1}{2}, -\tfrac{1}{2}) + f(-\tfrac{1}{2}, \tfrac{1}{2})f^*(-\tfrac{1}{2}, -\tfrac{1}{2})\right]$$

$$= i\frac{fg^* - f^*g}{|f|^2 + |g|^2}. \tag{9.34a}$$

From equns. (9.29) and (9.32)

$$T(\theta) = T_x = T_z = \frac{|f|^2 - |g|^2}{|f|^2 + |g|^2}, \tag{9.34b}$$

$$T_y = 1, \tag{9.34c}$$

and

$$U(\theta) = U_{xz} = U_{zx} = \frac{fg^* + f^*g}{|f|^2 + |g|^2}. \tag{9.34d}$$

In this case one must measure four independent observables for a complete experiment, namely the absolute differential cross section, the Sherman function S, and two polarisation components to yield T and U.

The case of elastic scattering from targets with arbitrary angular momentum $j_i = j_0$ is more complicated. Bartschat (1989) considered this case in detail and showed how, in general, six independent parameters are needed to describe completely the scattering process, time reversal invariance leading to the following relations between the observables.

$$S_A = S_P = S \tag{9.35a}$$

and

$$U_{xz} - U_{zx} = \tan\theta(T_z - T_x). \tag{9.35b}$$

9.2.3 Inelastic scattering from $j_0 = 0$ to $j_i = 0$

This case is a generalisation of the elastic scattering case. Taking into account the parities of the initial and final states one readily obtains

$$S_A = \Pi_0\Pi_i S_P, \tag{9.36a}$$
$$T_y = \Pi_0\Pi_i, \tag{9.36b}$$
$$T_x = \Pi_0\Pi_i T_z, \tag{9.36c}$$
$$U_{xz} = \Pi_0\Pi_i U_{zx}, \tag{9.36d}$$

showing that the process is again determined by four observables. Thus for the excitation of a $^3P_0^o$ state from a ground $^1S_0^e$ state (as in He or Hg), $S_A = -S_P$.

From equn. (9.30) we see that the y component of final polarisation for a $0 \to 0$ transition is given by

$$P'_y = \frac{S_P + \Pi_0\Pi_i P_y}{1 + \Pi_0\Pi_i S_P P_y}, \tag{9.37}$$

which gives for a totally polarised beam with $P_y = 1$ (using $\Pi_0^2\Pi_i^2 = 1$)

$$P'_y = \Pi_0\Pi_i, \tag{9.38}$$

giving a complete reversal of polarisation for a $^1S_0^e \to {}^3P_0^o$ transition.

9.2.4 Pure exchange scattering from spin 1/2 targets

Applying the above formalism for pure exchange scattering from spin 1/2 targets such as the alkali-metal atoms, we obtain for elastic scattering

$$\sigma_u(\theta) = \tfrac{1}{4}\Big[|f(\tfrac{1}{2}\tfrac{1}{2},\tfrac{1}{2}\tfrac{1}{2})|^2 + |f(\tfrac{1}{2}-\tfrac{1}{2},\tfrac{1}{2}-\tfrac{1}{2})|^2 + |f(\tfrac{1}{2}-\tfrac{1}{2},-\tfrac{1}{2}\tfrac{1}{2})|^2$$
$$+ |f(-\tfrac{1}{2}\tfrac{1}{2},\tfrac{1}{2}-\tfrac{1}{2})|^2 + |f(-\tfrac{1}{2}\tfrac{1}{2},-\tfrac{1}{2}\tfrac{1}{2})|^2 + |f(-\tfrac{1}{2}-\tfrac{1}{2},-\tfrac{1}{2}-\tfrac{1}{2})|^2\Big]. \tag{9.39}$$

Of the 16 terms in (9.18) for $j_i = j_0 = 1/2$, 10 can be omitted because they violate conservation of total spin angular momentum on the basis of pure exchange scattering, leaving only the above six terms. Recalling the definitions in section 9.1.2 of the direct and exchange amplitudes f and g (equns. (9.4a) and (9.4b) respectively) one has

$$f = f(\tfrac{1}{2}-\tfrac{1}{2},\tfrac{1}{2}-\tfrac{1}{2}),$$
$$g = f(\tfrac{1}{2}-\tfrac{1}{2},-\tfrac{1}{2}\tfrac{1}{2}),$$
$$f - g = f(\tfrac{1}{2}\tfrac{1}{2},\tfrac{1}{2}\tfrac{1}{2}). \tag{9.40}$$

The last relationship follows from the fact that both the direct and exchange processes contribute to the amplitude for

$$e \uparrow + A \uparrow \rightarrow e \uparrow + A \uparrow,$$

and therefore one must have a coherent superposition of the amplitudes. With these definitions one readily obtains equn. (9.6) for σ_u.

9.3 One-electron atoms

Scattering from alkali-metal atoms is understood as the three-body problem of two electrons interacting with an inert core. The electron—core potentials are frozen-core Hartree—Fock potentials with core polarisation being represented by a further potential (5.82).

The coupled-channels-optical calculation gives generally-good agreement for cross sections although, in the test case of sodium, experiments for larger-angle differential cross sections disagree with each other so strongly that they do not really test the calculations. In particular, inclusion of the ionisation continuum is not critically tested.

For electron scattering on lighter alkali-metals, spin asymmetry is due to the Pauli exclusion principle, not to relativistic effects. It tests the relationship between direct and exchange elements of the calculation. Since it is a ratio it is easier to measure accurately than the differential cross section, which varies over many orders of magnitude in the case of sodium.

9.3.1 Lithium

The theoretical treatment of asymmetry is rather generally tested for lithium by the energy-dependent measurements at three angles of Baum *et al.* (1986) for the ground ($2s$) state and Baum *et al.* (1989) for the first-excited ($2p$) state.

Fig. 9.2 compares the experiments with the coupled-channels-optical calculation of Bray, Fursa and McCarthy (1993). Here P space consisted of 13 channels, polarisation potentials for ionisation being calculated for all couplings in the first eight channels. Full convergence is achieved with these truncations. The effect of the continuum is seen by comparing the full calculation with a 13-state coupled-channels calculation. The continuum has a significant effect after the ionisation threshold at 5.4 eV and brings the calculations into good agreement with experiment.

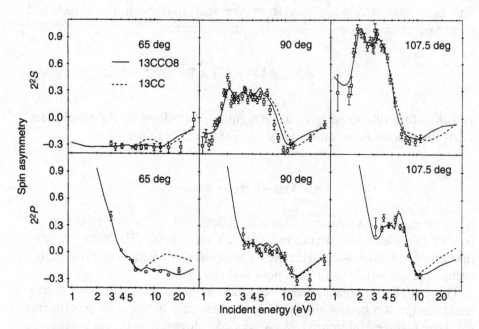

Fig. 9.2. Spin asymmetries for electron scattering to the 2s and 2p states of lithium plotted against incident energy for the indicated scattering angles (Bray et al., 1993). 2s experiment, Baum et al. (1986); 2p experiment, Baum et al. (1989); full curve, coupled channels optical; broken curve, 13-state coupled channels.

9.3.2 Sodium

The investigation of sodium as a critical test of the theoretical treatment of scattering is given a new dimension by the spin-dependent measurements of Kelley et al. (1992) in elastic and superelastic scattering experiments with polarised electrons on the polarised 3s and laser-excited 3p states. Not only have asymmetries been measured for these states, but spin-dependent observations of the magnetic substate parameter $L_\perp$ have been made for the 3p state.

The most comprehensive set of data is available at a 3s-channel energy of 20 eV. The experimental and theoretical situations are summarised in fig. 9.3. Asymmetries are shown for the 3s and 3p channels. Singlet and triplet differential cross sections are obtained from spin-independent measurements by using these asymmetries. Also shown are the singlet, triplet and spin-averaged values of $L_\perp$.

The coupled-channels-optical calculation converges at 15 channels in P space with polarisation potentials for the continuum included for all couplings in the first six channels. The effect of the inclusion of the continuum is shown by the 15-state coupled-channels calculation. The distorted-wave

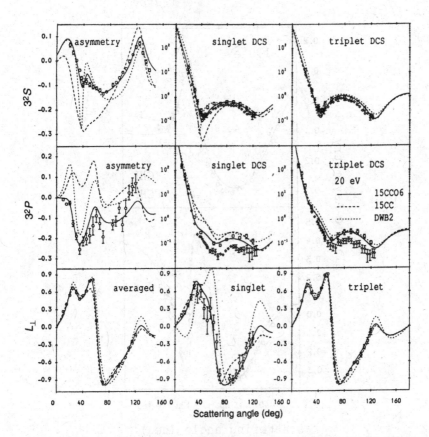

Fig. 9.3. Electron—sodium scattering at 20 eV in the 3s channel (Bray and McCarthy, 1992). Circles 3s, McClelland, Kelley and Celotta (1989); circles 3p, Kelley *et al.* (1992). Differential cross sections (multiplied by asymmetries) are: squares, Srivastava and Vušković (1980); diamonds, Lorentz and Miller (1991). Full curves, coupled channels optical; long-dashed curves, 15-state coupled channels; short-dashed curves, distorted-wave second Born (Madison *et al.*, 1992).

second Born calculation of Madison, Bartschat and McEachran (1992) includes the continuum in representing the second-order term. Comparing it with the full calculation tests the need for full coupling in P space.

The need for inclusion of the continuum or for full coupling is not very obvious for the triplet reactions, which dominate the spin-independent data. It is the asymmetry that provides the critical test of theory. Very good agreement with experiment is obtained by the full coupled-channels-optical calculation, but the other two calculations are qualitatively incorrect, even giving the opposite sign for the 3p asymmetry. These conclusions hold for experimental—theoretical comparisons at 1.0, 1.6, 4.1, 12.1 and 40 eV (Bray and McCarthy, 1992).

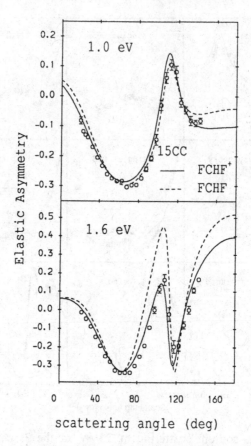

Fig. 9.4. Elastic asymmetry for electron—sodium scattering at 1.0 and 1.6 eV (Bray and McCarthy, 1992). Circles, Lorentz *et al.* (1991); full curves, 15-state coupled channels with core-polarisation in the bound states; broken curve, the same reaction calculation omitting core polarisation.

Essentially-complete agreement with experiment is achieved by the coupled-channels-optical calculation. We can therefore ask if scattering is so sensitive to the structure details in the calculation that it constitutes a sensitive probe for structure. The coupled-channels calculations in fig. 9.3 included the polarisation potential (5.82) in addition to the frozen-core Hartree—Fock potential. Fig. 9.4 shows that addition of the polarisation potential has a large effect on the elastic asymmetry at 1.6 eV, bringing it into agreement with experiment. However, in general the probe is not very sensitive to this level of detail.

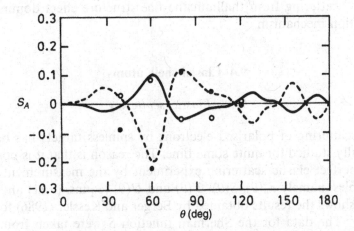

Fig. 9.5. Asymmetry function S_A for superelastic scattering from the $(3p)^2 P_{1/2}$ ($\bullet$ and - - -) and $(3p)^2 P_{3/2}$ (o and —) states of sodium at an incident electron energy of 12 eV. The theoretical curves are from a 4-state R-matrix calculation by Bartschat (1991a), and the experimental points are from Nickich *et al.* (1990).

9.3.3 *The fine-structure effect*

An example of the fine-structure effect, caused by the interplay of exchange scattering and atomic fine-structure splitting (section 9.1.3), is shown in fig. 9.5, where the measured asymmetry function S_A in superelastic scattering from the $3p^2 P^o_{1/2,3/2}$ states of sodium (Nickich *et al.*, 1990) is compared with a recent 4-state R-matrix calculation by Bartschat (1991a). In the pure fine-structure effect, which assumes that the total orbital angular momentum L and total spin S are separately conserved in the collision, transitions between fine-structure levels are described by purely algebraic recoupling techniques. This leads to some simple relations between polarisation parameters for such transitions (Bartschat, 1989). Thus the polarisation, asymmetry and rotation functions all vanish when averaged over all fine-structure states. As can be seen in fig. 9.5 the simple relationship

$$S_A(^2P^o_{1/2}) = -2S_A(^2P^o_{3/2})$$

is verified both theoretically and experimentally.

For heavier atoms the pure fine-structure effect is expected to break down due to relativistic effects. In the very heavy open-shell target atom thallium ($Z=81$) the ground-state atoms populate only one of the fine-structure levels, and the effect may be important at low energies. In an R-matrix calculation using magnetic potentials derived from the Dirac equation, Goerss, Nordbeck and Bartschat (1991) showed that for 5 eV scattering from thallium the fine-structure effect dominates the polarisation mechanism.

9.4 Closed-shell atoms

9.4.1 Elastic scattering

Elastic scattering of polarised electrons by spinless targets has been systematically studied for quite some time. The reason is that it is possible to do the perfect elastic scattering experiment by the measurement of four observables, namely $\sigma_u(\theta)$, $S(\theta)$, $T(\theta)$ and $U(\theta)$ (equns. (9.31) and (9.34)). Fig. 9.6 shows the results obtained by Berger and Kessler (1986) for xenon at 60 eV. The data for the Sherman function S were taken from Berger *et al.* (1982), and the data for σ_u, used in evaluating the moduli of the scattering amplitudes and their relative phases, from Register, Vuškovic and Trajmar (1986) and Williams and Crowe (1975). From equns. (9.33) and (9.34) it follows that

$$|f| = [\sigma_u(1 + T)/2]^{\frac{1}{2}} \tag{9.41a}$$

$$|g| = [\sigma_u(1 - T)/2]^{\frac{1}{2}} \tag{9.41b}$$

$$\gamma = \gamma_1 - \gamma_2 = \tan^{-1}(-S/U), \tag{9.41c}$$

where $f = f(\frac{1}{2},\frac{1}{2})$ and $g = f(-\frac{1}{2},\frac{1}{2})$ are the non-spin-flip (direct) and spin-flip amplitudes respectively, and γ_1 and γ_2 are their phases. Thus the relative phase γ can be obtained from S and U alone.

The measured values of S, T and U as well as the derived scattering amplitude parameters $|f|$, $|g|$, and γ are compared with the results of several calculations in fig. 9.6. Haberland, Fritsche and Noffke (1986) do their calculation within a Kohn–Sham type one-particle theory (Kohn and Sham, 1965) including exchange, treating the scattering process as an $(N+1)$-electron problem, which is solved self-consistently. Awe *et al.* (1983) in their relativistic calculation use energy-dependent equivalent-local exchange potentials with various local density approximation forms for exchange and correlation contributions. McEachran and Stauffer (1986) use a relativistic form of the Schrödinger equation, with the static and relativistic potentials derived from relativistic Hartree–Fock wavefunctions, while the polarisation potential is obtained from a non-relativistic polarised-orbital calculation. Exchange is included exactly for the large component of the scattered wavefunction. The agreement between theory and experiment is quite good, although there are a number of discrepancies. These are reduced at higher energies (Berger and Kessler, 1986).

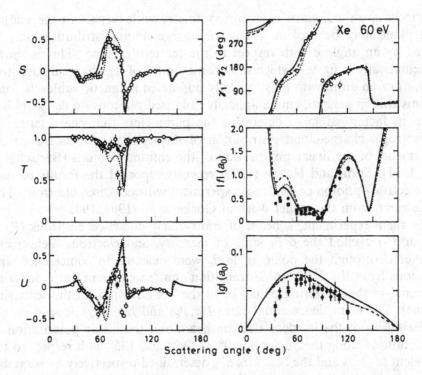

Fig. 9.6. Angular distribution of S, T and U and the derived moduli and relative phases of the scattering parameters for elastic scattering from Xe at 60 eV. Experiment: o, Berger and Kessler (1986); ▪, Möllenkamp *et al.* (1984) and Wübker, Möllenkamp and Kessler (1982). Absolute measured differential cross sections used in the derivation of the moduli $|f|$ and $|g|$: •, Register *et al.* (1986); ▪, Williams and Crowe (1975). Theory: ——, McEachran and Stauffer (1986); ·······, Haberland *et al.* (1986); - - -, Awe *et al.* (1983).

The data show that the differential cross section is dominated by the direct scattering amplitude f; the modulus for the spin-flip amplitude g is in general an order of magnitude smaller. Thus the spin–orbit interaction has only a small influence on the cross section, which is mainly influenced by the Coulomb interaction, exchange, and charge-cloud polarisation.

9.4.2 Inelastic scattering: electron–photon coincidences

As discussed in the last chapter, electron–photon (e, e′γ) measurements yield much more information on the scattering process than simple inelastic differential cross section measurements. In particular the population of magnetic sublevels can be obtained, which can be visualised by the corresponding charge-cloud probability distribution (fig. 8.1). The set of parameters discussed in the last chapter must be enlarged when polarised electrons are used, since reflection symmetry with respect to the scattering plane may be broken. Thus the charge-cloud distribution may be tilted by an angle ϵ with respect to the scattering plane. This is shown schematically in fig. 9.7, which shows an example of the charge distribution $\rho(\theta, \phi)$ of an atom with anistropically populated magnetic sublevels. Such atoms will, in general, emit elliptically-polarised photons on de-excitation. It is in fact possible to determine the parameters that characterise the anisotropic charge-cloud distribution of the radiating atoms from measurements of the linear polarisation of the emitted photons (Bartschat *et al.*, 1981). Sohn and Hanne (1992) recently reported the results of such an electron–photon coincidence experiment with polarised electrons. This follows on from the earlier work of Goeke *et al.* (1988, 1989).

In their experiment, a beam of transversely-polarised electrons ($P_e = P_y$ or P_x) excited the 6^3P_1 state of mercury, and electrons inelastically scattered through the polar angle θ_e were detected in coincidence with photons from the $6^3P_1 - 6^1S_0$ transition emitted either in the y direction (normal to the scattering plane) or in the $-x$ direction (in the scattering plane). The coincidence count rates $I(\alpha, P_e)$ and $I(\sigma^\pm, P_e)$ depend on the polarisation of the incident electron beam and the linear polarisation of the photons along the angles $\alpha = 0°, 45°, 90°$ and $135°$ with respect to the incident (z) axis and the helicities $\sigma^\pm$, determined respectively by rotatable linear polarisation and circular polarisation filters. It is then possible to derive six Stokes parameters or light polarisation components ($P_1 \cdots P_6$), and six spin up–down asymmetry parameters $A(\alpha, P_e)$ depending on the emitted photon direction and angles α and helicities $\sigma^\pm$. For light emitted in the y direction, we define (see equn. (8.30))

$$P_1 = \frac{I(0°) - I(90°)}{I(0°) + I(90°)} \tag{9.42a}$$

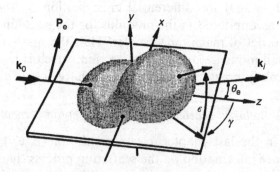

Fig. 9.7. Schematic of the angular dependence of a charge-cloud distribution of an atomic state excited by polarised electrons. The tilt (ϵ) out of the scattering plane must by parity conservation be zero if $P_x = 0$.

$$P_2 = \frac{I(45°) - I(135°)}{I(45°) + I(90°)} \tag{9.42b}$$

$$P_3 = \frac{I(\sigma^-) - I(\sigma^+)}{I(\sigma^-) + I(\sigma^+)}, \tag{9.42c}$$

which are the Stokes parameters for unpolarised electrons. For light emitted in the x-direction we have

$$P_4 = \frac{I(0°) - I(90°)}{I(0°) + I(90°)} \tag{9.42d}$$

$$P_5 = \frac{I(45°, +P_x) - I(135°, -P_x)}{I(45°, +P_x) + I(135°, -P_x)} \tag{9.42e}$$

$$P_6 = \frac{I(\sigma^-, +P_x) - I(\sigma^+, +P_x)}{I(\sigma^-, +P_x) + I(\sigma^+, +P_x)}. \tag{9.42f}$$

P_4 is again a Stokes parameter for unpolarised electrons. P_5 and P_6 must be zero if $P_x = 0$. This can be seen by taking a mirror reflection in the scattering plane. Nonzero values of P_5 and P_6 would violate parity conservation if $P_x = 0$, since the electron beam geometry would remain unchanged unless there is an x-component of P_e. Similarly the tilt (ϵ) in the charge-cloud distribution out of the scattering plane indicated in fig. 9.7 must vanish unless $P_x \neq 0$.

The spin up–down asymmetries are defined by

$$A(\alpha)_{x,y} = \frac{I(\alpha, +P_y) - I(\alpha, -P_y)}{I(\alpha, +P_y) - I(\alpha, -P_y)}. \tag{9.43}$$

Due to parity conservation $A(\alpha)_x = 0$ for $\alpha = 45°$ and $135°$. These parameters can be related to the state multipoles $\langle T_q^k(j) \rangle$ (8.22) describing the atomic state, which depend on the electron polarisation components as follows (Bartschat *et al.*, 1981)

$$\langle T_q^k(j)^+ \rangle = \langle T_q^k(j) \rangle^u + \langle T_q^k(\alpha) \rangle^{P_x} P_x + \langle T_q^k(j) \rangle^{P_y} P_y$$
$$+ \langle T_q^k(j) \rangle^{P_z} P_z. \tag{9.44}$$

Here we have simplified the state multipole notation (8.22) to show only the angular momentum of the excited state.

Since the absolute differential cross section for scattering by unpolarised electrons was not determined by Sohn and Hanne, they analysed their results using normalised state multipoles defined by

$$T_{kq} = \frac{\langle T_q^k(1)^+ \rangle}{\langle T_0^0(1)^+ \rangle^u}, \tag{9.45}$$

where $j_0 (= s_0 = \ell_0) = 0$ and $j_i = j = 1$. The observables of their experiment can be written in terms of 11 normalised state multipoles, namely

$$\text{Im}\,T_{11}^{u}, T_{20}^{u}, \text{Re}\,T_{21}^{u}, \text{Re}\,T_{22}^{u} \quad \text{for } \mathbf{P}_e = 0 \tag{9.46a}$$

$$T_{00}^{P_y}, T_{20}^{P_y}, \text{Re}\,T_{21}^{P_y}, \text{Re}\,T_{22}^{P_y} \quad \text{for } \mathbf{P}_e = P_y \tag{9.46b}$$

$$\text{Re}\,T_{11}^{P_x}, \text{Im}\,T_{21}^{P_x}, \text{Im}\,T_{22}^{P_x} \quad \text{for } \mathbf{P}_e = P_x. \tag{9.46c}$$

Except for $\text{Im}\,T_{22}^{P_x}$ these observables can be derived from the measured Stokes and asymmetry parameters (9.42, 9.43) as shown for instance by Sohn and Hanne (1992).

The angular distribution of the charge cloud of an atomic state (normalised to $\langle T_0^0(j)^u\rangle$) is given by (Blum, 1985; Sohn and Hanne, 1992)

$$\rho(\theta,\phi) = \sum_{kqmm'} (-1)^{j-m}\langle j j m' - m|kq\rangle\, T_{kq}(j)\, Y_{jm'}(\theta,\phi)\, Y_{jm}^{*}(\theta,\phi), \tag{9.47}$$

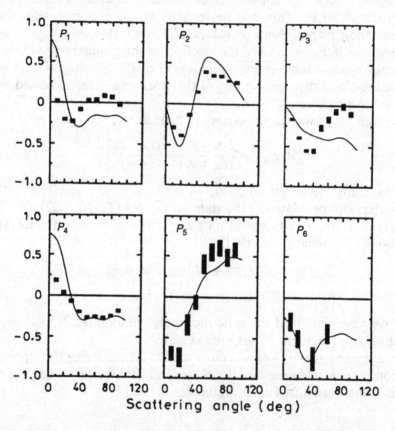

Fig. 9.8. Stokes parameters for the $\text{Hg}(6^3P_1 - 6^1S_0)$ transition after electron-impact excitation at E_0=8 eV. $P_1 - P_4$ are for unpolarised electrons and P_5 and P_6 are normalised to $P_e = P_x = 1$. Experiment, P_1, P_4 (Goeke *et al.*, 1989), P_2, P_3, P_4, P_5 (Sohn and Hanne, 1992); theory, Bartschat (1991b).

where $\langle jjm'\ -m|kq\rangle$ is a Clebsch–Gordan coefficient. Since the normalised state multipoles depend on P_e (9.44, 9.45), the charge-cloud distribution depends on P_e. Using the intermediate coupling scheme to describe the $Hg^*(6s6p6^3P_1)$ state, Sohn and Hanne show that its normalised charge-cloud distribution is given by

$$\rho(\theta,\phi) = \frac{\sqrt{3}}{4\pi}T_{00} - \frac{3}{4\pi}(\beta^2 - \frac{\alpha^2}{2})\left[\frac{1}{\sqrt{6}}T_{20}(3\cos^2\theta - 1) + \mathrm{Re}\,T_{22}\sin^2\theta\cos2\phi\right.$$
$$\left. - \mathrm{Re}\,T_{21}\sin2\theta\cos\phi + \mathrm{Im}\,T_{21}\sin2\theta\sin\phi - \mathrm{Im}\,T_{22}\sin^2\theta\sin2\phi\right], \quad (9.48)$$

where $\alpha\ (= 0.985)$ and $\beta\ (= 0.171)$ are the intermediate coupling coefficients in the expansion of the excited-state wave function in terms of pure *LS*-coupling states involving the $6s$ and $6p$ orbitals

$$\phi(6^3P_1) = \alpha\phi_{LS}(6^3P_1) + \beta\phi_{LS}(6^1P_1). \quad (9.49)$$

In fig. 9.8 the experimental results for the Stokes parameters $P_2, P_3, P_5,$ P_6 (Sohn and Hanne, 1992) and P_1 and P_4 (Goeke *et al.*, 1989) are shown together with theoretical results of a relativistic *R*-matrix calculation by Bartschat (1991*b*). The parameters P_5, P_6 and the asymmetry parameters (fig. 9.9) are normalised to $P_e = 1$. The large values of these latter spin-dependent parameters show the importance of exchange and the spin–orbit interaction in the excitation of $Hg^*(6^3P_1)$ by electron impact.

Fig. 9.10 shows the normalised state multipoles derived from the Stokes and asymmetry parameters compared with the results of the *R*-matrix

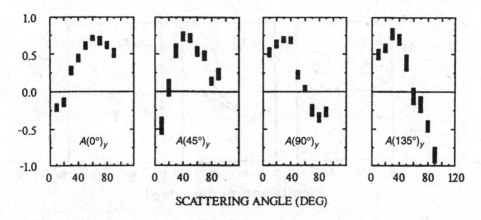

Fig. 9.9. Spin up–down asymmetries $A(\alpha)_y$ normalised to $P_e = P_y = 1$ (from Hanne, 1992).

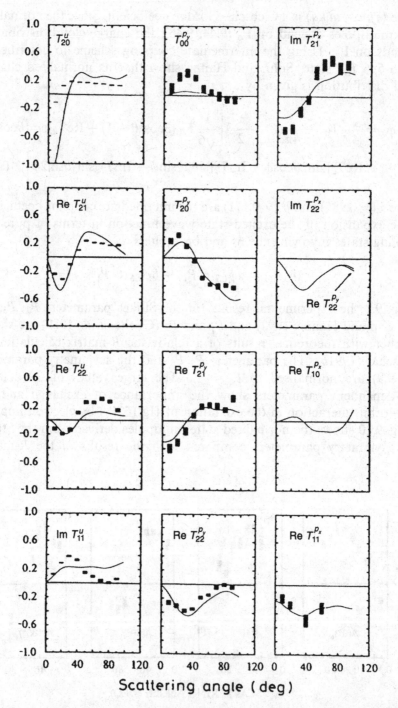

Fig. 9.10. Normalised state multipoles T_{kq} plotted against scattering angle for electron-impact excitation of $Hg^*(6^3P_1)$ at 8 eV (from Sohn and Hanne, 1992). Curve: relativistic R-matrix calculation (Bartschat, 1991b).

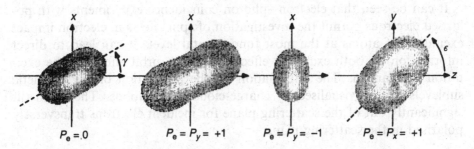

$P_e = 0$ $P_e = P_y = +1$ $P_e = P_y = -1$ $P_e = P_x = +1$

Fig. 9.11. Normalised charge-cloud distribution of the excited $Hg^*(6^3P_1)$ state after collision with 8 eV electrons scattered through 20°. For $P_e = 0$ and $P_e = P_y = \pm 1$ the view is perpendicular to the scattering plane (y direction), and for $P_e = P_x = \pm 1$ the view is in the scattering plane from the $-x$ direction (fig. 9.7) (from Hanne, 1992).

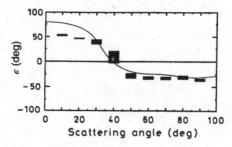

Fig. 9.12. Plot of the tilt angle ϵ in the $y - z$ plane vs scattering angle θ_e compared with the R-matrix theory of Bartschat (1991b).

calculation. Surprisingly the overall agreement between theory and experiment is better for the multipoles that depend on the electron polarisation than those for unpolarised electrons.

Sohn and Hanne (1992) also show some examples of the charge cloud distribution of $Hg^*(6^3P_1)$ after excitation by unpolarised and polarised electrons derived using equn. (9.48). As can be seen in fig. 9.11, the relative size of the charge cloud distribution can be very different for $P_y = +1$ and $P_y = -1$, illustrating a spin up–down asymmetry. The alignment angle γ and shape can also be very different. For unpolarised electrons the shape is just the average of those with $P_y = +1$ and $P_y = -1$. The tilt of the charge cloud out of the scattering plane projected on the $z - y$ plane, indicated by the alignment angle ϵ in fig. 9.11, is also quite large and is significantly different from 0° or 90° over the entire range of scattering angles (fig. 9.12). Sohn and Hanne (1992) show that such a tilt out of the scattering plane would not be allowed if the whole collision system could be described in pure LS coupling.

It can be seen that electron—photon coincidence experiments with polarised electrons permit the investigation of spin effects in electron impact excitation of atoms at the most fundamental level. It can lead to direct information on both exchange effects and spin—orbit effects in the excitation mechanism. The information on the population of the magnetic sublevels can be visualised by charge-cloud distributions. These can tilt significantly out of the scattering plane for incident electrons transversely polarised in the scattering plane.

10

Ionisation

Electron—atom collisions that ionise the target provide a very interesting diversity of phenomena. The reason for this is that a three-body final state allows a wide range of kinematic regions to be investigated. Different kinematic regions depend sensitively on different aspects of the description of the collision.

Up to now there has been no calculation of differential cross sections by a method that is generally valid. We use a formulation due to Konovalov (1993). Understanding of ionisation has advanced by an iterative process involving experiments and calculations that emphasise different aspects of the reaction. Kinematic regions have been found that are completely understood in the sense that absolute differential cross sections in detailed agreement with experiment can be calculated. These form the basis of a structure probe, electron momentum spectroscopy, that is extremely sensitive to one-electron and electron-correlation properties of the target ground state and observed states of the residual ion. It forms a test of unprecedented scope and sensitivity for structure calculations that is described in chapter 11.

Other kinematic regions require a complete description of the collision, which may be facilitated by including the boundary condition for the three charged particles in the final state. This is nontrivial because there is no separation distance at which the Coulomb forces in the three-body system are strictly negligible. The pioneering experiments of Ehrhardt *et al.* (1969) are of this type.

An accurate description of ionisation channels is essential in a theory of scattering, even to low-lying discrete states at low incident energy. The first test of such a description is provided by the total ionisation cross section and asymmetry. Very convincing evidence that our present understanding of collisions is on the right track is the complete agreement of the convergent-close-coupling method with experimental measurements of these quantities for hydrogen.

In this chapter we treat ionisation as a three-body problem in which one target electron is knocked out from a one-electron orbital and the remainder of the target atom acts like an inert third body. In chapter 11 we will see that this approximation can be quite accurately realised in a wide range of kinematic situations. Electron—hydrogen ionisation is of course a true three-body problem.

In a kinematically-complete ionisation experiment for an incident beam of momentum k_0, the differential cross section is normally measured for a range of a single variable determining the momenta k_f and k_s of the faster and slower final-state electrons. The kinematic variables are the kinetic energies E_f, E_s, the polar angles θ_f, θ_s, measured from $\hat{k}_0$, and the relative azimuthal angle

$$\phi = \phi_f - \phi_s + \pi. \tag{10.1}$$

The measurement determines the separation energy ϵ_α from the kinetic energies

$$\epsilon_\alpha = E_0 - E_f - E_s. \tag{10.2}$$

This is the negative of the energy eigenvalue of the state of the residual ion. Its relationship to the orbital energy in the one-electron model of the target will be left to chapter 11. Here we assume that ϵ_α is the negative of the orbital energy eigenvalue. There is a discrete set of values of ϵ_α up to the second-ionisation continuum and resonance structure for low energies in the continuum. The differential cross section is recorded for each value of ϵ_α or summed for ion states that cannot be resolved.

The differential cross section (6.60) is sometimes called the *triple differential cross section* because it is differential in two solid angles and one energy. In the absence of spin analysis it provides the most-detailed information about the ionisation mechanism, but it is impracticable to study it over the full kinematic range available to a three-body final state. It is more informative to study it as a function of one variable in restricted kinematic regions.

The double differential cross section involves integration over the solid angle of either the slow or fast electron. These cases are sometimes called *primary* and *secondary* respectively. It enables a wide kinematic region to be investigated at the expense of losing information due to integration. These considerations apply even more to the single differential cross section in which the integration is over both solid angles.

The measurement of the total ionisation cross section as a function of total energy gives an important overall check on theoretical methods for describing a collision. Total ionisation cross section experiments have also been performed with spin analysis, yielding the total ionisation asymmetry.

Two types of kinematic range have been most-commonly observed in kinematically-complete experiments. In coplanar asymmetric kinematics $\phi = 0$, $E_f \gg E_s$, θ_f is fixed at a value less than about 30° and θ_s is varied. An important subregion is known as the Bethe ridge. Here we are close to the billiard-ball kinematics of a free two-electron collision, for which the recoil momentum $\mathbf{p}$ of the ion, given by

$$\mathbf{p} = \mathbf{k}_0 - \mathbf{k}_f - \mathbf{k}_s, \tag{10.3}$$

is zero. This region is defined by a condition that is common to the billiard-ball collision

$$k_s = |\mathbf{k}_0 - \mathbf{k}_f|. \tag{10.4}$$

Note that this is a conditon on $\mathbf{k}_0$ and $\mathbf{k}_f$. It is not violated by varying θ_s.

In noncoplanar-symmetric kinematics $E_f = E_s$, $\theta_f = \theta_s = 45°$ and ϕ is varied. For small values of ϕ this is again close to billiard-ball kinematics. Both noncoplanar-symmetric and coplanar-asymmetric ranges fix the momentum transfer K, which is conventionally defined by

$$\mathbf{K} = \mathbf{k}_0 - \mathbf{k}_f. \tag{10.5}$$

10.1 Formulation of the three-body ionisation problem

The differential cross section for ionisation is given by (6.60). To formulate the T-matrix element we partition the total Hamiltonian H into a channel Hamiltonian K and a short-range potential V and use the distorted-wave representation (6.77). The three-body model is defined as follows.

$$H = K_1 + K_2 + v_1 + v_2 + v_3, \tag{10.6}$$
$$K = (K_1 + U_1) + (K_2 + v_2), \tag{10.7}$$
$$V = v_1 + v_3 - U_1. \tag{10.8}$$

The electron with coordinate—spin variables $x_i(i = 1, 2)$ has kinetic energy K_i and electron—ion potential v_i. The electron—electron potential is v_3. The electron—ion potential is given to a good approximation by the frozen-core Hartree—Fock potential, which is nonlocal. The distorting potential U_1 is uncharged and acts in the space x_1. State vectors will not be explicitly antisymmetrised in the electron coordinates for the formal discussion. The coordinates x_1 and x_2 are assigned to the fast or incident and slow or bound electrons respectively. Exchange amplitudes are calculated from direct amplitudes by reversing the roles of $\mathbf{k}_f$ and $\mathbf{k}_s$.

The channel Hamiltonian K (10.7) is separable in the electron coordinates. We define the following one-electron states.

$$[E_0 - K_1 - U_1]|\chi^{(+)}(\mathbf{k}_0)\rangle = 0, \tag{10.9a}$$
$$[E_f - K_1 - U_1]|\chi^{(-)}(\mathbf{k}_f)\rangle = 0, \tag{10.9b}$$

$$[E_s - K_2 - v_2]|\chi^{(-)}(\mathbf{k}_s)\rangle = 0, \tag{10.10}$$

$$[\epsilon_\alpha - K_2 - v_2]|\alpha\rangle = 0. \tag{10.11}$$

The distorted-wave integral equation for the full collision state, corresponding to (6.81), is

$$|\Psi_\alpha^{(+)}(\mathbf{k}_0)\rangle = |\alpha\chi^{(+)}(\mathbf{k}_0)\rangle + \frac{1}{E^{(+)} - K}V|\Psi_\alpha^{(+)}(\mathbf{k}_0)\rangle, \tag{10.12}$$

and the unsymmetrised T-matrix element is

$$\langle\mathbf{k}_f\mathbf{k}_s|T|\alpha\mathbf{k}_0\rangle = \langle\chi^{(-)}(\mathbf{k}_f)\chi^{(-)}(\mathbf{k}_s)|V|\Psi_\alpha^{(+)}(\mathbf{k}_0)\rangle. \tag{10.13}$$

The T-matrix element obtained by time-reversing the arguments of chapter 6 is

$$\langle\mathbf{k}_f\mathbf{k}_s|T|\alpha\mathbf{k}_0\rangle = \langle\Psi^{(-)}(\mathbf{k}_f,\mathbf{k}_s)|V|\alpha\chi^{(+)}(\mathbf{k}_0)\rangle. \tag{10.14}$$

Here $|\Psi^{(-)}(\mathbf{k}_f,\mathbf{k}_s)\rangle$ is the solution of the Schrödinger equation for the final state. Its boundary condition describes three charged particles separated by large distances. The Coulomb potentials acting between each pair of particles can never be strictly neglected, so the boundary condition is not simple. It was first given by Rosenberg (1973), based on unpublished work by Redmond. It was first used explicitly in an ionisation calculation by Brauner, Briggs and Klar (1989).

With the appropriate definition of the coordinate–spin variables x_i, and using the spin wave functions (3.79), the asymptotic form of the coordinate–spin representation of the collision state $|\Psi^{(-)}(\mathbf{k}_f,\mathbf{k}_s)\rangle$ is

$$\langle x_1 x_2|\Psi^{(-)}(\mathbf{k}_f,\mathbf{k}_s)\rangle \to (2\pi)^{-3}e^{i\mathbf{k}_f\cdot\mathbf{r}_1}\chi_{v_f}^{1/2}(\sigma_1)e^{i\mathbf{k}_s\cdot\mathbf{r}_2}\chi_{v_s}^{1/2}(\sigma_2)e^{i\Phi}. \tag{10.15}$$

The asymptotic phase Φ is given in terms of the relative momentum $\mathbf{k}_i$ and relative position $\mathbf{r}_i$ of each pair i by

$$\Phi = \Sigma_i\phi_i, \tag{10.16}$$

where

$$\phi_i = \eta_i\ln(k_i r_i + \mathbf{k}_i\cdot\mathbf{r}_i). \tag{10.17}$$

The pair Coulomb parameter η_i is given by (4.61). The asymptotic form of the Coulomb wave (4.84) with time-reversed boundary conditions for a charged pair i (Schiff, 1955) is

$$\psi_{\eta_i}^{(-)}(\mathbf{k}_i,\mathbf{r}_i) \to e^{i\mathbf{k}_i\cdot\mathbf{r}_i}e^{i\eta_i\ln(k_i r_i + \mathbf{k}_i\cdot\mathbf{r}_i)}, r_i \to \infty. \tag{10.18}$$

The asymptotic form of the three-body wave function for the final state is therefore the product of the asymptotic pair wave functions, with allowance for the fact that there are only two independent momenta in the centre-of-mass system, which is the reference frame with the ion stationary in the approximation that the kinetic energy of the ion can be neglected.

In the stationary-ion approximation the relative coordinate and momentum of each electron–ion pair are $\mathbf{r}_i$ and $\mathbf{k}_i$ ($i = 1,2$). The relative coordinate and momentum of the electron–electron pair are, in atomic units,

$$\mathbf{r} = \mathbf{r}_1 - \mathbf{r}_2,$$
$$\mathbf{k} = \tfrac{1}{2}(\mathbf{k}_1 - \mathbf{k}_2), \tag{10.19}$$

and the coordinate and momentum of the electron–electron centre of mass are

$$\mathbf{R} = \tfrac{1}{2}(\mathbf{r}_1 + \mathbf{r}_2),$$
$$\mathbf{K} = \mathbf{k}_1 + \mathbf{k}_2. \tag{10.20}$$

The corresponding kinetic energy operators are related by

$$\tfrac{1}{2}(\nabla_1^2 + \nabla_2^2) = \tfrac{1}{4}\nabla_R^2 + \nabla^2. \tag{10.21}$$

The formulation of the ionisation problem proceeds by defining an auxiliary Schrödinger equation for the final state.

$$[E^{(-)} - K - U]|\Phi^{(-)}(\mathbf{k}_f, \mathbf{k}_s)\rangle = 0. \tag{10.22}$$

The auxiliary state $|\Phi^{(-)}(\mathbf{k}_f, \mathbf{k}_s)\rangle$ will be defined conveniently. The corresponding potential U differs from V by a potential U'.

$$U + U' = V. \tag{10.23}$$

The integral equation formally satisfied by $\langle\Phi^{(-)}(\mathbf{k}_f, \mathbf{k}_s)|$ is

$$\langle\Phi^{(-)}(\mathbf{k}_f, \mathbf{k}_s)| = \langle\chi^{(-)}(\mathbf{k}_f)\chi^{(-)}(\mathbf{k}_s)| + \langle\Phi^{(-)}(\mathbf{k}_f, \mathbf{k}_s)|U\frac{1}{E^{(+)} - K}. \tag{10.24}$$

By substituting for the bra vector of (10.13) using (10.24) and by using (10.12) and the definitions (10.6–10.8,10.22,10.23) we obtain the following rearranged form for the unsymmetrised T-matrix element.

$$\langle\mathbf{k}_f\mathbf{k}_s|T|\alpha\mathbf{k}_0\rangle = \langle\Phi^{(-)}(\mathbf{k}_f, \mathbf{k}_s)|H - E|\Psi_\alpha^{(+)}(\mathbf{k}_0) - \alpha\chi^{(+)}(\mathbf{k}_0)\rangle$$
$$+ \langle\Phi^{(-)}(\mathbf{k}_f, \mathbf{k}_s)|V|\alpha\chi^{(+)}(\mathbf{k}_0)\rangle. \tag{10.25}$$

The Hamiltonian H operates on the bra vector of (10.25).

The first term of (10.25) may be considered as a correction to the second term, which may be minimised by an optimum choice of either, or preferably both, of two criteria. First we may choose U_1 so that $|\alpha\chi^{(+)}(\mathbf{k}_0)\rangle$ is a good approximation to $|\Psi_\alpha^{(+)}(\mathbf{k}_0)\rangle$. Second we may choose $|\Phi^{(-)}(\mathbf{k}_f, \mathbf{k}_s)\rangle$ to approximate $|\Psi^{(-)}(\mathbf{k}_f, \mathbf{k}_s)\rangle$ closely. Note that the initial-state boundary condition ensures the vanishing of the integrand of the correction term in the asymptotic region. It is therefore not necessary to satisfy the final-state boundary condition.

The auxiliary state $|\Phi^{(-)}(\mathbf{k}_f, \mathbf{k}_s)\rangle$ can be chosen so as to exhibit the final-state correlation explicitly. Note that the choice $U = v_3$ reduces (10.25) to (10.14). A useful choice is

$$|\Phi^{(-)}(\mathbf{k}_f, \mathbf{k}_s)\rangle = |\bar{\chi}^{(-)}(\mathbf{k}_f)\chi^{(-)}(\mathbf{k}_s)\phi_\eta^{(-)}(\mathbf{k}_f - \mathbf{k}_s)\rangle, \qquad (10.26)$$

where the final-state correlation function $\phi_\eta^{(-)}(\mathbf{k}', \mathbf{r})$ is given by

$$\phi_\eta^{(-)}(\mathbf{k}', \mathbf{r}) = \psi_\eta^{(-)}(\mathbf{k}', \mathbf{r})e^{-i\mathbf{k}'\cdot\mathbf{r}} \qquad (10.27)$$

and $\bar{\chi}^{(-)}(\mathbf{k}_f)$ is calculated in the charged potential v_1. This choice satisfies the boundary condition (10.15) for three charged bodies.

The approximations to be discussed all treat at least one two-body pair interaction fully. Different kinematic regions depend differently on the amount of detail necessary in the treatment of particular pair interactions. Some success in isolated cases has been achieved by calculations based on low-order terms of the Born series. They are not considered here.

10.1.1 Multichannel approximation

With the choice $U_1 = v_1$, $U = 0$, (10.25) reduces to (10.13). Curran and Walters (1987) and Curran, Whelan and Walters (1991) have approximated $|\Psi_\alpha^{(+)}(\mathbf{k}_0)\rangle$ in the case of hydrogen by the multichannel wave function obtained from the pseudostate calculation of van Wyngaarden and Walters (1986), which achieves very good results for scattering data as we have seen in section 8.2.

10.1.2 The distorted-wave Born approximation

The weak-coupling approximation for the collision state in (10.13) involves neglecting the possibility of exciting the target, except perhaps by including excitations through an optical potential. The approximation is

$$|\Psi_\alpha^{(+)}(\mathbf{k}_0)\rangle = |\alpha\chi^{(+)}(\mathbf{k}_0)\rangle, \qquad (10.28)$$

where $|\chi^{(+)}(\mathbf{k}_0)\rangle$ is a distorted wave calculated in the potential U_1 of (10.9a). A simple choice of U_1 is justified in section 6.10.

$$U_1 = \langle\alpha|v_1 + v_3|\alpha\rangle. \qquad (10.29)$$

The T-matrix element in the distorted-wave Born approximation is

$$\langle\mathbf{k}_f\mathbf{k}_s|T|\alpha\mathbf{k}_0\rangle = A_S\langle\chi^{(-)}(\mathbf{k}_f)\chi^{(-)}(\mathbf{k}_s)|v_3|\alpha\chi^{(+)}(\mathbf{k}_0)\rangle. \qquad (10.30)$$

The potentials for the separable bra state may be chosen by the second criterion for (10.25).

This approximation sets the standard for ionisation calculations. In many cases it gives at least a good semiquantitative description of cross sections. An example is shown in fig. 10.1 for the $3p$ orbital of argon in coplanar asymmetric kinematics, $E_0 = 1000$ eV, $E_s = 120$ eV.

The wide applicability of (10.30) justifies showing its computational form. Formally (10.30) is a potential matrix element (6.88) in the distorted-wave representation for a three-body collision with the bound orbital $|i\rangle$ replaced by the continuum orbital (distorted wave) $|\chi^{(-)}(\mathbf{k}_s)\rangle$. The direct matrix element is written in a form analogous to (7.62) using the distorted-wave form (4.58) of (7.45) for the continuum orbitals, (7.49) for the bound orbital $|\alpha\rangle$ and (7.60) for the electron–target potential. Note that the term $1/r_0$ in (7.60) vanishes if we require $|\chi^{(-)}(\mathbf{k}_s)\rangle$ to be orthogonal to $|\alpha\rangle$. This requirement is implicit in (10.10,10.11) and is normally imposed in implementing (10.30).

The algebra used in obtaining the direct potential matrix element (7.67) gives the following LS-coupling expression

$$\langle \mathbf{k}_f \mathbf{k}_s | T | \alpha \mathbf{k}_0 \rangle$$

$$= (2\pi)^{-9/2}(4\pi)^{5/2}(k_f k_s k_0)^{-1} \sum_{L'L''M''\lambda} (-1)^{M''} \begin{pmatrix} \ell & \lambda & L'' \\ 0 & 0 & 0 \end{pmatrix}$$

$$\times \begin{pmatrix} \ell & \lambda & L'' \\ m & M''-m & -M'' \end{pmatrix}$$

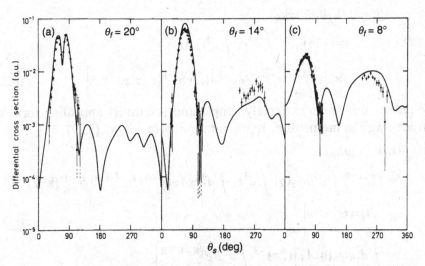

Fig. 10.1. Coplanar-asymmetric ionisation from the $3p$ orbital of argon at $E_0 = 1000$ eV, $E_s = 120$ eV (Avaldi et al., 1989). The fast and slow electrons are respectively indicated by the subscripts f and s on the diagram. (a) $\theta_f = 20°$ (Bethe ridge), (b) $\theta_f = 14°$, (c) $\theta_f = 8°$. Full curve, distorted-wave Born approximation.

$$\times \Sigma_L \begin{pmatrix} L' & \lambda & L \\ 0 & 0 & 0 \end{pmatrix} \begin{pmatrix} L' & \lambda & L \\ m-M'' & M'' -m & 0 \end{pmatrix}$$

$$\times R^{(\lambda)}_{L'L''L\alpha}(k_f,k_s,k_0)Y_{L'm-M''}(\hat{\mathbf{k}}_f)Y_{L''M''}(\hat{\mathbf{k}}_s)$$

$$+ \text{exchange amplitude,} \tag{10.31}$$

where the orbital quantum numbers of the state $|\alpha\rangle$ are ℓ,m and

$$R^{(\lambda)}_{L'L''L\alpha}(k_f,k_s,k_0) = i^{L-L'-L''}\exp[i(\sigma_{L'}+\sigma_{L''})]\hat{L}'\hat{L}''\hat{\ell}\hat{L}^2$$

$$\times \int dr_1 \int dr_2 u_{L'}(k_f,r_1)u_L(k_0,r_1)\frac{r_<^\lambda}{r_>^{\lambda+1}}u_{L''}(k_s,r_2)u_\alpha(r_2). \tag{10.32}$$

10.1.3 The impulse approximation

The simplest way of including the full interaction of the two final-state electrons is to use the impulse approximation. In its simplest plane-wave form this approximation is obtained from (10.14) by neglecting v_1 and v_2 in the definition of the collision state $|\Psi^{(-)}(\mathbf{k}_f,\mathbf{k}_s)\rangle$. It retains the two-electron function $\phi_\eta^{(-)}(\mathbf{k}',\mathbf{r})$. In the spirit of this approximation it replaces $|\chi^{(+)}(\mathbf{k}_0)\rangle$ with a plane wave. We expect the plane-wave impulse approximation to describe kinematic regions where the two-electron collision dominates the reaction mechanism such as the higher-energy billiard-ball range.

The T-matrix element in the plane-wave impulse approximation is

$$\langle \mathbf{k}_f\mathbf{k}_s|T|\alpha\mathbf{k}_0\rangle$$

$$= (2\pi)^{-9/2}\int d^3r_1 \int d^3r_2 A_S$$

$$\times \left[e^{-i\mathbf{k}_f\cdot\mathbf{r}_1}e^{-i\mathbf{k}_s\cdot\mathbf{r}_2}\phi_\eta^{(-)*}(\mathbf{k}',\mathbf{r})r^{-1}\langle\mathbf{r}_2|\alpha\rangle e^{-i\mathbf{k}_0\cdot\mathbf{r}_1} \right]. \tag{10.33}$$

We reduce this to an extremely simple and intuitively-appealing form by introducing the momentum representation $\phi_\alpha(\mathbf{q})$ of $|\alpha\rangle$ (3.29).

$$\langle \mathbf{k}_f\mathbf{k}_s|T|\alpha\mathbf{k}_0\rangle$$

$$= (2\pi)^{-6}\int d^3q\phi_\alpha(\mathbf{q})\int d^3r_1 \int d^3r_2 A_S\left[e^{-i\mathbf{k}_f\cdot\mathbf{r}_1}e^{-i\mathbf{k}_s\cdot\mathbf{r}_2}\phi_\eta^{(-)*}(\mathbf{k}',\mathbf{r})\right.$$

$$\times \left. r^{-1}e^{i\mathbf{q}\cdot\mathbf{r}_2}e^{i\mathbf{k}_0\cdot\mathbf{r}_1}\right]$$

$$= \int d^3q\phi_\alpha(\mathbf{q})A_S\left[(2\pi)^{-3}\int d^3R e^{i(\mathbf{K}-\mathbf{K}')\cdot\mathbf{R}}\right]$$

$$\times \left[(2\pi)^{-3}\int d^3r\psi_\eta^{(-)*}(\mathbf{k}',\mathbf{r})r^{-1}e^{i\mathbf{k}\cdot\mathbf{r}}\right]$$

$$= A_S\langle\mathbf{k}'|t_\eta|\mathbf{k}\rangle\phi_\alpha(-\mathbf{p}), \tag{10.34}$$

where $\mathbf{p}$ is the observed ion recoil momentum (10.3) and t_η is the two-electron T matrix.

In the derivation (10.34) we have used (10.27), the transformation (10.19–10.21) from electron coordinates and momenta to relative and centre-of-mass coordinates and momenta, the definition corresponding to the time reversal of (4.112) for the half-on-shell T matrix, and the representation reciprocal to (3.30) of $\delta(\mathbf{K}' - \mathbf{K})$.

In this approximation the (e,2e) amplitude factorises into the anti-symmetrised product of amplitudes for simultaneously finding the target electron with momentum $-\mathbf{p}$ and knocking it out. The momenta $\mathbf{k}', \mathbf{k}$ and $\mathbf{p}$ are all directly observed in the experiment and η is the Coulomb parameter of $\mathbf{k}'$.

$$\mathbf{k} = \tfrac{1}{2}(\mathbf{k}_0 + \mathbf{p}), \quad \mathbf{k}' = \tfrac{1}{2}(\mathbf{k}_f - \mathbf{k}_s), \quad \eta = 1/k'. \tag{10.35}$$

The factorisation is characteristic also of the plane-wave Born approximation, which is (10.30) with distorted waves replaced by plane waves. Here the two-electron T-matrix element is replaced by the two-electron potential matrix element (3.41).

$$\langle \mathbf{k}'|v_3|\mathbf{k}\rangle = (2\pi^2|\mathbf{k} - \mathbf{k}'|^2)^{-1}. \tag{10.36}$$

Ford (1964) has obtained the half-on-shell Coulomb T-matrix element as the limit of the T-matrix element for the screened potential $e^{-\lambda r}/r$ as $\lambda \to 0$. It is

$$\langle \mathbf{k}'|t_\eta|\mathbf{k}\rangle = \langle \psi_\eta^{(-)}(\mathbf{k}')|v_3|\mathbf{k}\rangle$$
$$= \lim_{\lambda \to 0} C(\eta)\left[\frac{k^2 - k'^2}{|\mathbf{k} - \mathbf{k}'|^2}\right]^{i\eta} e^{i\delta_0}(2\pi^2|\mathbf{k} - \mathbf{k}'|^2)^{-1}, \tag{10.37}$$

where

$$C(\eta) = e^{-\pi\eta/2}|\Gamma(1 + i\eta)| = \left[\frac{2\pi\eta}{e^{2\pi\eta} - 1}\right]^{1/2},$$
$$\delta_0 = \sigma_0 - \eta\ln(2k'/\lambda),$$
$$\sigma_0 = \arg\Gamma(1 + i\eta),$$
$$\mathbf{k} \neq \mathbf{k}'. \tag{10.38}$$

The divergent phase vanishes when the absolute square of (10.37) is taken for the differential cross section (6.60), which has the form

$$\frac{d^5\sigma}{d\Omega_f d\Omega_s dE_s} = (2\pi)^4 \frac{k_f k_s}{k_0} f_{ee} \frac{1}{2j_0 + 1} \sum_m |\phi_\alpha(-\mathbf{p})|^2. \tag{10.39}$$

Here j_0 is the total angular momentum of the ground state and m is the magnetic quantum number of the orbital $|\alpha\rangle$. The spin-averaged two-electron collision factor is

$$f_{ee} = \sum_{av} |A_S \langle \mathbf{k}'|t_\eta|\mathbf{k}\rangle|^2$$

$$= \frac{1}{(2\pi^2)^2} \frac{2\pi\eta}{e^{2\pi\eta} - 1} \left[\frac{1}{|\mathbf{k}_0 - \mathbf{k}_f|^4} + \frac{1}{|\mathbf{k}_0 - \mathbf{k}_s|^4} - \frac{1}{|\mathbf{k}_0 - \mathbf{k}_f|^2} \frac{1}{|\mathbf{k}_0 - \mathbf{k}_s|^2} \right.$$

$$\left. \times \cos\left(\eta \ln \frac{|\mathbf{k}_0 - \mathbf{k}_s|^2}{|\mathbf{k}_0 - \mathbf{k}_f|^2} \right) \right]. \tag{10.40}$$

The plane-wave impulse approximation is very successful in high-energy cases where the distortion of the incident and outgoing electron waves can be neglected. There is an experimental test of the validity of this condition. The expression (10.34) enables an apparent momentum-space wave function $\phi_\alpha(-\mathbf{p})$ to be extracted from an experiment. Distortion is negligible if an experiment at higher energy gives the same apparent wave function.

The approximation is the basis of the probe for the probability of finding an electron of momentum $\mathbf{p}$ discussed in section 3.2. In noncoplanar-symmetric and Bethe ridge kinematics f_{ee} is essentially constant for a range of p up to several atomic units, so the approximation gives a direct estimate of $|\phi_\alpha(-\mathbf{p})|^2$. Fig. 3.1 shows that for hydrogen this corresponds

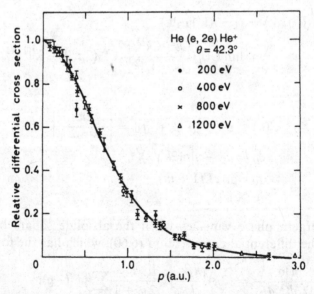

Fig. 10.2. Noncoplanar-symmetric ionisation of helium at the indicated total energies E (McCarthy and Weigold, 1976). Curve, plane-wave impulse approximation.

to the squared momentum-space wave function at several values of the total energy

$$E = E_f + E_s, \qquad (10.41)$$

at least within a normalisation factor which is not measured in the experiment.

Fig. 10.2 shows that the plane-wave impulse approximation is as good for relative helium differential cross sections at different energies as it is for hydrogen. Here $\phi_{1s}(-\mathbf{p})$ is the Hartree–Fock orbital. For helium there is an absolute experiment by van Wingerden *et al.* (1979) for $\phi = 0$ in symmetric kinematics at different total energies. Fig. 10.3 shows that the plane-wave impulse approximation using the Ford T-matrix element is consistent with the experiment.

There is another significance to this result. The Coulomb T matrix is analytically very difficult (Chen and Chen, 1972). For example, it is not uniformly convergent to the half shell as the half-shell value is approached from above or below. The Ford version is the only one that correctly describes the experiment (McCarthy and Roberts, 1987).

In many cases distortion of the continuum-electron wave functions is significant. The (e,2e) T-matrix element in the distorted-wave impulse

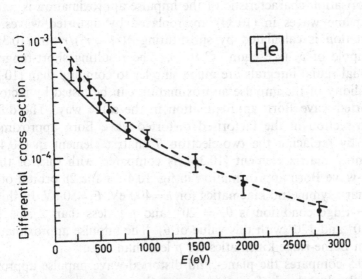

Fig. 10.3. The differential cross section for electron–helium ionisation at $\phi = 0$ in symmetric kinematics, plotted against total energy (van Wingerden *et al.*, 1979). Full curve, distorted-wave impulse approximation; broken curve, plane-wave impulse approximation. From McCarthy and Weigold (1988).

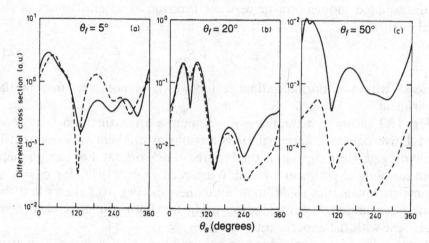

Fig. 10.4. Factorisation test of the distorted-wave Born approximation for coplanar asymmetric ionisation from the $2p$ orbital of neon at $E = 400$ eV, $E_s = 50$ eV (Madison, McCarthy and Zhang, 1989). The fast and slow electrons are respectively indicated by the subscripts f and s on the diagram. The Bethe ridge condition is $\theta_f = 20°$. Full curve, unfactorised; broken curve, factorised.

approximation is

$$\langle \mathbf{k}_f \mathbf{k}_s | T | \alpha \mathbf{k}_0 \rangle = A_S \langle \mathbf{k}' | t_\eta | \mathbf{k} \rangle \langle \chi^{(-)}(\mathbf{k}_f) \chi^{(-)}(\mathbf{k}_s) | \alpha \chi^{(+)}(\mathbf{k}_0) \rangle. \qquad (10.42)$$

The factorisation characteristic of the impulse approximation is retained, but the plane waves in (10.34) are replaced by distorted waves. The approximation is calculated by substituting $\delta(r_1 - r_2)/r_1^2$ in (10.32) for the multipole of v_3 (see equn. (3.102)). The resulting short-range one-dimensional radial integrals are much simpler to compute than (10.32).

The validity of the impulse approximation can be tested by factorising the distorted-wave Born approximation in the same way. The differential cross section in the factorised distorted-wave Born approximation, obtained by replacing the two-electron T-matrix element in (10.42) by the potential matrix element (10.36), is compared with that of the full distorted-wave Born approximation in fig. 10.4 for the $2p$ orbital of neon in coplanar-asymmetric kinematics for $E=400$ eV, $E_s=50$ eV. In this case the Bethe-ridge condition is $\theta_f = 20°$, and p is less than 2 a.u. for θ_s between $0°$ and $120°$ with this value of θ_f. The impulse approximation is verified in Bethe-ridge kinematics for p less than 2 a.u.

Fig. 10.5 compares the plane- and distorted-wave impulse approximations for the $3p$ orbital of argon in noncoplanar-symmetric kinematics at $E=1500$ eV. Here distortion makes a difference beyond $p=1.5$ a.u. The experiment is described excellently (within an unmeasured normalisation) by the distorted-wave impulse approximation. Figs. 10.3 and 10.5 support

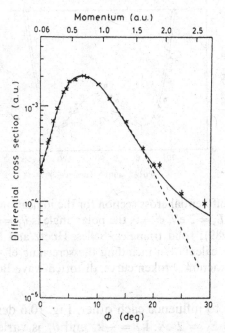

Fig. 10.5. Differential cross section for the 1500 eV noncoplanar-symmetric ionisation from the $3p$ orbital of argon (McCarthy *et al.*, 1989). Full curve, distorted-wave impulse approximation × 0.95; broken curve, plane-wave impulse approximation × 0.83.

the conclusion reached by experience with many different reactions that the distorted-wave impulse approximation describes ionisation within experimental error in noncoplanar symmetric kinematics for $E > 1000$ eV and $p < 2.5$ a.u. There is evidence that this is true also for Bethe-ridge kinematics.

10.1.4 *Final-state interaction*

The inclusion of the two-electron interaction $\phi_\eta^{(-)}(\mathbf{k}', \mathbf{r})$ in approximating (10.25) is difficult in spherical polar coordinates because it is not centred on the ion. This difficulty has been circumvented by Pan and Starace (1991) and Jones, Madison and Srivastava (1992) within the computational framework of the distorted-wave Born approximation. The potential v_3 is accounted for as a screening potential in the calculation of the final-state distorted waves in a way that depends on the final-state kinematic variables. We do not describe the calculations, which differ from (10.30) in details such as the inclusion of a term due to the nonorthogonality of $|\chi^{(-)}(\mathbf{k}_s)\rangle$ and $|\alpha\rangle$. The methods are applied to a kinematic region where the final-state energy is only a few electron volts and the final-state

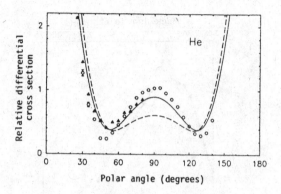

Fig. 10.6.   Relative differential cross section for the ionisation of helium (Pan and Starace, 1991). $E_f = E_s = 2$ eV, $\phi = 0$, the polar angle is $\theta_f = \pi + \theta_s$. Open circles, Schlemmer *et al.* (1989); solid triangles, Selles, Huetz and Mazeau (1987); full curve, distorted-wave calculation including the screening effect of the final-state electron-electron interaction; broken curve, distorted-wave Born approximation.

electrons have time to influence each other. Fig. 10.6 describes the case of helium where $E_f = E_s = 2$ eV, $\mathbf{k}_f = -\mathbf{k}_s$ and θ_f is varied. Inclusion of v_3 in the calculation of the distorted waves takes the calculated cross section in the direction of the experiment and illustrates the possible usefulness of the final-state interaction in approximating (10.25).

Brauner, Briggs and Klar (1989) have performed the first calculation of differential cross sections that uses an approximation in which the boundary condition (10.15) is explicitly satisfied. The target was hydrogen. Their calculation may be considered in terms of (10.25). The first term is omitted. The second term is evaluated with the choice (10.26) for the auxiliary state $|\Phi^{(-)}(\mathbf{k}_f, \mathbf{k}_s)\rangle$ and $U_1 = 0$.

Fig. 10.7 compares the calculations of Brauner *et al.* with the distorted-wave Born approximation and the approximation to (10.13) of Curran and Walters (1987) for a coplanar-asymmetric experiment on hydrogen at $E_0 = 150$ eV. No calculation yields fully-quantitative agreement with experiment in the peak for small values of p, but all describe the relative shape. The cross section that is observed at much larger p is not well described by the distorted-wave Born approximation, but the other two calculations predict the trends better.

10.2 Inner-shell ionisation

Most ionisation experiments have concentrated on the valence electrons of the target. However, experiments and calculations for the $1s$ shell of neon and the $n=2$ shells of argon have been reported by Zhang *et al.* (1992). Fig. 10.8 shows an example. Here the $2p$ orbital of argon

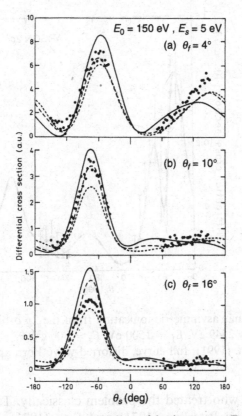

Fig. 10.7. Coplanar-asymmetric ionisation of hydrogen at $E_0 = 150$ eV, $E_s = 5$ eV. The fast and slow electrons are respectively indicated by the subscripts f and s on the diagram. Experimental data, Klar *et al.* (1987); full curve, distorted-wave Born approximation (McCarthy and Zhang, 1990); long-dashed curve, pseudostate approximation to (10.13) (Curran and Walters, 1987); short-dashed curve, interacting final state (Brauner *et al.*, 1989). From McCarthy and Zhang (1990).

is observed in coplanar-asymmetric kinematics (Bickert *et al.* 1991) with $E_0 = 2549$ eV, $E_f = 1500$ eV, $E_s = 800$ eV, $\theta_f = 33.8°$. The distorted-wave Born approximation with the Hartree—Fock orbital achieves excellent shape agreement with the relative differential cross section.

10.3 Ionisation near threshold

A kinematic region that is extremely difficult for quantum calculations is just above the ionisation threshold. Near threshold the two slow continuum electrons moving in the field of a positive ion are strongly correlated and suitable approximations are difficult to evaluate. The main features in the asymptotic region were first established theoretically by

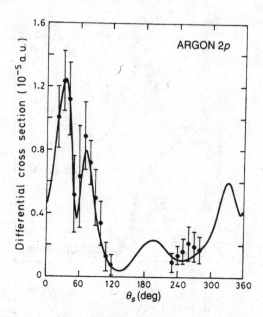

Fig. 10.8. Coplanar asymmetric ionisation from the $2p$ orbital of argon (Zhang *et al.*, 1992). $E_0 = 2549$ eV, $E_f = 1500$ eV, $E_s = 800$ eV, $\theta_f = 33.8°$. Experimental data, Bickert *et al.* (1991); full curve, distorted-wave Born approximation.

Wannier (1953) who treated the problem classically. The Wannier theory was confirmed by Peterkop (1971) and Rau (1971) using semiclassical methods. This earlier work, as well as more recent work on near-threshold ionisation, have all emphasised the role of radial and angular correlations in the final two-electron state. In the Wannier theory, and its semiclassical extensions, the details of the collision process and the structure of the target play no role, since only the asymptotic region is considered.

Close to threshold the system must be viewed as consisting of a correlated pair of electrons attached to the ion. It is convenient to introduce the hyperspherical coordinates

$$R = (r_f^2 + r_s^2)^{1/2}, \tag{10.43a}$$

$$\theta_{fs} = \cos^{-1}(\hat{\mathbf{r}}_f \cdot \hat{\mathbf{r}}_s) \tag{10.43b}$$

and

$$\alpha = \tan^{-1} r_s / r_f, \tag{10.43c}$$

where θ_{fs}, the angle between the radial directions of the two electrons, describes their angular correlation and the angle α describes the radial correlation of the two electrons. The Coulomb energy expressed in terms of these coordinates is

$$V = \frac{C(\alpha, \theta_{fs})}{R}, \tag{10.44a}$$

where

$$C(\alpha, \theta_{fs}) = \frac{-Z}{\cos\alpha} - \frac{Z}{\sin\alpha} + \frac{1}{(1 - \sin2\alpha\cos\theta_{fs})^{1/2}}. \tag{10.44b}$$

The dependence of C on the two angles α and θ_{fs} is shown in fig. 10.9. C is symmetric about the point $\alpha = \pi/4$, $\theta_{fs} = \pi$ (i.e. $\mathbf{r}_f = -\mathbf{r}_s$), which is often called the Wannier point.

The Wannier point is a saddle point with C (i.e. V) increasing in the θ_{fs} direction but decreasing in the α direction. The angle θ_{fs} therefore tends to converge to the value π (i.e. $\hat{\mathbf{r}}_f = -\hat{\mathbf{r}}_s$) as the system evolves, but the angle α will tend to diverge from $\pi/4$ (i.e. $r_f = r_s$). If this divergence is great enough the system will fall into the valleys at $\alpha = 0$ or $\pi/2$ (i.e. $r_s = 0$), which will leave one electron bound to the ion while the other electron goes free (i.e. excitation of the atom). Double escape (ionisation) can only occur if the system stays close to the Wannier point as R increases to the point at which the two electrons are essentially independent. This boundary between the 'Coulomb' region, in which the coupling between the potential and kinetic energies is decisive, and the asymptotic region, where the kinetic energy dominates, is known as the 'Wannier radius', R_W. It scales as E^{-1} where E is the total energy of the two free electrons (Fano and Rau, 1986). Therefore since it takes a long time for R to reach the

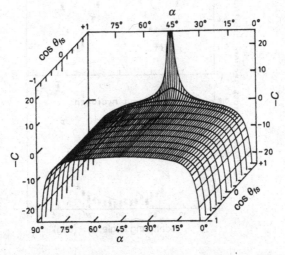

Fig. 10.9. The dependence of the potential function C (10.44) on α and θ_{fs} (Fano and Lin, 1975). C is symmetrical about the plane through $\theta_{fs} = \pi$.

value R_W when E is small, the ionisation cross section is necessarily very small when E is small.

The Hamiltonian has radial (K_R) and angular (K_α, K_θ) kinetic energy operators in addition to the potential V (10.44). By treating these on par with $V(R, \theta_{fs}, \alpha)$ and by assuming an initial quasi-ergodic distribution in phase space of the escape trajectories as they enter the Coulomb zone, Wannier was able to show that at threshold (small E) the total ionisation cross section was dominated by the instability in the escape trajectories and was given by

$$\sigma_I \propto E^n, \tag{10.45}$$

where

$$n = \tfrac{1}{4}\left[\left(\frac{100Z - 9}{4Z - 1}\right)^{1/2} - 1\right]. \tag{10.46}$$

This is known as the Wannier threshold law and for $Z = 1, 2, 3$, n has the values 1.127, 1.056, and 1.036 respectively. The deviation of n from the expected value of unity arises from the instability in the escape configuration discussed above.

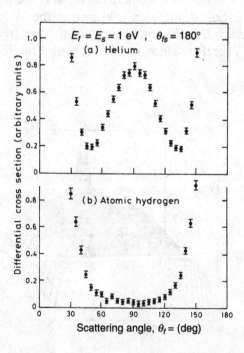

Fig. 10.10.   The (e,2e) differential cross section for He and H for $E_f = E_s = 1$ eV and $\theta_{fs} = \pi$. From Rösel *et al.* (1989).

Total ionisation cross section measurements are in excellent agreement with the Wannier threshold law (10.45) with $n = 1.127$ (e.g. Read, 1985). Differential measurements are difficult close to threshold because of the low energy of the emitted electrons and the high energy resolution required in the incident beam. Cvejanović and Read (1974), with a coincidence time-of-flight technique, were nevertheless able to show for electron impact ionisation of helium at $E = 0.37$ and 0.60 eV that the angular correlation did indeed have a maximum at $\theta_{fs} = \pi$ with a width $\Delta\theta_{fs} \sim \theta_0 E^{1/4}$, and that the energy-partitioning probability was indeed uniform to within about 5%. These results are all in agreement with the extended Wannier theory. The value of θ_0 has recently been investigated by the Paris group (e.g. Selles *et al.*, 1987) using coplanar geometry, and by the Manchester group (Hawley-Jones *et al.*, 1992) using the perpendicular-plane geometry ($\theta_f = \theta_s = \pi$) with $E_f = E_s = E/2$. At $E = 1$ eV, Hawley-Jones *et al.* found $\theta_0 = 1.30 \pm 0.04$, and an empirical energy dependence up to $E = 6$ eV of $\Delta\theta_{fs} = 70E^{0.16}$ where E is in electron volts and the angle is in degrees. The deviation from the expected exponent of $1/4$ is due to d-wave components in the cross section in addition to the dominant s-wave components (p-wave terms do not contribute to the perpendicular plane differential cross section since $\theta_f = \theta_s = \pi/2$).

The Wannier model and its semiclassical extensions only treat the long-range Coulomb forces, and they therefore cannot distinguish between different systems with the same Coulomb asymptotic behaviour. Ionisation from atomic hydrogen and helium are two such systems, the case of helium differing from hydrogen only in the inner region due to the role of polarisation forces and screening of the ion. In a very nice experiment Rösel *et al.* (1989) measured the coplanar (e,2e) cross sections for atomic hydrogen and helium a few electron volts above threshold with $E_f = E_s = E/2$ and θ_{fs} fixed. Their results for $E_f = E_s = 1$eV and $\theta_{fs} = \pi$ are shown in fig. 10.10 as a function of θ_f. The pronounced maximum at $\theta_f = 90°$ in helium is missing in the hydrogen case. This difference, which must be due to short range interactions, shows the limitations of the Wannier-type theories. A full description of the cross section close to threshold obviously requires a theory which takes into account the full collision process (see fig. 10.6).

10.4 Excitation of autoionising resonances

The study of resonances in chapter 8 involved cross sections for elastic and inelastic scattering that are affected by resonances in the electron–target compound system. A new dimension in the study of resonances and the ionisation mechanism is provided by kinematically-complete experiments that can be considered as the excitation of autoionising resonances of the target system by collision with a fast electron. The scattering reaction that observes the same resonances is electron scattering by the ion. The reaction is an extremely-sensitive test of theoretical methods, since it depends simultaneously on two types of amplitude, one for direct ionisation and one for resonance scattering.

The experiment of Lower and Weigold (1990) on a helium target employed coplanar-asymmetric geometry with E_0 and θ_f fixed. The He^+ ion was left in its ground state. The energy E_s was varied through the region 32–37 eV, which includes the lowest autoionising resonances of helium in the manifolds ^1S, ^1P, ^3P, ^1D. In the corresponding scattering reaction they are resonances below the $n=2$ excitation threshold of the He^+ ion at 54.4 eV. The independent-particle interpretation of the analogous resonances in electron scattering by the H atom is given in section 8.2.5.

The differential cross section for resonant ionisation has been calculated by McCarthy and Shang (1993). The approximation treats the T-matrix element as a coherent superposition of two amplitudes. One describes direct ionisation and is analogous to the amplitudes of section 10.1. The other contains the momentum $\mathbf{k}_s$ in a resonant amplitude that has an entirely different structure. One can find values of θ_s for which the direct amplitude is small (see fig. 10.1 for an analogous reaction). Here the resonant amplitude dominates the T-matrix element. At angles θ_s where the direct amplitude is large, interference between the direct and resonant amplitudes is observed.

Resonant amplitudes are calculated by scattering methods that are very well understood (section 8.2.5). We can therefore consider the resonant ionisation reaction as an extremely-sensitive test of the approximation used for direct ionisation, since it depends on the magnitude and phase of the direct amplitude.

10.4.1 Calculation of the amplitude

The amplitude for resonant ionisation is written in a distorted-wave formalism as

$$\langle \mathbf{k}_f \mathbf{k}_s 0_+ | T | 0 \mathbf{k}_0 \rangle \equiv A_{S'} \langle \chi^{(-)}(\mathbf{k}_f) \Psi_{0S}^{(-)}(\mathbf{k}_s) | \tau | 0 \chi^{(+)}(\mathbf{k}_0) \rangle, \tag{10.47}$$

where the collision operator τ is to be approximated. The operator $A_{S'}$ antisymmetrises the amplitude in the coordinates of the two continuum electrons in spin state S'. The ground state of helium is $|0\rangle$. The incident and fast electrons are represented by distorted waves (10.9), but the resonant interaction of the slow electron with the He^+ ion must be treated in more detail. The state $|\Psi_{0S}^{(-)}(\mathbf{k}_s)\rangle$ of the resonant system describes an electron of momentum $\mathbf{k}_s$ leaving the ion in its ground state $|0_+\rangle$. The electron—ion spin quantum number is S. The state is the time reversal of the electron—ion collision state $|\Psi_{0S}^{(+)}(\mathbf{k}_s)\rangle$, which satisfies the distorted-wave Lippmann—Schwinger equation (6.81). In the notation of the present chapter this is

$$|\Psi_{0S}^{(+)}(\mathbf{k}_s)\rangle = |0_+\psi^{(+)}(\mathbf{k}_s)\rangle$$
$$+ \frac{1}{E_s^{(+)} + \epsilon_0 - K_2 - H_+ - U}(v_2 - U)|\Psi_{0S}^{(+)}(\mathbf{k}_s)\rangle. \quad (10.48)$$

The distorted wave $|\psi^{(+)}(\mathbf{k}_s)\rangle$ is calculated in the Coulomb potential U of the electron—ion system. The Schrödinger equation of the ion is

$$(\epsilon_i - H_+)|i_+\rangle = 0. \quad (10.49)$$

The notation i_+ includes the He$^+$ continuum.

We use the time reversal of (10.48) and introduce the unit operator in electron—ion channel space, using a notation that implies the inclusion of electron—ion bound states. The unsymmetrised reaction amplitude of (10.47) becomes

$$\langle \chi^{(-)}(\mathbf{k}_f)\Psi_{0S}^{(-)}(\mathbf{k}_s)|\tau|0\chi^{(+)}(\mathbf{k}_0)\rangle$$
$$= \langle \chi^{(-)}(\mathbf{k}_f)\psi^{(-)}(\mathbf{k}_s)0_+|\tau|0\chi^{(+)}(\mathbf{k}_0)\rangle + \Sigma_i \int d^3k$$
$$\times \langle \Psi_{0S}^{(-)}(\mathbf{k}_s)|v_2 - U|i_+\psi^{(+)}(\mathbf{k})\rangle \frac{1}{E_S^{(-)} + \epsilon_0 - \epsilon_i - \frac{1}{2}k^2}$$
$$\times \langle \chi^{(-)}(\mathbf{k}_f)\psi^{(+)}(\mathbf{k})i_+|\tau|0\chi^{(+)}(\mathbf{k}_0)\rangle. \quad (10.50)$$

Practical approximations for τ are binary-encounter operators such as v_3, which do not depend on the ion coordinates. The target and ion are therefore represented by the overlap $\langle i_+|0\rangle$, which is shown by experiments described in section 11.1.2 to be small unless $|i_+\rangle$ is the ground state $|0_+\rangle$. These approximations were made by McCarthy and Shang (1993). They represented the final-state interaction in the long-range on-shell direct amplitude of (10.50) by the factor $C(\eta)e^{i\sigma_0}$ (10.38). This factor was omitted from the resonant amplitude on the basis of the semiclassical picture of the two electrons emerging at different times.

The first amplitude in the integrand of the second term of (10.50) is a half-on-shell element of the time-reversed distorted-wave T matrix $T_+^\dagger$ for the electron—ion collision (6.87). The approximation calculated for resonant ionisation is

$$\langle \chi^{(-)}(\mathbf{k}_f)\Psi_{0S}^{(-)}(\mathbf{k}_s)|\tau|0\chi^{(+)}(\mathbf{k}_0)\rangle$$
$$= C(\eta)e^{i\sigma_0}\langle \chi^{(-)}(\mathbf{k}_f)\psi^{(-)}(\mathbf{k}_s)0_+|v_3|0\chi^{(+)}(\mathbf{k}_0)\rangle$$
$$+ \int d^3k \langle \mathbf{k}_s^{(-)}0_+|T_+^\dagger|0_+\mathbf{k}^{(+)}\rangle \frac{1}{E_s^{(-)} - \frac{1}{2}k^2}$$
$$\times \langle \chi^{(-)}(\mathbf{k}_f)\psi^{(+)}(\mathbf{k})0_+|v_3|0\chi^{(+)}(\mathbf{k}_0)\rangle. \qquad (10.51)$$

The direct amplitudes involving v_3 are analogous to the distorted-wave Born approximation and are calculated by (10.31). The T-matrix element in the second amplitude of (10.51), which has the observed resonances, is calculated by solving the problem of electron scattering on He^+. The solution consists of half-on-shell T-matrix elements at the quadrature points for the scattering integral equations (6.87). The same points are used for the k integration of (10.51).

10.4.2 Comparison of theory and experiment

The electron–ion T-matrix elements were calculated in the coupled-channels approximation (6.87) using the lowest six states of He^+ (McCarthy and Shang, 1993). The four lowest resonances are compared with experiment in table 10.1. The calculation predicts the resonances very closely, so that we may consider the reaction as a test of the amplitudes used to describe direction ionisation. Fig. 10.11 shows the example $E_0=100$ eV, $\theta_f=13°$. The direct-resonant interference is strong at $\theta_s=42°$, but the influence of the direct amplitude is much reduced at $\theta_s = 24°$. Semiquan-

Table 10.1. *Comparison of energies (eV) and widths ($10^{-3}eV$) of experimental and calculated resonances below the $n=2$ threshold in the compound system of an electron and a He^+ ion. EXP, van den Brink et al. (1989); CC, nine coupled channels (McCarthy and Shang, 1993)*

	EXP		CC	
Symmetry	E_s	Γ	E_s	Γ
^1S	33.24	138±15	33.28	128
^3P	33.72	8	33.72	8
^1D	35.32	72±18	35.40	72
^1P	35.555	38±2	35.57	35

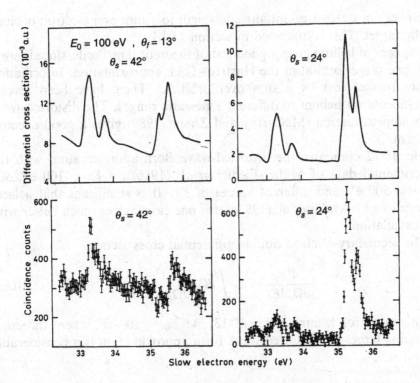

Fig. 10.11. Resonant ionisation of helium at $E_0 = 100\,\text{eV}$, $\theta_f = 13°$. Experimental data, Lower and Weigold (1990); full curve, calculation of (10.51) by McCarthy and Shang (1993). From McCarthy and Shang (1993).

titative agreement between the calculation and the experiment is achieved in both cases.

10.5 Integrated cross sections

Experiments that integrate over some of the kinematic variables of the differential cross section observe a wide kinematic range at the expense of losing information because of the integration.

10.5.1 Double differential cross section

The primary-electron double differential cross section is

$$\frac{d^3\sigma}{d\Omega_f dE_f} = \Sigma_i \int d\Omega_s \frac{d^5\sigma_i}{d\Omega_f d\Omega_s dE_f}. \qquad (10.52)$$

Here we have used a discrete notation i for the states of the residual ion, which implicitly includes states above the second-ionisation threshold. The sum over ion states is essentially equivalent to a sum over electron orbitals of the target. This is discussed in section 11.1.1.

The case of helium gives a good test of theoretical methods, since there is only one target orbital in the Hartree–Fock approximation. Information is not further lost by a sum over orbitals. There have been several experiments on helium in different kinematic ranges. The distorted-wave Born approximation (McCarthy and Zhang, 1989) gives a good account of them.

Fig. 10.12 compares the distorted-wave Born approximation with the experimental data of Müller-Fiedler *et al.* (1986) for $E_0 = 100$ eV, 300 eV and 500 eV and different values of E_f. It is significant that a later experiment by Avaldi *et al.* (1987a), for one case, agrees much better with the calculation.

The secondary-electron double-differential cross section

$$\frac{d^3\sigma}{d\Omega_s dE_s} = \Sigma_i \int d\Omega_f \frac{d^5\sigma_i}{d\Omega_f d\Omega_s dE_s} \tag{10.53}$$

is illustrated for helium in fig. 10.13. At $E_0 = 100$ eV, where the cross section is large, the distorted-wave Born approximation is a considerable

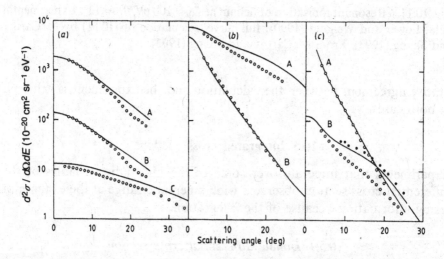

Fig. 10.12. Primary-electron double differential cross section for electron–helium ionisation. Experimental data are due to Müller-Fiedler *et al.* (1986) (open circles) and Avaldi *et al.* (1987a) (full circles). Full curves, distorted-wave Born approximation (McCarthy and Zhang, 1989). Cases illustrated are (a) $E_0 = 100$ eV, $E_f = 73.4$ eV(A), 71.4 eV(B), 55.4 eV(C); (b) $E_0 = 300$ eV, $E_f = 235.4$ eV (cross section multiplied by 100) (A), 271.4 eV(B); (c) $E_0 = 500$ eV, $E_f = 471.4$ eV(A), 435.4 eV(B). From McCarthy and Zhang (1989).

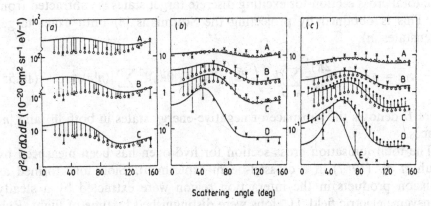

Scattering angle (deg)

Fig. 10.13. Secondary-electron double-differential cross section for electron-helium ionisation. Experimental data and curves are as for fig. 10.12 with Opal, Beaty and Peterson, (1972) (crosses). The value of E_s for an experimental point is indicated by a vertical line joining it to the corresponding curve. Cases illustrated are (a) $E_0 = 100$ eV, $E_s = 4$ eV(A), 10 eV(B), 20 eV(C); (b) $E_0 = 300$ eV, $E_s = 4$ eV(A), 20 eV(B), 40 eV(C), 100 eV(D); (c) $E_0 = 500$ eV, $E_s = 4$ eV(A), 20 eV(B), 40 eV(C), 102 eV(D), 205 eV(E). From McCarthy and Zhang (1989).

over-estimate, suggesting that this kinematic region could be responsible for an over-estimate of the total ionisation cross section. In higher-energy cases, where experimental data have been obtained by more than one group, the disagreement between the distorted-wave Born approximation and any one experiment is less than the disagreement between experiments.

10.5.2 Total ionisation cross section

The total ionisation cross section is a very important quantity in the study of electron–atom collisions. Not only does it give an overall test of theoretical methods for ionisation, but it is an essential check on the treatment of the complete set of target states in a calculation of scattering.

The convergent-close-coupling method (Bray and Stelbovics, 1993) is a complete calculation of scattering to low-lying channels in the prototype case of hydrogen. It represents the complete set of target states by a Sturmian expansion (5.55) of dimension M and calculates T-matrix elements by solving the coupled Lippmann–Schwinger equations (6.73). The total ionisation cross section is

$$\sigma_I = \tfrac{1}{4}\Sigma_S(2S + 1)\sigma_{IS}, \tag{10.54}$$

where σ_{IS} is the total ionisation cross section for total electron spin S. In each spin state the total ionisation cross section is calculated as follows. The total cross section σ_{TS} is calculated by the optical theorem (6.47).

The total cross section for exciting discrete target states is subtracted from it. This is obtained by projecting the Sturmians $|\bar{n}\rangle$ onto exact target eigenstates $|n\rangle$.

$$\sigma_{IS} = \sigma_{TS} - \frac{(2\pi)^4}{k_0} \sum_{\bar{n} \in D} k_{\bar{n}} \int d\hat{\mathbf{k}}_{\bar{n}} |\langle \mathbf{k}_{\bar{n}}\bar{n}| T_S |0\mathbf{k}_0\rangle|^2 \sum_{n \in D} |\langle n|\bar{n}\rangle|^2. \qquad (10.55)$$

Here D denotes the subspace of negative-energy states in both $|\bar{n}\rangle$ and $|n\rangle$ spaces.

The total ionisation cross section for hydrogen has been measured by Shah *et al.* (1987) in a crossed-beam experiment. Slow ions formed as collision products in the interaction region were extracted by a steady transverse electric field. H^+ ions were distinguished by time of flight. Relative cross sections were normalised to previously-measured cross sections for hydrogen ionisation by protons of the same velocity. The proton cross sections were normalised to the Born approximation at 1500 keV.

Fig. 10.14 shows that the convergent-close-coupling method describes the total ionisation cross section within experimental error for the whole energy range above total energy $E = 4$ eV. Just above threshold it underestimates the cross section by up to 30%.

Other scattering calculations that account for the complete target space can also be tested. The method (10.55) can be used for the pseudostate

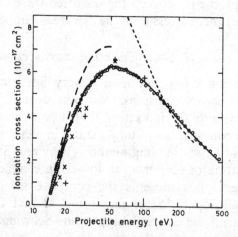

Fig. 10.14. Total ionisation cross section for hydrogen. Experimental data, Shah *et al.* (1987); full curve, convergent close coupling (Bray and Stelbovics, 1992b); plus signs, coupled channels optical (Bray *et al.*, 1991c), crosses, pseudostate method (Callaway and Oza, 1979); long-dashed curve, intermediate-energy R-matrix (Scholz *et al.*, 1990); short-dashed curve, distorted-wave Born approximation.

calculation of Callaway and Oza (1979). The total ionisation cross section for the coupled-channels-optical method is calculated by subtracting the sum of integrated cross sections for the converged P space from the total cross section calculated with a polarisation potential that describes the ionisation space. The results of both these calculations are shown in fig. 10.14. The coupled-channels-optical method (Bray *et al.* 1991c) is quite accurate for $E_0 > 40$ eV, but the results illustrated in chapters 8 and 9 show that discrepancies are insignificant over the whole energy range from the point of view of scattering calculations. The intermediate-energy *R*-matrix method in the implementation of Scholz, Walters and Burke (1990) is less accurate.

The distorted-wave Born approximation for ionisation considerably overestimates the total ionisation cross section for hydrogen below about 150 eV. This is a good indication of its lower limit of validity.

It is useful to test approximations for the total ionisation cross section of helium, since it is a common target for the scattering and ionisation reactions treated in chapters 8, 10 and 11. Fig. 10.15 compares the data reported as the experimental average by de Heer and Jansen (1977) with the distorted-wave Born approximation and the coupled-channels-optical calculation using the equivalent-local polarisation potential. Cross sections

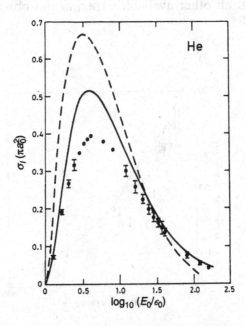

Fig. 10.15. Total ionisation cross section for helium. Experimental data, de Heer and Jansen (1977); full curve, coupled channels optical (equivalent local) (McCarthy and Stelbovics, 1983a); broken curve, distorted-wave Born approximation.

are overestimated by both approximations below 400 eV. The discrepancy at lower energies for the coupled-channels-optical method is about 20%.

10.6 Total ionisation asymmetry

The total ionisation asymmetry is defined by

$$A_I = \frac{\sigma_{I0} - \sigma_{I1}}{\sigma_I}. \tag{10.56}$$

It has been measured for hydrogen (Fletcher *et al.*, 1985; Crowe *et al.*, 1990) and for lithium, sodium and potassium (Baum *et al.*, 1985) at incident energies from threshold to several hundred electron volts. The data were obtained by ionisation of polarised target atoms by polarised electrons. The relative total ionisation cross sections for parallel and antiparallel spins were determined by counting the ions, regardless of the kinematics of the final-state electrons.

The experimental data for hydrogen are compared with calculations in fig. 10.16. Both the convergent-close-coupling and coupled-channels-optical methods come close to complete agreement with experiment. The total ionisation cross section is a more severe test of theory, since it is an absolute quantity, whereas the asymmetry is a ratio. However, the correct prediction of the asymmetry reinforces the conclusion, reached by comparison with all other available experimental observables, that these methods are valid.

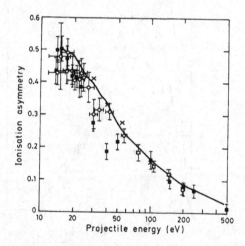

Fig. 10.16. Total ionisation asymmetry for hydrogen. Open circles, Fletcher *et al.* (1985); squares, Crowe *et al.* (1990); full curve, convergent close coupling (Bray and Stelbovics, 1992*b*); crosses, coupled channels optical (Bray *et al.*, 1991*c*).

11

Electron momentum spectroscopy

Some atomic bound states have simple structure in the sense that a straightforward calculation obtains correct energy levels. In some cases optical oscillator strengths probe further detail. Collision theory has reached the stage where experimental observables for electron collisions involving such states can be calculated within experimental error. Observables whose calculation is sensitive to structure details constitute a probe for structure which verifies the details in more-difficult cases.

Scattering experiments are usually not very sensitive to structure. On the other hand the differential cross section for ionisation in a kinematic region where the plane-wave impulse approximation is valid gives a direct representation (10.31) of the structure of simple targets in the form of the momentum-space orbital of a target electron.

Electron momentum spectroscopy (McCarthy and Weigold, 1991) is based on ionisation experiments at incident energies of the order of 1000 eV, where the plane-wave impulse approximation is roughly valid. The differential cross section is measured for each ion state over a range of ion recoil momentum p from about 0 to 2.5 a.u. Noncoplanar-symmetric kinematics is the usual mode. In such experiments the distorted-wave impulse approximation turns out to be a sufficiently-refined theory. Checks of this based on a generally-valid sum rule will be described.

The reaction depends as much on the observed state $|f\rangle$ of the residual ion as on the ground state $|0\rangle$ of the target. Not only the single-particle structure but electron correlations in each state are sensitively probed in different circumstances.

11.1 Basic theory

The T-matrix element for ionisation of an atom from the ground state $|0\rangle$ to an ion state $|f\rangle$ is understood by successively-more-detailed approximations that are capable of direct experimental verification. The first is the

binary-encounter approximation, which assumes that the collision operator T does not depend explicitly on the coordinates of the electrons in the
ion so that it commutes with the ion state. Introducing a complete set of
plane-wave states $|\mathbf{q}\rangle$ for the knocked-out electron the binary-encounter
approximation is written as

$$\langle \mathbf{k}_f \mathbf{k}_s f | T | 0 \mathbf{k}_0 \rangle = \int d^3 q \langle \mathbf{k}_f \mathbf{k}_s | T | \mathbf{q} \mathbf{k}_0 \rangle \langle \mathbf{q} f | 0 \rangle. \tag{11.1}$$

The kinematic variables are defined in chapter 10.

The approximation involves neglecting exchange terms for electrons of
the ion in orbitals $|\mu\rangle$. These terms have factors such as $\langle \mathbf{k} | \mu \rangle$, where $\mathbf{k}$ is
the momentum of one of the external electrons. Such terms are essentially
zero for $k > 4$ a.u. in the case of valence electrons. This gives 400 eV as the
minimum total energy for symmetric ionisation. Correspondingly-higher
energies are needed to probe inner-shell structure.

The amplitude $\langle \mathbf{k}_f \mathbf{k}_s | T | \mathbf{q} \mathbf{k}_0 \rangle$ may be considered as a probe amplitude
for the momentum-space structure amplitude $\langle \mathbf{q} f | 0 \rangle$. It is the structure
amplitude that is essentially probed by the reaction. Increasingly-refined
forms may be used for the probe amplitude. The plane-wave impulse
approximation is

$$\langle \mathbf{k}_f \mathbf{k}_s | T | \mathbf{q} \mathbf{k}_0 \rangle = A_S \langle \mathbf{k}' | t_\eta | \mathbf{k} \rangle \delta(\mathbf{q} - \mathbf{k}_f - \mathbf{k}_s + \mathbf{k}_0), \tag{11.2}$$

where the kinematic variables are defined by (10.35). The distorted-wave
impulse approximation is

$$\langle \mathbf{k}_f \mathbf{k}_s | T | \mathbf{q} \mathbf{k}_0 \rangle = A_S \langle \mathbf{k}' | t_\eta | \mathbf{k} \rangle \langle \chi^{(-)}(\mathbf{k}_f) \chi^{(-)}(\mathbf{k}_s) | \mathbf{q} \chi^{(+)}(\mathbf{k}_0) \rangle. \tag{11.3}$$

Further refinements would be given by comparing (11.1) with (10.13) or
(10.25), but (11.3) is all that is necessary in the selected kinematic region.

Electron momentum spectroscopy can therefore be considered in terms
of $\langle \mathbf{q} f | 0 \rangle$. For the one-electron model of the target

$$\langle \mathbf{q} f | 0 \rangle = \phi_\alpha(\mathbf{q}) \tag{11.4}$$

and the reaction is a direct probe of the momentum-space orbital.

In most experiments the target is not oriented. The differential cross
section is averaged over the solid angle $\hat{\mathbf{p}}$. In the distorted-wave impulse
approximation it is

$$\frac{d^5 \sigma}{d\Omega_f d\Omega_s dE_s} = (2\pi)^4 \frac{k_f k_s}{k_0} f_{ee} \frac{1}{4\pi}$$

$$\times \int d\hat{\mathbf{p}} \Sigma_m |\langle \chi^{(-)}(\mathbf{k}_f) \chi^{(-)}(\mathbf{k}_s) f | 0 \chi^{(+)}(\mathbf{k}_0) \rangle|^2, \tag{11.5}$$

where the symmetry of the ion state $|f\rangle$ is described by the quantum
numbers ℓ, j, m. In the plane-wave impulse approximation for the one-

electron model, where

$$\langle \mathbf{p}\sigma|\alpha\rangle = u_\alpha(p)\langle -\hat{\mathbf{p}}\sigma|\ell jm\rangle, \tag{11.6}$$

this reduces to

$$\frac{d^5\sigma}{d\Omega_f d\Omega_s dE_s} = (2\pi)^4 \frac{k_f k_s}{k_0} f_{ee} \frac{2j+1}{4\pi} |u_\alpha(p)|^2. \tag{11.7}$$

Target—ion structure is usually observed as a function of the separation energy ϵ, which is varied over a range relevant to the shells to be studied by changing E_0, keeping the total energy E constant.

$$E_0 = E + \epsilon. \tag{11.8}$$

For valence structure this range is up to about 50 eV. The momentum distribution is observed for each resolved cross-section peak corresponding to an ion eigenvalue $-\epsilon_f$. In order to characterise the observation of the target—ion structure we choose a quantity that is as independent as possible of the probe characteristics such as total energy. In conditions where the plane-wave impulse approximation is valid we consider the reaction as a perfect probe for the energy—momentum spectral function

$$S(\epsilon, q) = \int d\hat{\mathbf{q}} |\langle f|a(\mathbf{q})|0\rangle|^2 \delta(\epsilon - \epsilon_f). \tag{11.9}$$

From now on we write the structure amplitude more explicitly in second-quantised notation.

$$\langle \mathbf{q}f|0\rangle \equiv \langle f|a(\mathbf{q})|0\rangle. \tag{11.10}$$

The operator $a(\mathbf{q})$ annihilates an electron of momentum $\mathbf{q}$ in the target ground state.

In practice transfer of momentum to the ion by the mechanism described by distortion renders the probe imperfect. Experience has shown that the distorted-wave impulse approximation is sufficient to describe the distorted spectral function

$$S_D(\epsilon, q, E) = \frac{1}{2j+1} \Sigma_m \int d\hat{\mathbf{q}} |\langle \chi^{(-)}(\mathbf{k}_f)\chi^{(-)}(\mathbf{k}_s)f|0\chi^{(+)}(\mathbf{k}_0)\rangle|^2 \delta(\epsilon - \epsilon_f). \tag{11.11}$$

To a very good approximation S_D is independent of E_0 in noncoplanar-symmetric kinematics over the range of ϵ_f relevant to valence electrons, but it does change somewhat for large changes of E. We write the distorted-wave differential cross section as

$$\frac{d^5\sigma}{d\Omega_f d\Omega_s dE_s} = (2\pi)^4 \frac{k_f k_s}{k_0} f_{ee} \frac{2j+1}{4\pi} S_D(\epsilon, p, E). \tag{11.12}$$

11.1.1 The weak-coupling approximation

To understand the structure amplitude (11.10) we first consider the configuration-interaction expansion (section 5.6) of the target and ion states. For algebraic simplicity we choose the same orbital basis set $|\alpha\rangle$ for both states. By analogy with the hydrogen atom this must be a set of target orbitals, generated as described in chapter 5 by procedures to which we give the generic name *Hartree–Fock*.

In general the observed ion states can be grouped into symmetry manifolds, characterised by the quantum numbers ℓ, j. We consider each symmetry manifold separately. The configuration-interaction basis for the target consists of symmetry configurations $|r\rangle$, which are linear combinations, with symmetry ℓ, j, of determinants formed from the set of orbitals $|\alpha\rangle$.

The representation that we choose for expanding the ion states $|f\rangle$ is not the configuration-interaction representation but the weak-coupling representation. Here the orthonormal basis states $|j\rangle$ are linear combinations of configurations formed by annihilating an electron in a target eigenstate. The weak-coupling states are identical to the symmetry configurations if the target Hartree–Fock configuration is the only one contributing significantly to the target ground state. The approximation made by ignoring other contributions (target correlations) is the target Hartree–Fock approximation. The weak-coupling representation of the structure amplitude is

$$\langle f|a(\mathbf{q})|0\rangle = \Sigma_j \langle f|j\rangle\langle j|a(\mathbf{q})|0\rangle. \tag{11.13}$$

We now introduce the very important concept of a one-hole state. This is a state $|i\rangle$ of the weak-coupling basis $|j\rangle$ that is formed by annihilating an electron in the target ground state $|0\rangle$. Clearly $\langle j|a(\mathbf{q})|0\rangle$ in (11.13) is zero if $|j\rangle$ is not a one-hole state. A one-hole state $|\alpha\rangle$ is denoted by the orbital of the annihilated electron. In the context of the many-electron ion space this notation cannot be confused with that of the orbital.

The weak-coupling approximation is made by taking only the leading term $|\alpha\rangle$ of the expansion (11.13) in one-hole states. Its validity can be experimentally verified. In the overwhelming majority of cases it turns out to be valid. In this approximation

$$\langle f|a(\mathbf{q})|0\rangle = \langle f|\alpha\rangle\langle \alpha|a(\mathbf{q})|0\rangle. \tag{11.14}$$

The approximation generalises the target Hartree–Fock approximation to cases where target correlations are not negligible.

The spectral function (11.9) in the weak-coupling approximation is proportional to the square of the one-hole structure amplitude $\langle \alpha|a(\mathbf{q})|0\rangle$, averaged over the angles of $\mathbf{q}$. This amplitude is the momentum-space orbital $\phi_\alpha(\mathbf{q})$ for an uncorrelated ground state, and very close to it for weak coupling. The experiment determines an experimental orbital $\psi_\alpha(\mathbf{q})$, defined by

$$\psi_\alpha(\mathbf{q}) = \langle \alpha | a(\mathbf{q}) | 0 \rangle. \tag{11.15}$$

If the experiment requires the distorted-wave formulation (11.11), the observed momentum profile is distorted. It is still possible to extract $\psi_\alpha(\mathbf{q})$ by a statistical fitting procedure.

It is clear from (11.1,11.10,11.14) that the differential cross section in the weak-coupling binary-encounter approximation is proportional to the spectroscopic factor $S_f(\alpha)$, defined by

$$S_f(\alpha) = |\langle f | \alpha \rangle|^2. \tag{11.16}$$

We are now in a position to consider the consequences of the approximations that can be experimentally verified.

The manifold momentum profile

For a given total energy E we can identify a manifold of ion states $|f\rangle$ that all have a momentum profile of the same shape, given by (11.11,11.12,11.14). The shape is characteristic of an orbital $|\alpha\rangle$ of the target. The manifold is characterised not only by the symmetry, but by the set of quantum numbers α that includes a principal quantum number. We call it the orbital manifold α.

The spectroscopic sum rule

The sum of spectroscopic factors for the whole manifold α is

$$\Sigma_f S_f(\alpha) = \Sigma_f \langle \alpha | f \rangle \langle f | \alpha \rangle = 1. \tag{11.17}$$

The experimental orbital

The experiment observes a distorted spectral function (11.11) characterised by the experimental orbital $\psi_\alpha(\mathbf{q})$ (11.15). The summed cross sections for the α manifold are characterised by the normalised orbital $\psi_\alpha(\mathbf{q})$, in view of (11.17). We say that the orbital is split among the states $|f\rangle$ of the manifold by the ion correlations. The weight of each state in the manifold sum is $S_f(\alpha)$. The differential cross sections for the states of the manifold are in the ratios of these weights.

The experimental orbital energy

The orbital energy is the expectation value for the one-hole state $|\alpha\rangle$ of the ion Hamiltonian H_I.

$$\epsilon_\alpha = \langle \alpha | H_I | \alpha \rangle = \Sigma_f \langle \alpha | f \rangle \epsilon_f \langle f | \alpha \rangle = \Sigma_f S_f(\alpha) \epsilon_f. \tag{11.18}$$

It is the centroid of the energies of the states $|f\rangle$ of the manifold α, where the weights are the spectroscopic factors.

The existence of a common momentum profile for the manifold α confirms the weak-coupling binary-encounter approximation. Within these approximations we must make further approximations to calculate differential cross sections. For the probe amplitude of (11.1) we may make, for example, the distorted-wave impulse approximation (11.3). This enables us to identify a normalised experimental orbital for the manifold. If normalised experimental orbitals are used to calculate the differential cross sections for two different manifolds within experimental error this confirms the whole approximation to this stage. An orbital approximation for the target structure (such as Hartree–Fock or Dirac–Fock) is confirmed if the experimental orbital energy agrees with the calculated orbital energy and if it correctly predicts differential cross sections.

The spectroscopic factors are critical quantities in determining the accuracy of a configuration-interaction calculation of the structure of the ion. The ion state $|f\rangle$ is written in the weak-coupling representation as

$$|f\rangle = \Sigma_j |j\rangle \langle j|f\rangle. \tag{11.19}$$

The spectroscopic factor (11.16) is the absolute square of the coefficient of the one-hole state $|\alpha\rangle$ in this expansion. Determination of one coefficient in each of several eigenstates of H_I in the representation is a strong constraint on the calculation. Furthermore the determination of spectroscopic factors depends only on the validity of the weak-coupling

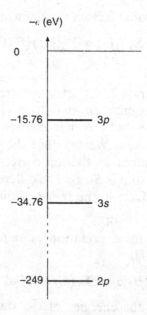

Fig. 11.1. Orbital energy levels of argon.

binary-encounter approximation and the identification of enough states of the one-hole manifold essentially to exhaust the sum rule (11.17). It does not depend on the validity of a model for the probe amplitude. The ratios of the spectroscopic factors are the ratios of the corresponding differential cross sections and they are normalised by the sum rule.

The valence structure of argon provides a complete illustration of the application of electron momentum spectroscopy to correlations in the ion. The Hartree–Fock single-electron level diagram of fig. 11.1 illustrates the values of the separation energy ϵ to be expected on the basis of the independent-electron model. The experimental situation is illustrated in fig. 11.2 by the first experiment in the field (Weigold, Hood and Teubner, 1973). The noncoplanar-symmetric differential cross section at 10° is plotted against E_0 for $E=400$ eV. There is an ion state at 15.76 eV, as predicted by Hartree–Fock, but there are at least two further states rather than the predicted one.

Fig. 11.3 illustrates the relative momentum profile of the 15.76 eV state in a later experiment at $E=1200$ eV, compared with the plane-wave impulse approximation with orbitals calculated by three different methods. The sensitivity of the reaction to the structure calculations is graphically illustrated. A single Slater-type orbital (4.38) with a variationally-determined exponent provides the worst agreement with experiment. The Hartree–Fock–Slater approximation (Herman and Skillman, 1963), in which exchange is represented by an equivalent-local potential, also disagrees. The Hartree–Fock orbital agrees within experimental error.

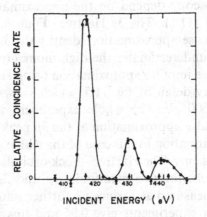

Fig. 11.2. Relative differential cross section at $\phi = 10°$ for the 400 eV noncoplanar-symmetric ionisation of argon (Weigold *et al.*, 1973). The arrows indicate known energy levels of Ar^+.

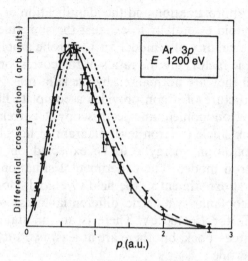

Fig. 11.3. The 1200 eV noncoplanar-symmetric momentum profile for the 15.76 eV state of Ar^+ (McCarthy and Weigold, 1988). Plane-wave impulse approximation curves are calculated with $3p$ orbitals. Full curve, Hartree—Fock (Clementi and Roetti, 1984); long-dashed curve, Hartree—Fock—Slater (Herman and Skillman, 1963); short-dashed curve, minimal variational basis.

Fig. 11.4 illustrates the momentum profiles of the other ion states observed in a later experiment with better energy resolution than that of fig. 11.2. All these states have momentum profiles of essentially the same shape. They are thus identified as states of the same orbital manifold, for which the experiment obeys the criterion for the validity of the weak-coupling binary-encounter approximation. Details of electron momentum spectroscopy depend on the approximation adopted for the probe amplitude of (11.1). The $3s$ Hartree—Fock momentum profiles in the plane-wave impulse approximation identify the $3s$ manifold. However, the approximation underestimates the high-momentum profile.

The distorted-wave impulse approximation using Hartree—Fock orbitals is confirmed in every detail by fig. 11.5, which shows momentum profiles for argon at $E=1500$ eV. The whole experiment is normalised to the distorted-wave impulse approximation at the $3p$ peak. It represents the remainder of the confirmation in this case of the whole procedure of electron momentum spectroscopy. The Hartree—Fock orbitals give complete agreement with experiment for two manifolds, $3p$ and $3s$. The spectroscopic factor $S_{15.76}(3p)$ is measured as 1, since no further states of the $3p$ manifold are identified. Later experiments give 0.95 and this is the value used for normalisation. The approximation describes the momentum-profile shape for the first member of the $3s$ manifold at 29.3 eV within experimental error. The shape for the manifold sum of cross sections agrees and its

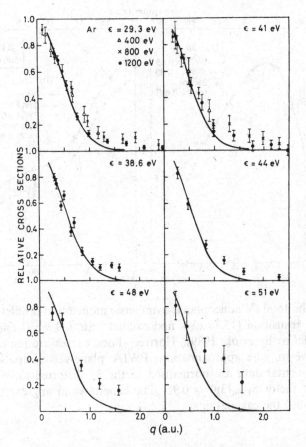

Fig. 11.4. Noncoplanar-symmetric momentum profiles at the indicated energies for the ionisation of argon to some more-strongly excited ion states above the ion ground state (Weigold and McCarthy, 1978). Full curve, plane-wave impulse approximation for the Hartree–Fock 3s orbital.

magnitude exactly exhausts the spectroscopic sum rule for $3s$ in comparison with $3p$. The plane-wave impulse approximation is distinctly less valid. Its profile shape agrees with experiment up to about 1 a.u., but $\Sigma_f S_f(3s)$ falls short at 0.76. The plane-wave impulse approximation must be multiplied by 0.83 to agree with the distorted-wave approximation at the $3p$ peak. This is because it neglects refraction in the continuum-electron wave functions. The present experimental spectroscopic factors are shown in table 11.1. The $3s$ orbital energy is $\epsilon_{3s} = 35.2 \pm 0.2$ eV, which is close to the Hartree–Fock value 34.76.

The spectroscopic information obtained for argon is firstly about orbitals. The concept of the experimental orbital is verified. Its shape and its energy agree closely with the Hartree–Fock calculations in both the $3p$

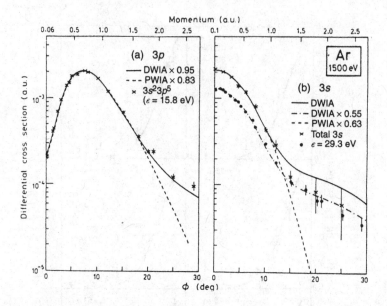

Fig. 11.5.   The 1500 eV noncoplanar-symmetric momentum profiles for the argon ground-state transition (15.76 eV), first excited state (29.3 eV) and the total $3s$ manifold (McCarthy *et al.*, 1989). Hartree—Fock curves are indicated: DWIA, distorted-wave impulse approximation; PWIA, plane-wave impulse approximation. Experimental data are normalised to the $3p$ distorted-wave curve with a spectroscopic factor $S_{15.76}(3p) = 0.95$. The experimental angular resolution has been folded into the calculations.

and $3s$ cases. Next the $3s$ spectroscopic factors of table 11.1 test structure calculations with correlated electrons. Configuration-interaction calculations of the target-ion overlap (11.10) by Mitroy, Amos and Morrison (1984) and Hibbert and Hanson (1987) show semiquantitative agreement with experiment. Amusia and Kheifets (1985, 1990) obtained much closer agreement using the perturbation theory of section 5.7.

Perturbation theory considers the residual electron—electron interaction as a perturbation on the independent-particle model for the ion, which splits the one-hole configuration among eigenstates of the one-hole manifold. A measure of the strength of the perturbation is the ratio of the standard deviation of the distribution of eigenstates, weighted by the spectroscopic factors, to the orbital energy. For the $3p$ and $3s$ manifolds of argon the ratios are 0 and 0.18 respectively. For this reason we consider the perturbation to be weak.

Table 11.1. *Spectroscopic factors for the 3s manifold of argon. EXP, McCarthy et al. (1989). The error in the last figure is given in parentheses. Target–ion, configuration interaction in the target and ion. Ion, configuration interaction in the ion only. Pert, perturbation theory*

Dominant ion state configuration	ϵ(eV)	EXP	Mitroy et al. (1984)		Hibbert and Hansen (1987)	Amusia and Kheifets (1985)
			Target–ion	Ion	Ion	Pert
$3s3p^6$	29.24	0.55(1)	0.649	0.600	0.618	0.553
$3p^44s$	36.50	0.02(1)	0.013	0.006	0.006	–
$3p^43d$	38.58	0.16(1)	0.161	0.142	0.112	0.199
$3p^44d$	41.21	0.08(1)	0.083	0.075	0.057	0.107
$3p^45d$	42.65				0.021	0.042
		} 0.08(1)	0.081	0.095		
$3p^46d$	43.40				0.009	0.021
$Ar^{++} + e$		0.12(1)	0.013	0.08	0.18	0.076

11.1.2 Ground-state correlations

Ion states that do not obey the weak-coupling approximation give information about electron correlations in the target ground state. The weak-coupling representation (11.13) of such states $|f\rangle$ has significant contributions only from basis states $|j\rangle$ for which the hole j is in an excited target configuration but not in the Hartree–Fock configuration. The structure amplitude is very sensitive to the coefficients $\langle f|j\rangle$ of excited configurations, i.e. to correlations.

A simple example is the $2s$ state of the helium ion. It has a small overlap with the $1s$ Hartree–Fock orbital of helium, since the Hartree–Fock potential of helium is not the same as the Coulomb potential of the helium ion. However, it has a large overlap with helium configurations that contain a $2s$ orbital. The $2s$ orbital is not occupied in the Hartree–Fock configuration.

Fig. 11.6 shows the noncoplanar-symmetric differential cross sections at 1200 eV for the $1s$ state and the unresolved $n=2$ states, normalised to theory for the low-momentum $1s$ points. Here the structure amplitude is calculated from the overlap of a converged configuration-interaction representation of helium (McCarthy and Mitroy, 1986) with the observed helium ion state. The distorted-wave impulse approximation describes the $1s$ momentum profile accurately. The summed $n=2$ profile does not have the shape expected on the basis of the weak-coupling approximation (long-dashed curve). Its shape and magnitude are given quite well by

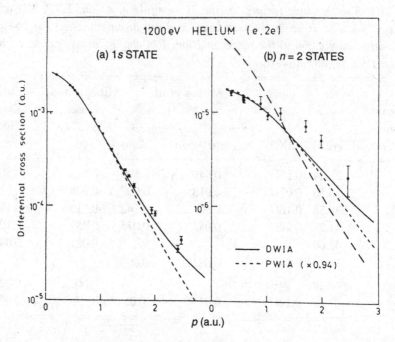

Fig. 11.6.   The 1200 eV noncoplanar-symmetric momentum profiles for the ground-state ($n = 1$) and summed $n = 2$ transitions in helium (Cook *et al.*, 1984). Curves indicated: DWIA, distorted-wave impulse approximation; PWIA, plane-wave impulse approximation. The curves are calculated using a converged configuration-interaction expansion (McCarthy and Mitroy, 1986) for the helium ground state. The long-dashed curve is the distorted-wave impulse approximation for the Hartree—Fock ground state. Experimental data are normalised to the $1s$ curve at low momentum. From McCarthy and Weigold (1991).

the calculation. The plane-wave impulse approximation again underestimates the high-momentum profile and is not quite accurate for relative magnitudes.

11.2 Examples of structure information

The principles derived and illustrated in section 11.1 show that electron momentum spectroscopy gives information about orbitals, about orbital manifolds that are split by electron correlations in the ion, and about correlations in the target ground state. We give examples of the kind of information that is obtained.

11.2.1 Argon

The dominant features of the argon ion spectrum observed in an ionisation reaction have illustrated the electron momentum spectroscopy of the $3p$

and 3s manifolds. However, argon is also an example of the detail that can be achieved by observing ion states with very small cross sections. Above the lowest energy 29.3 eV of the 3s manifold, ion states have been observed whose symmetry is known from photon spectroscopy to be $^2D^e$ and $^2P^o$. We adopt the convention of characterising a state by its leading configuration.

The $^2D^e$ states can only be observed if the argon ground state has excited configurations with an occupied d orbital. These states do not obey the weak-coupling approximation. The most prominent $^2D^e$ transition is to the $3s^23p^4(^1D)4s\ ^2D^e$ state at 34.20 eV. It is shown in fig. 11.7(a). The narrow peak at about $p=0.25$ a.u. is best described by the Hartree–Fock 4d orbital, which is diffuse in coordinate space. This peak could be

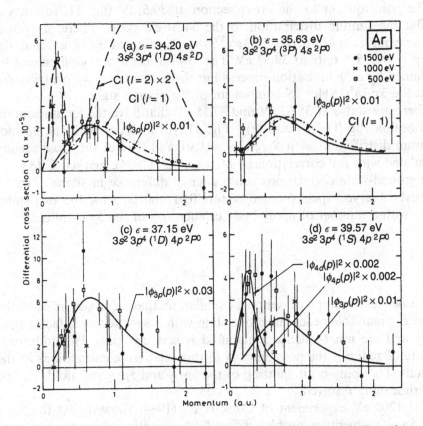

Fig. 11.7. Noncoplanar-symmetric momentum profiles at the indicated energies for the indicated transitions in argon, compared with calculated profiles (McCarthy *et al.*, 1989). Experimental data are normalised to the distorted-wave impulse approximation for the summed 3s manifold. Calculations are indicated by the square of a Hartree–Fock orbital multiplied by a spectroscopic factor. Configuration-interaction curves (CI) are described in the text.

accounted for by a 0.4% admixture of the configuration $3s^2 3p^4(^1D)4s4d\ ^1S$ in the ground state.

Mitroy *et al.* (1984) carried out an extensive configuration-interaction calculation of the structure amplitude $\langle qf|0 \rangle$ for correlated target and ion states. The long-dashed curve in fig. 11.7(a) shows their momentum distribution multiplied by 2. They found that the dominant contribution came from the pseudo-orbital $\overline{3d}$, calculated by the natural-orbital transformation. Pseudo-orbitals are localised to the same part of space as the occupied $3s$ and $3p$ Hartree–Fock orbitals and therefore contribute to the cross section at much higher momenta than the diffuse Hartree–Fock $3d$ and $4d$ orbitals. The measurements show that the $4d$ orbital has a larger weight than is calculated by Mitroy *et al.*, who overestimate the $\overline{3d}$ component.

The contribution to the cross section at 39.5 eV (fig. 11.7(d)) has a similar momentum distribution to the 34.20 eV state. There are possible contributions from the $3s^2 3p^4(^3P)4d\ ^2D^e$ state at 39.64 eV and the $3s^2 3p^4(^1S)4p\ ^2P^o$ state at 39.57 eV. The low-momentum region cannot be explained by a $3p$ ionisation process but there could be a $4p$ contribution from the $3s^2 3p^4(^1S)4p^2\ ^1S$ component of the ground state.

There are states at 35.63 eV and 37.15 eV that have the $3p$ momentum distribution (fig. 11.7(b) and (c)). Fig. 11.7(b) includes both the $3p$ momentum distribution with $S_{35.63}(3p) = 0.01$ and 0.67 of the cross section calculated with full correlation by Mitroy *et al.* (1984), marked CI($\ell = 1$). The ground-state correlations cause a small difference in shape. The respective observed spectroscopic factors 0.01 and 0.03 for the two states agree with a number of many-body calculations of the $3p$ manifold.

11.2.2 Xenon

The valence structure of xenon is similar to that of argon in that the valence p manifolds each have one state with a spectroscopic factor near unity and the inner-valence s manifold is severely split. The additional feature of xenon is the possibility of testing relativistic calculations of the orbitals. The spin–orbit splitting of the $5p_{3/2}$ and $5p_{1/2}$ manifolds can be experimentally resolved.

The 1200 eV experiment of Cook *et al.* (1984) showed that the $5p_{3/2}$ and $5p_{1/2}$ momentum profiles differed significantly. They are not consistent with nonrelativistic Hartree–Fock orbitals but can be described within experimental error by the distorted-wave impulse approximation using Dirac–Fock orbitals. The $5p_{3/2} : 5p_{1/2}$ branching ratio is shown in fig. 11.8, where it is compared with the distorted-wave impulse approximation using relativistic and nonrelativistic orbitals. The $5p_{3/2}$ orbital

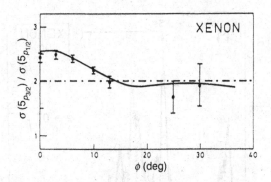

Fig. 11.8. The ratio of $5p_{3/2}$ to $5p_{1/2}$ noncoplanar-symmetric differential cross sections for xenon at 1000 eV (Cook *et al.*, 1986). Full curve, distorted-wave impulse approximation with Dirac–Fock orbitals; broken line, nonrelativistic branching ratio. From McCarthy and Weigold (1988).

has significantly more low-momentum components than the $5p_{1/2}$ orbital. The approximation is accurate over the observed momentum range, 0.1 to 2.2 a.u.

The spectroscopy of the $5p$ and $5s$ manifolds has been observed in great detail. Fig. 11.9 shows the separation energy spectra for $E=500$ eV at $\phi = 0°$ ($p = 0.16$) and $10°$ ($p = 0.6$). The $5s$ transitions are much stronger at low momentum than higher momentum and the $5p$ transitions are the opposite. The $5p_{3/2}$ and $5p_{1/2}$ manifolds are dominated by single transitions. Cook *et al.* (1986) obtained a spectroscopic factor 0.98 for the ground-state transition. A large-basis configuration-interaction calculation reported by these authors gave spectroscopic factors 0.980 and 0.983 at $p=0.5$ a.u. for the lowest $5p_{3/2}$ and $5p_{1/2}$ states respectively.

The $5s$ manifold shows great complexity. For the lowest state $S_{23.4}(5s)$ $= 0.37$. This value is considerably lower than many structure calculations predict, but the perturbation calculation of Kheifets and Amusia (1992) obtains 0.384. The orbital energy ϵ_{5s} (11.18) is 27.6 ± 0.3eV, which is to be compared to the Dirac–Fock value 27.49 eV. The Hartree–Fock value is 25.70 eV. The criterion for the strength of the perturbation, given by the ratio of the standard deviation to the mean of the $5s$ manifold is 0.18. The ratios $S_{29.1}(5s) : S_{23.4}(5s)$ and $S_{23.4}(5s) : \Sigma_f S_f(5s)$ are compared at different momenta in fig. 11.10. The condition for the validity of the weak-coupling binary-encounter approximation is completely satisfied within experimental error.

The complete understanding of the reaction is summarised by fig. 11.11, in which the 1000 eV $5p$ and $5s$ manifold sums (normalised at the $5p$ peak) are compared with the distorted- and plane-wave impulse approximations

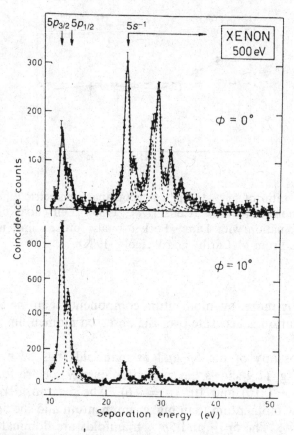

Fig. 11.9.   The 500 eV noncoplanar-symmetric valence energy spectra for xenon at $\phi = 0°$ and $10°$ (Cook *et al.*, 1986). The full curve is obtained by fitting peaks (broken curves) of the known experimental resolution function at known energy levels of Xe$^+$. From McCarthy and Weigold (1991).

for different momenta. It is necessary to use distorted waves to obtain profile shapes and relative magnitudes within experimental error, but plane waves predict the profile shape below $p=1$ a.u.

11.2.3 Lead

Lead illustrates relativistic orbitals and correlations in both target and ion states. The Dirac–Fock configuration has two electrons in each of the $6p_{1/2}$ and $6s_{1/2}$ orbitals. Frost, Mitroy and Weigold (1986) found that the strong transition to the $6p$ manifold has two components, described as $6p_{3/2}$ and $6p_{1/2}$ by a two-configuration calculation for the ground state, $6p_{3/2}^2$ and $6p_{1/2}^2$. Since the splitting is due to ground-state correlations the weak-coupling approximation is not obeyed and the branching ratio

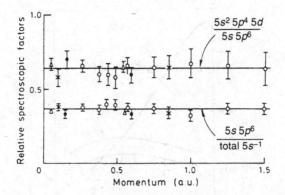

Fig. 11.10. Spectroscopic factors for the noncoplanar-symmetric ionisation of xenon to the $5s5p^6$ ^2S ($\epsilon=23.4$ eV) ion state (lower) and the ratio of spectroscopic factors for the $5s^25p^45d$ ^2S ($\epsilon=29.1$ eV) and 23.4 eV states (upper), plotted against recoil momentum p. Full circles, 500 eV; open triangles, 1000 eV; open circles, 1200 eV; crosses, 2000 eV. From McCarthy and Weigold (1991).

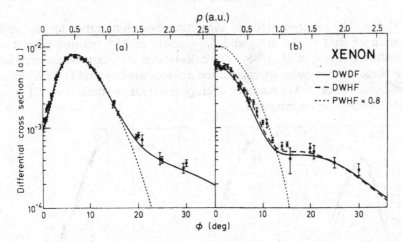

Fig. 11.11. The 1000 eV noncoplanar-symmetric momentum profiles for the summed (a) $5p$ and (b) $5s$ manifolds of xenon (McCarthy and Weigold, 1991). Distorted- and plane-wave impulse approximations are indicated respectively by DW and PW. Dirac–Fock and Hartree–Fock orbitals are indicated respectively by DF and HF. The experimental angular resolution has been folded into the calculation. The experimental data are normalised at the peak of the $5p$ profile.

$R(q)$ is momentum-dependent. Both the extended-average-level and the optimal-level multiconfiguration Dirac–Fock methods were tried. The results are illustrated in fig. 11.12. The sensitivity of electron momentum spectroscopy is shown by the fact that the former calculation is completely

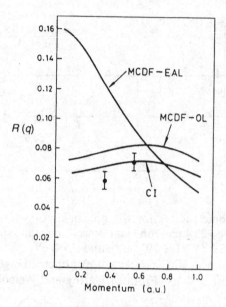

Fig. 11.12.   The experimental and theoretical branching ratios for the 1000 eV ionisation of lead to the $6p_{3/2}$ and $6p_{1/2}$ states of Pb^+, plotted against recoil momentum p (Frost *et al.*, 1986). The calculations with target-state correlations in the plane wave impulse approximation are indicated by: MCDF, multiconfiguration Dirac–Fock; EAL, extended average level; OL, optimal level. CI indicates ion-state configuration interaction.

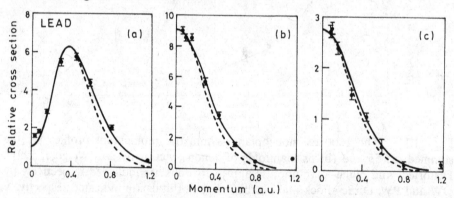

Fig. 11.13.   The 1000 eV noncoplanar-symmetric momentum profiles for lead (Frost *et al.*, 1986). Curves show the plane-wave impulse approximation. The experiment is normalised at the peak of the $6p$-manifold profile (a). The 14.6 eV and 18.4 eV states of the $6s$ manifold are indicated by (b) and (c). Spectroscopic factors are given in table 11.2. For (a), (b) and (c) respectively the Hartree–Fock calculation (broken curve) is normalised to multiconfiguration Dirac–Fock (solid curve) by factors 0.82, 0.70 and 0.64.

ruled out by the experiment. Agreement requires the optimal-level method, with ion correlations included in the structure amplitude $\langle \mathbf{q}f|0\rangle$.

Fig. 11.13(a) shows the summed momentum profiles for the states of the 6p manifold at 7.4 eV and 9.2 eV. Figs. 11.13(b) and (c) describe states that are identified by the plane-wave impulse approximation with the Dirac–Fock orbital as belonging to the 6s manifold. Since the valence states of lead are diffuse in coordinate space most of the momentum profile is within the 1 a.u. limit of validity of the plane-wave impulse approximation for the profile shape. The experiment agrees with the Dirac–Fock profile but rules out the nonrelativistic Hartree–Fock method.

The spectroscopic factors for the 6p and 6s manifolds are compared in table 11.2 with the relativistic calculations of Frost *et al.* (1986) that include target and ion correlations. Fig. 11.12 shows that states of the 6s manifold obey the weak-coupling approximation, so that their spectroscopic factors are momentum-independent.

11.3 Excited and oriented target states

Excited target atoms can be prepared in well-defined states by optical pumping with a tunable laser. Specific magnetic substates are excited by polarised light. Momentum distributions are observed for these states by electron momentum spectroscopy.

We consider sodium atoms pumped by σ^+ circularly-polarised laser light tuned to the $3^2S_{1/2}(F=2) \longleftrightarrow 3^2P_{3/2}(F'=3)$ transition. The quantum

Table 11.2. *Eigenvalues and spectroscopic factors for observed states of Pb$^+$ (Frost et al., 1986). The spectroscopic factors for the 6p manifold are evaluated at p=0.35 a.u. (upper) and p=0.55 a.u. (lower)*

	Experiment		Theory	
	ϵ_f (eV)	S_f	ϵ_f (eV)	S_f
6p manifold	7.42	0.944±0.01	7.10	0.907
		0.933±0.01		0.901
	9.16	0.056±0.006	8.76	0.062
		0.067±0.007		0.065
6s manifold	14.59	0.762±0.008	13.79	0.729
	18.35	0.227±0.005	18.34	0.214
	20.34	0.011±0.003	19.97	0.036

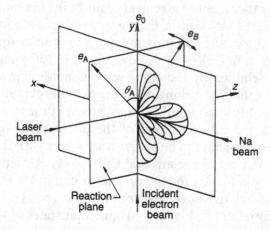

Fig. 11.14. Schematic diagram of the laser-excited noncoplanar-symmetric ion-isation of sodium (Zheng *et al.*, 1990). The $3p_x$ and $3p_y$ charge clouds are indicated. Final-state polar angles are θ_A and θ_B. From McCarthy and Weigold (1991).

numbers F and F' refer to the hyperfine transition structure due to the nuclear magnetic moment. Because of the high resolution of the laser it is important to select a particular hyperfine transition. For the purpose of *LS*-coupling atomic structure we consider only the orbital quantum numbers n, ℓ and m of the valence electron. The magnetic substate is $3p_1$.

The geometry of the experiment is shown in fig. 11.14. The axis of quantisation $\hat{\mathbf{z}}$ is the direction of the light beam. The atomic sodium beam is incident in the $\hat{\mathbf{x}}$ direction and the electron beam in the $\hat{\mathbf{y}}$ direction. The schematic diagram also shows the $3p_x$ and $3p_y$ lobes of the target charge cloud for the $3p_1$ substate. The scattering plane is the zy plane so that the component p_x of the recoil momentum **p** is observed in noncoplanar-symmetric kinematics. Because of the finite angular resolution of the spectrometer the components p_y and p_z are of the order of 0.06 a.u. rather than zero.

The momentum distributions for the $3s$ ground state and the $3p_1$ state are shown in fig. 11.15. They are compared with the momentum distri-butions calculated using Hartree—Fock orbitals and folding in the exper-imental momentum resolution function. Because the $3s$ and $3p$ orbitals are very diffuse in coordinate space the momentum profile is well within the $p=1$ limit of validity of the plane-wave impulse approximation.

The $3s$ momentum distribution confirms the validity of the approxi-mations. The experiment confirms the $3p_1$ Hartree—Fock momentum distribution and eliminates the distribution for the sum over magnetic sub-states that would apply to target atoms that are not oriented. With perfect

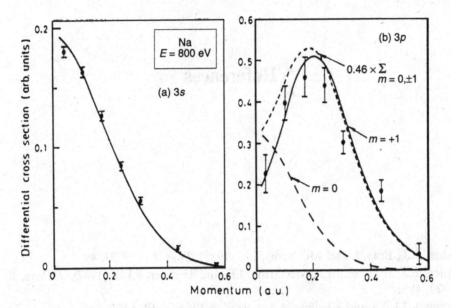

Fig. 11.15. The 800 eV noncoplanar-symmetric momentum profiles for the laser-assisted ionisation of sodium (Zheng *et al.*, 1990). Hartree—Fock curves for the indicated states are calculated in the plane-wave impulse approximation. From McCarthy and Weigold (1991).

angular resolution the $3p_0$ component would be entirely in the $\hat{z}$ direction and therefore would not be observed in the experiment. However in the present experiment it would increase the $p{=}0$ momentum components beyond the limits of statistical errors.

References

Allen, L.J., Bray, I. and McCarthy, I.E. (1988). *Phys. Rev. A* **37**, 49.

Allen, L.J., Brunger, M.J., McCarthy, I.E. and Teubner, P.J.O. (1987). *J. Phys. B* **20**, 4861.

Amusia, M. Ya and Kheifets, A.S. (1985). *J. Phys. B* **18**, L679.

Amusia, M. Ya and Kheifets, A.S. (1990). Unpublished.

Andersen, N., Gallagher, J.W. and Hertel, I. (1988). *Phys. Rep.* **165**, 1.

Anderson, E., Bai, Z., Bischof, C., Demmel, J., Dongarra, J., DuCroz, J., Greenbaum, A., Hammarling, S., McKenney, A., Ostrouchov, S. and Sorensen, D. (1992). *LAPACK Users' Guide* (Society for Industrial and Applied Mathematics, Philadelphia).

Andrick, D. and Bitsch, A. (1975). *J.Phys.B* **8**, 393.

Antosiewicz, H.A. (1973). In *Handbook of Mathematical Functions*, ed. M. Abramowitz and I.A. Stegun (Dover, New York).

Avaldi, L., Camilloni, R., Fainelli, E. and Stefani, G. (1987a). *Nuovo Cimento D* **19**, 97.

Avaldi, L., Camilloni, R., Fainelli, E., Stefani, G., Franz, A., Klar, H. and McCarthy, I.E. (1987b). *J. Phys. B* **20**, 5827.

Avaldi, L., Stefani, G., McCarthy, I.E. and Zhang, X. (1989). *J. Phys. B* **22**, 3079.

Awe, B., Kemper, F., Rosicky, F. and Feder, R. (1983). *J. Phys. B* **16**, 603.

Barnett, A.R., Feng, D.H., Steed, J.W. and Goldfarb, L.J.B. (1974). *Comp. Phys. Comm.* **8**, 377.

Bartschat, K. (1989). *Phys. Rep.* **180**, 1.

Bartschat, K. (1991a). *J. Phys. B* **24**, 4615.

Bartschat, K. (1991b). Unpublished.

Bartschat, K., Blum, K., Hanne, G.F. and Kessler, J. (1981). *J. Phys. B* **14**, 3761.

Bartschat, K. and Madison, D.H. (1988). *J. Phys. B* **21**, 2621.

Baum, G., Caldwell, C.D. and Schröder, W. (1980). *Appl. Phys.* **21**, 121.

Baum, G., Frost, L., Raith, W. and Sillmen, U. (1989). *J. Phys. B* **22**, 1677.

Baum, G., Moede, M., Raith, W. and Schröder, W. (1985). *J. Phys. B* **18**, 531.

Baum, G., Moede, M., Raith, W. and Sillmen, U. (1986). *Phys. Rev. Lett.* **57**, 1855.

Baum, G., Raith, W. and Schröder, W. (1988). *J. Phys. B* **21**, L501.

Beaty, E.C., Hesselbacher, K.H., Hong, S.P. and Moore, J.H. (1977). *J. Phys. B* **10**, 611.

Bederson, B. (1968). *Meths. Exp. Phys.* **7A**, 67.

Bederson, B. (1969). *Comm. At. Mol. Phys.* **1**, 41, 65.

Bederson, B. (1970). *Comm. At. Mol. Phys.* **2**, 160.

Bederson, B. (1973). In *Atomic Physics 3*, ed. S.J. Smith and G.K. Walters. (Plenum, New York).

Bederson, B. and Kieffer, L.J. (1971). *Rev. Mod. Phys* **43**, 601.

Beijers, J.P.M., Madison, D.H., van Eck, J. and Heideman, H.G.M. (1987). *J. Phys. B* **20**, 167.

Berger, O. and Kessler, J. (1986). *J. Phys. B* **19**, 3539.

Berger, O., Wübker, W., Möllenkamp, R. and Kessler, J. (1982). *J. Phys. B* **15**, 2473.

Bergmann, K. (1988). In *Atomic and Molecular Beam Methods Vol.I*, ed. G. Scoles (Oxford University Press, Oxford).

Bickert, P., Hink, W., Dal Cappello, C. and Lahmam-Bennani, A. (1991). *J. Phys. B* **24**, 4603.

Blaauw, H.J., Wagenaar, R.W., Barends, D.H. and de Heer, F.J. (1980). *J. Phys. B* **13**, 359.

Bloch, C. (1957). *Nucl. Phys* **4**, 503.

Blum, K. (1981). *Density Matrix Theory and Applications* (Plenum, New York).

Blum, K. and Kleinpoppen, H. (1979). *Phys. Rep.* **52**, 203.

Blum, K. (1985). NATO ASI Series B, *Phys.* **134**, 103.

Bonn, J., Huber, G., Kluger, H.J., Otten, E.W. and Lode, D. (1975). *Z. Phys. A* **272**, 375.

Born, M. and Wolf, E. (1970). *Principles of Optics* (Pergamon, Oxford).

Brauner, M., Briggs, J.S. and Klar, H. (1989). *J. Phys. B* **22**, 2265.

Bray, I., Fursa, D.V. and McCarthy, I.E. (1993). *Phys. Rev. A* **47**, 1101.

Bray, I., Konovalov, D.A. and McCarthy, I.E. (1991*a*). *Phys. Rev. A* **43**, 1301.

Bray, I., Konovalov, D.A. and McCarthy, I.E. (1991*b*). *Phys. Rev. A* **43**, 5878.

Bray, I., Konovalov, D.A. and McCarthy, I.E. (1991*c*). *Phys. Rev. A* **44**, 5586.

Bray, I., Konovalov, D.A. and McCarthy, I.E. (1991*d*). *Phys. Rev. A* **44**, 7179.

Bray, I., Madison, D.H. and McCarthy, I.E. (1990). *Phys. Rev. A* **41**, 5916.

Bray, I. and McCarthy, I.E. (1992). *Phys. Rev. A* **46**, .

Bray, I., McCarthy, I.E., Mitroy, J. and Ratnavelu, K. (1989). *Phys. Rev. A* **39**, 4998.

Bray, I. and Stelbovics, A.T. (1992*a*). *Phys. Rev. Lett.* **61**, 53.

Bray, I. and Stelbovics, A.T. (1992*b*). *Phys. Rev. A* **46**, 6995.

Bray, I. and Stelbovics, A.T. (1993). *Phys. Rev. Lett.* **70**, 746.

Breit, G. and Wigner, E.P. (1936). *Phys. Rev.* **49**, 519; *ibid*, 642.

Brink, D.M. and Satchler, G.R. (1971). *Angular Momentum* (Clarendon, Oxford).

Brinkmann, R.T. and Trajmar, S. (1981). *J.Phys.E* **14**, 245.

Brown, G.E. (1952). *Phil. Mag.* **43**, 467.

Brunger, M.J., McCarthy, I.E., Ratnavelu, K., Teubner, P.J.O., Weigold, A.M., Zhou, Y. and Allen, L.J. (1990). *J. Phys. B* **23**, 1325.

Brunger, M.J., Riley, J.L., Scholten, R.E. and Teubner, P.J.O. (1988). *J. Phys. B* **21**, 1639.

Buckman, S.J. and Teubner, P.J.O. (1979). *J. Phys. B* **12**, 1741.

Buckman, S.J., Gulley, R.J., Moghbelalhossein, M. and Bennett, S.J. (1993). *Meas. Sci. Technol.* **4**, 1143.

Burke, P.G. and Mitchell, J.F. (1973). *J. Phys. B* **6**, 320.

Burke, P.G., Noble, C.J. and Scott, M.P. (1987). *Proc. R. Soc. Lond. A* **410**, 289.

Burke, P.G. and Robb, W.D. (1975). *Adv. At. Mol. Phys.* **11**, 144.

Bussert, W. (1986). *Z. Phys. D* **1**, 321.

Byron, F.W. Jr., Joachain, C.J. and Potvliege, R.M. (1982). *J. Phys. B* **15**, 3916.

Callaway, J. (1978). *Phys. Rep.* **45**, 89.

Callaway, J. (1982). *Phys. Rev. A* **26**, 199.

Callaway, J. and Oza, D.H. (1979). *Phys. Lett.* **72A**, 207.

Callaway, J. and Unnikrishnan, K. (1989). *Phys. Rev. A* **40**, 1660.

Callaway, J. and Williams, J.F. (1976). *Phys. Rev. A* **12**, 2312.

Campbell, D.M. Hermann, C., Rampel, G. and Owen, R. (1985). *J. Phys. E* **18**, 664.

Cartwright, D.C., Csanak, G., Trajmar, S. and Register, D.F. (1992). *Phys. Rev. A* **45**, 1602.

Chan, M., Crowe, D.M., Lubell, M.S., Tang, F.C., Vasilakis, A., Mulligan, F.J. and Slevin, J. (1988). *Z. Phys. D* **10**, 393.

Chen, J.C.Y. and Chen, A.C. (1972). *Adv. At. Mol. Phys.* **8**, 71.

Chernysheva, L.V., Cherepkov, N.A. and Radojevic, V. (1976). *Comp. Phys. Comm.* **11**, 57.

Clementi, E. and Roetti, C. (1974). *At. Dat. Nucl. Dat. Tab.* **14**, 177.

Cohen, E.R. and Taylor, B.N. (1987). *Rev. Mod. Phys.* **59**, 1121.

Coleman, P.G. (1979). *J. Phys. E: Sci. Instr.* **12**, 590.

Conroy, H. (1967). *J. Chem. Phys.* **47**, 5307.

Cook, J.P.D., McCarthy, I.E., Mitroy, J. and Weigold, E. (1986). *Phys. Rev. A* **33**, 211.

Cook, J.P.D., McCarthy, I.E., Stelbovics, A.T. and Weigold, E. (1984). *J. Phys. B* **17**, 2339.

Crowe, D.M., Guo, X.Q., Lubell, M.S., Slevin, J. and Eminyan, M. (1990). *J. Phys. B* **23**, L325.

Curran, E.P. and Walters, H.R.J. (1987). *J. Phys. B* **20**, 337.

Curran, E.P., Whelan, C. and Walters, H.R.J. (1991). *J. Phys. B* **24**, L19.

Cvejanović, S. and Read, F.H. (1974). *J. Phys. B* **10**, 1841.

de Heer, F.J. and Jansen, R.H.J. (1977). *J. Phys. B* **10**, 3741.

Dehmelt, H.G. and Jefferts, K.B. (1962). *Phys. Rev.* **12S**, 1318.

Dirac, P.A.M. (1928). *Proc. R. Soc. Lond. A* **117**, 610.

Dirac, P.A.M. (1958). *The Principles of Quantum Mechanics (4th edition)* (Clarendon, Oxford).

Dreves, W., Jänsch, H., Koch, E. and Fick, D. (1983). *Phys. Rev. Lett.* **50**, 1759.

Dreves, W., Kamke, W., Broermann, W. and Fick, D. (1981). *Z. Phys. A* **303**, 203.

Düren, R. (1988). In *Atomic and Molecular Beam Methods Vol.II*, ed. G. Scoles (Oxford University Press, Oxford).

Ehrhardt, H., Schulz, M., Tekaat, T. and Willmann, K. (1969). *Phys. Rev. Lett.* **22**, 89.

Eminyan, M., MacAdam, K.B., Slevin, J., Standage, M.C. and Kleinpoppen, H. (1974). *J. Phys. B* **7**, 1519.

Enemark, E.A. and Gallagher, A. (1972). *Phys. Rev. A* **6**, 162.

Ertmer, W., Blatt, R., Hall, J.L. and Zhu, M. (1985). *Phys. Rev. Lett.* **54**, 985.

Fano, U. (1969). *Phys. Rev.* **178**, 131.

Fano, U. and Lin, C.D. (1975). *Atomic Phys.* **4**, 47.

Fano, U. and Macek, J.H. (1973). *Rev. Mod. Phys.* **45**, 553.

Fano, U. and Rau, A.R.P. (1986). *Atomic Collisions and Spectra* (Academic, Orlando).

Farrell, P.M., MacGillivray, W.R. and Standage, M.C. (1988). *Phys. Rev. A* **37**, 4240.

Feneuille, G. and Jacquinot, P. (1981). *Adv. At. Mol. Phys.* **17**, 99.

Ferch, J., Granitza, B., Masche, C. and Raith, W. (1985). *J. Phys. B* **18**, 967.

Feshbach, H. (1962). *Ann. Phys. (N.Y.)* **19**, 287.

Fischer, A. and Hertel, I.V. (1982). *Z. Phys. A* **304**, 103.

Fletcher, G.D., Alguard, M.J., Gay, T.J., Hughes, V.W., Wainwright, P.F., Lubell, M.S. and Raith, W. (1985). *Phys. Rev. A* **31**, 2854.

Fletcher, G.D., Gay, T.J. and Lubell, M.S. (1986). *Phys. Rev. A* **34**, 911.

Fock, V. (1930). *Z. Phys.* **61**, 126; *ibid.* **62**, 795.

Fon, W.C., Aggarwal, K.M. and Ratnavelu, K. (1992). *J. Phys. B* **25**, 2625.

Ford, W.F. (1964). *Phys. Rev. B* **133**, 1616.

Fraser, G.W. and Mathieson, E. (1981). *Nucl. Instr. Meth.* **180**, 597.

Froese-Fischer, C. (1975). *Can. J. Phys.* **53**, 184, 338.

Froese-Fischer, C. (1977). *The Hartree–Fock Method for Atoms* (Wiley, New York).

Froese-Fischer, C. (1978). *Comp. Phys. Comm.* **14**, 145.

Froese-Fischer, C. (1979). *J. Opt. Soc. Am.* **69**, 118.

Frost, L., Mitroy, J. and Weigold, E. (1986). *J. Phys. B* **19**, 4063.

Fry, E.S. and Williams, W.L. (1969). *Rev. Sci. Instr.* **40**, 141.

Garwin, E.L., Pierce, D.T. and Siegmann, H.C. (1974). *Helv. Phys. Acta* **47**, 393.

Gell-Mann, M. and Goldberger, M.L. (1953). *Phys. Rev.* **91**, 398.

Gerlach, W. and Stern, O. (1924). *Ann. Phys.* **74**, 673.

Gibson, J.R. and Dolder, K.T. (1969). *J. Phys. B* **2**, 741.

Goeke, J., Hanne, G.F., Kessler, J. and Wolcke, A. (1983). *Phys. Rev. Lett.* **51**, 2273.

Goeke, J., Hanne, G.F. and Kessler, J. (1988). *Phys. Rev. Lett.* **61**, 58.

Goeke, J., Hanne, G.F. and Kessler, J. (1989). *J. Phys. B* **22,** 1075.

Goerss, H-J., Nordbeck, R-P. and Bartschat, K. (1991). *J. Phys. B* **24,** 2833.

Gorunganthu, R.R. and Bonham, R.A. (1986). *Phys. Rev. A* **34,** 103.

Granitza, B., Guo, X., Hurn, J., Shen, Y. and Weigold E. (1993). (Abstracts XVIII ICPEAC, Aarhus, Denmark, p201).

Grant, I.P. (1970). *Adv. Phys.* **19,** 747.

Grant, I.P., McKenzie, B.J., Norrington, P.H., Mayers, D.F. and Pyper, N.C. (1980). *Comp. Phys. Comm.* **21,** 207.

Gray, L.G., Hart, M.W., Dunning, F.B. and Walters, G.K. (1984). *Rev. Sci. Instr.* **55,** 88.

Haberland, R., Fritsche, L. and Noffke, J. (1986). *Phys. Rev. A* **33,** 2305.

Hanne, G.F. (1976). *J. Phys. B* **9,** 805.

Hanne, G.F. (1983). *Phys. Rep.* **95,** 95.

Hanne, G.F. (1992). In *Correlations and Polarization in Electronic and Atomic Collisions and (e,2e) Reactions,* ed. P.J.O. Teubner and E. Weigold (IOP, Bristol).

Hanne, G.F., Szmytkowski, Cz. and van der Wiel, M. (1982). *J. Phys. B* **15,** L109.

Happer, W. (1972). *Rev. Mod. Phys.* **44,** 169.

Harting, E. and Read, F.H. (1976). *Electrostatic Lenses* (Elsevier, Amsterdam).

Hartree, D.R. (1927). *Proc.Camb.Phil.Soc.* **24,** 89, 111.

Hawkes, P.W. and Kasper, E. (1988). *Principles of·Electron Optics* (Academic, N.Y.).

Hawley-Jones, T.J., Read, F.H., Cvejanović, S., Hammond, P. and King, G.C. (1992). *J. Phys. B* **25,** 2393.

Heck, L. and Williams, J.F. (1987). *J. Phys. B* **20,** 2871.

Heddle, D.W.O. and Keesing, R.G.W. (1968). *Adv. At. Mol. Phys.* **4,** 267.

Heinzmann, U. (1987). *Physica Scripta T* **17,** 77.

Heller, E.J., Reinhardt, W.P. and Yamani, H.A. (1973). *J. Comp. Phys.* **13,** 536.

Herman, F. and Skillman, S. (1963). *Atomic Structure Calculations* (Prentice-Hall, New York).

Hermann, H.W. and Hertel, I.V. (1980). *J. Phys. B* **13,** 4285.

Hermann, H.W. and Hertel, I.V. (1982). *Comm. At. Mol. Phys.* **12,** 61.

Hertel, I.V. and Stoll, W. (1974). *J. Phys. B* **7,** 570,583.

Hertel, I.V. and Stoll, W. (1978). *Adv. Atom. Molec. Phys.* **13,** 113.

Hertel, I.V., Schmidt, H., Bähring, A. and Meyer, E. (1985). *Rep. Prog. Phys.* **48,** 375.

Hertel, I.V., Kelley, M.H. and McClelland, J.J. (1987). *Z. Phys. D* **6,** 163.

Hibbert, A. (1975). *Comp. Phys. Comm.* **9,** 141.

Hibbert, A. and Hansen, J.E. (1987). *J. Phys. B* **20,** L245.

Hodge, L.A., Moravec, T.J., Dunning, F.B. and Walters, G.K. (1979). *Rev. Sci. Instr.* **50,** 5.

Hood, S.T., Weigold, E. and Dixon, A.J. (1979). *J. Phys. B* **12,** 621.

Hughes, V.W., Long, R.L., Jr., Lubell, M.S., Posner, M. and Raith, W. (1972). *Phys. Rev. A* **5,** 195.

Huxley, L.G.H. and Crompton, R.W. (1974). *The Diffusion and Drift of Electrons in Gases* (Wiley, New York).

Iannotta, S. (1988). In *Atomic and Molecular Beam Methods Vol.II*, ed. G. Scoles (Oxford University Press, Oxford).

Imhof, R.E. and Read, F.R. (1977). *Rep. Prog. Phys.* **40**, 1.

Jones, S., Madison, D.H. and Srivastava, M.K. (1992). *J. Phys. B* **25**, 1899.

Jung, K., Müller-Fiedler, R., Schlemmer, P., Ehrhardt, M. and Klar, H. (1985). *J. Phys. B* **16**, 2955.

Karstensen, F. and Schneider, M. (1975). *Z. Phys. A* **273**, 321.

Kastler, A. (1950). *J. Phys. Radium* **11**, 225.

Kelley, M.H., McClelland, J.J., Lorentz, S.R., Scholten, R.E. and Celotta, R.J. (1992). In *Correlations and Polarization in Electronic and Atomic Collisions and (e,2e) Reactions*, ed. P.J.O. Teubner and E. Weigold (IOP, Bristol).

Kelly, F.M. and Mathur, M.S. (1980). *Can. J. Phys.* **58**, 1980.

Kennerly, R.E. and Bonham, R.A. (1978). *Phys. Rev. A* **17**, 1844.

Kessler, J. (1985). *Polarized Electrons (2nd edition)* (Springer-Verlag, Berlin).

Kessler, J. (1991). *Adv. At. Mol. Phys.* **27**, 81.

Kheifets, A.S. and Amusia, M.Ya. (1992). *Phys. Rev. A* **46**, 1261.

Kim, Y.K. (1983). *Phys. Rev. A* **18**, 656.

Kingston, A.E. and Walters, H.R.J. (1980). *J. Phys. B* **13**, 4633.

Klar, H., Roy, A.C., Schlemmer, P., Jung, K. and Ehrhardt, H. (1987). *J. Phys. B* **20**, 821.

Kohn, W. and Sham, L.J. (1965). *Phys. Rev.* **140A**, 1133.

Konovalov, D.A. (1993). Private communication.

Kuize, R.J., Wu, Z. and Happler, W. (1988). *Adv. At. Molec. Phys.* **24**, 223.

Kumar, K. (1984). *Phys. Rep.* **112**, 319.

Kuyatt, C.E. and Simpson, J.A. (1967). *Rev. Sci. Instrum.* **38**, 103.

Kwan, C.K., Kauppila, W.E., Lukaszew, R.A., Parikh, S.P., Stein, T.S., Wan, Y.J. and Dababneh, M.S. (1991). *Phys. Rev. A* **44**, 1620.

Lahmam-Bennani, A., Duguet, A., Wellenstein, H.F. and Roualt, M. (1980). *J. Chem. Phys.* **72**, 6398.

Lahmam-Bennani, A., Cherid, M. and Duguet, A. (1987). *J. Phys. B* **20**, 2531.

Lahmam-Bennani, A., Wellenstein, H.F., Duguet, A. and Lecas, M. (1985). *Rev. Sci. Instr* **56**, 43.

Lanczos, C. (1950). *J. Res. NBS* **45**, 255.

Land, J.E. and Raith, W.R. (1974). *Phys. Rev. A* **9**, 1592.

Lassetre, E.N., Skeberle, A. and Dillon, M.A. (1969). *J. Chem. Phys.* **50**, 1829.

Leckey, R.C.G. (1987). *J. Electron Spectrosc.* **43**, 183.

Leep, D. and Gallagher, A. (1976). *Phys. Rev. A* **13**, 148.

Leung, K.T. and Brion, C.E. (1985). *J. Electron Spectrosc.* **35**, 327.

Liljeby, L., Lindgard, A., Mannervik, S., Veje, E. and Jelenkovic, B. (1980). *Physica Scripta* **21**, 805.

Lohmann, B. and Weigold, E. (1981). *Phys. Lett.* **86A**, 139.

Lorentz, S.R. and Miller, T.M. (1991). Unpublished.

Lorentz, S.R., Scholten, R.E., McClelland, J.J., Kelley, M.H. and Celotta, R.J. (1991). *Phys. Rev. Lett.* **67**, 761.

Löwdin, P-O. (1955). *Phys. Rev.* **97**, 1474.

Lower, J., McCarthy, I.E. and Weigold, E. (1987). *J. Phys. B* **20**, 4571.

Lower, J. and Weigold, E. (1989). *J. Phys. E* **22**, 421.

Lower, J. and Weigold, E. (1990). *J. Phys. B* **23**, 2819.

Luke, Y.L. (1973). In *Handbook of Mathematical Functions,* ed. M. Abramowitz and I.A. Stegun (Dover, New York).

Lundin, L., Engman, B., Hilke, J. and Martinson, I. (1973). *Physica Scripta* **8**, 274.

Lurio, A. (1964). *Phys. Rev. A* **136**, 376.

Macek, J.H. and Hertel, I.V. (1974). *J. Phys. B* **7**, 2173.

Madison, D.H., Bartschat, K. and McEachran, R.P. (1992). *J.Phys.B* **25**, 5199.

Madison, D.H., Bray, I. and McCarthy, I.E. (1991). *J. Phys. B* **24**, 3861.

Madison, D.H. and Callaway, J. (1987). *J. Phys. B* **20**, 4197.

Madison, D.H., McCarthy, I.E. and Zhang, X. (1989). *J. Phys. B* **22**, 2041.

Madison, D.H., Winters, K.H. and Downing, S.L. (1989). *J. Phys. B* **22**, 1651.

Mahan, A.H., Gallagher, A. and Smith, S.J. (1976). *Phys. Rev. A* **13**, 156.

Martin, C., Jelensky, P., Lampton, M. and Malina, R.F. (1981). *Rev. Sci. Instr.* **52**, 1067.

Maruyama, T., Garwin, E.L., Prepost, R. and Zapalac, G.H. (1992). *Phys. Rev. B* **46**, 4261.

McAdams, R., Hollywood, M.T., Crowe, A. and Williams, J.F. (1980). *J. Phys. B* **13**, 3961.

McCarthy, I.E. and Mitroy, J. (1986). *Phys. Rev. A* **34**, 4426.

McCarthy, I.E., Pascual, R., Storer, P.J. and Weigold, E. (1989). *Phys. Rev. A* **40**, 3041.

McCarthy, I.E., Ratnavelu, K. and Zhou, Y. (1991). *J. Phys. B* **24**, 4431.

McCarthy, I.E. and Roberts, M.J. (1987). *J. Phys. B* **20**, L231.

McCarthy, I.E. and Shang, B. (1992). *Phys. Rev. A* **46**, 3959.

McCarthy, I.E. and Shang, B. (1993). *Phys. Rev. A* **47**, 4807.

McCarthy, I.E. and Stelbovics, A.T. (1983a). *Phys. Rev. A* **28**, 1322.

McCarthy, I.E. and Stelbovics, A.T. (1983b). *Phys. Rev. A* **28**, 2693.

McCarthy, I.E. and Weigold, E. (1976). *Phys. Rep.* **C27**, 275.

McCarthy, I.E. and Weigold, E. (1983). *Am. J. Phys.* **51**, 152.

McCarthy, I.E. and Weigold, E. (1988). *Rep. Prog. Phys.* **51**, 299.

McCarthy, I.E. and Weigold, E. (1991). *Rep. Prog. Phys.* **54**, 789.

McCarthy, I.E. and Zhang, X. (1989). *J. Phys. B* **22**, 2189.

McCarthy, I.E. and Zhang, X. (1990). *Aust. J. Phys.* **43**, 291.

McClelland, J.J., Buckman, S.J., Kelley, M.H. and Celotta, R.J. (1990). *J. Phys. B* **23**, L21.

McClelland, J.J. and Kelley, M.H. (1985). *Phys. Rev. A* **31**, 3704.

McClelland, J.J., Kelley, M.H. and Celotta, R.J. (1985). *Phys. Rev. Lett.* **55**, 688.

McClelland, J.J., Kelley, M.H. and Celotta, R.J. (1986). *Phys. Rev. Lett.* **56**, 1362.

McClelland, J.J., Kelley, M.H. and Celotta, R.J. (1987). *Phys. Rev. Lett.* **58**, 2198.

McClelland, J.J., Kelley, M.H. and Celotta, R.J. (1989). *Phys. Rev. A* **40**, 2321.

McConkey, J.W., van der Burgt, P.J.M. and Corr, J.J. (1992). In *Correlations and Polarization in Electronic and Atomic Collisions and (e,2e) Reactions*, ed. P.J.O. Teubner and E. Weigold (IOP, Bristol).

McEachran, R.P. and Stauffer, A.D. (1986). *J. Phys. B* **19**, 3523.

Merzbacher, E. (1970). *Quantum Mechanics (2nd edition)* (Wiley, New York).

Metcalf, H. and van der Straten, P. (1994). *Phys. Rep.* **244**, 203.

Mitchell, C.J. (1975). *J. Phys. B* **8**, 25.

Mitroy, J. (1983). Studies in atomic structure and (e,2e) reactions. University of Melbourne Ph.D. thesis.

Mitroy, J., Amos, K.A. and Morrison, I. (1979). *J. Phys. B* **12**, 1081.

Mitroy, J., Amos, K.A. and Morrison, I. (1984). *J. Phys. B* **17**, 1659.

Mitroy, J. and McCarthy, I.E. (1989). *J. Phys. B* **22**, 641.

Möllenkamp, R., Wübker, W., Berger, O., Jost, K. and Kessler, J. (1984). *J. Phys. B* **17**, 1107.

Moore, C.E. (1949). *Atomic Energy Levels (NBS circular No. 467, vol.1)* (U.S. government printing office, Washington).

Moore, J.H., Coplan, M.A., Skillman, L. Jr. and Brooks, E.D. III (1978). *Rev. Sci. Instr.* **49**, 463.

Morgan, L.A. and McDowell, M.R.C. (1977). *Comm. At. Mol. Phys.* **7**, 123.

Mott, N.F. (1929). *Proc. R. Soc. A* **124**, 425.

Müller-Fiedler, R., Jung, K. and Ehrhardt, H. (1986). *J. Phys. B* **19**, 1211.

Murray, A.J., Turton, B.C.H. and Read, F.H. (1992). *Rev. Sci. Inst.* **63**, 3349.

Nakanishi, T., Aoyagi, H., Horinaka, H., Kamiya, Y., Kato, T., Nakamura, S., Saka, T. and Tsubata, M. (1991). *Phys. Lett. A* **158**, 345.

Nickel, J.C., Imre, K., Register, D.F. and Trajmar, S. (1985). *J. Phys. B* **18**, 125.

Nickich, V., Hegemann, T., Barsch, M. and Hanne, G.F. (1990). *Z. Phys. D* **23**, 261.

Oberhettinger, F. (1973). In *Handbook of Mathematical Functions*, ed. M. Abramowitz and I.A. Stegun (Dover, New York).

Oda, N. (1975). *Rad. Res.* **64**, 80.

O'Malley, T.F., Spruch, L. and Rosenberg, L. (1961). *J. Math. Phys.* **2**, 491.

Opal, C.B., Beaty, E.C. and Peterson, W.K. (1972). *At. Data* **4**, 209.

Oza, D.H. and Callaway, J. (1983). *Phys. Rev. A* **27**, 2840.

Pan, C. and Starace, A.F. (1991). *Phys. Rev. Lett.* **67**, 185.

Parkes, W., Evans, K.D. and Mathieson, E. (1974). *Nucl. Instr. Meth.* **121**, 151.

Pathak, A., Kingston, A.E. and Berrington, K.A. (1988). *J. Phys. B* **21**, 939.

Pauly, H. (1988). In *Atomic and Molecular Beam Methods Vol.I*, ed. G. Scoles (Oxford University Press, Oxford).

Peterkop, R. (1971). *J. Phys. B* **4**, 513.

Pierce, D.T., Celotta, R.J., Wang, G.C., Unertl, W.N., Galip, A., Kuyatt, C.E. and Mielczarek, S.R. (1980). *Rev. Sci. Instr.* **51**, 478.

Poet, R. (1978). *J. Phys. B* **11**, 3081.

Poet, R. (1981). *J. Phys. B* **14**, 91.

Pollaczek, F. (1950). *Compt. Rend. Acad. Sci. (Paris)* **230**, 1563.

Racah, G. (1942). *Phys. Rev.* **62**, 438.

Raith, W.R. (1976). *Adv. At. Mol. Phys.* **12**, 281.

Ramsauer, C. (1921). *Ann. Phys.* **64**, 513; *ibid.* **66**, 546.

Ramsauer, C. and Kollath, R. (1932). *Ann. Phys.* **12**, 529, 837.

Ramsey, N.F. (1956). *Molecular Beams* (Clarendon Press, Oxford).

Rapp, D. and Englander-Golden, P. (1965). *J. Chem. Phys.* **43**, 1464.

Rau, A.R.P. (1971). *Phys. Rev. A* **4**, 207.

Read, F.H. (1985). In *Electron Impact Ionization,* ed. T.D. Mark and G.H. Dunn (Springer-Verlag, Berlin).

Read, F.H., Comer, J., Imhof, R.E., Brunt, J.N.H. and Harting, E. (1974). *J. Electron Spectrosc.* **4**, 293.

Register, D.F., Trajmar, S. and Srivastava, S.K. (1980). *Phys. Rev. A* **21**, 1134.

Register, D.F., Vušković, L. and Trajmar, S. (1986). *J. Phys. B* **19**, 1685.

Reuss, J. (1988). In *Atomic and Molecular Beam Methods Vol.I,* ed. G. Scoles (Oxford University Press, Oxford).

Riddle, T.W., Onellion, M., Dunning, F.B. and Walters, G.K. (1981). *Rev. Sci. Instr.* **52**, 797.

Robb, W.D. (1974). *J. Phys. B* **7**, 1006.

Roothaan, C.C.J. (1960). *Rev. Mod. Phys.* **32**, 179.

Roothaan, C.C.J. and Bagus, P. (1963). *Methods in Computational Physics, Vol. 2.* (Academic Press, New York).

Rösel, T., Bär, R., Jung, K. and Ehrhardt, H. (1989). *Proc. 2nd Eur. Conf. on (e,2e) Collisions and Related Problems* (Kaiserslautern, unpublished).

Rösel, T., Röder, J., Frost, L., Jung, K. and Ehrhardt, H. (1992). *J. Phys. B* **25**, 3859.

Rosenberg, L. (1973). *Phys. Rev. D* **8**, 1833.

Rotenberg, M. (1962). *Ann. Phys. (N.Y.)* **19**, 262.

Rotenberg, M., Bivins, R., Metropolis, N. and Wooten, J.K. Jr. (1959). *The 3-j and 6-j Symbols* (Technology Press, MIT Cambridge).

Rudd, M.E. (1991). *Phys. Rev. A* **44**, 1644.

Rutherford, E. (1911). *Phil. Mag.* **21**, 669.

Sams, W.N. and Kouri, D.J. (1969). *J. Chem. Phys.* **51**, 4809.

Schiff, L. (1955). *Quantum Mechanics (2nd edition)* (McGraw-Hill, New York).

Schlemmer, P., Rösel, T., Jung, K. and Ehrhardt, H. (1989). *Phys. Rev. Lett.* **63**, 252.

Schnetz, M. and Sandner, W. (1992). In *Correlations and Polarization in Electronic and Atomic Collisions and (e,2e) Reactions,* ed. P.J.O Teubner and E. Weigold (IOP, Bristol).

Scholz, T.T., Scott, M.P. and Burke, P.G. (1988). *J. Phys. B* **21**, L139.

Scholz, T.T., Walters, H.R.J. and Burke, P.G. (1990). *J. Phys. B* **23**, L467.

Scholz, T.T., Walters, H.R.J., Burke, P.G. and Scott, M.P. (1991). *J. Phys. B* **24**, 2097.

Scoles, G. (1988). In *Atomic and Molecular Beam Methods Vol.I,* ed. G. Scoles (Oxford University Press, Oxford).

Seaton, M.J. (1973). *Comp. Phys. Comm.* **6**, 245.

Seiler, G.J., Oberoi, R.S. and Callaway, J. (1971). *Phys. Rev. A* **3**, 2006.

Selles, P., Huetz, A. and Mazeau, J. (1987). *J. Phys. B* **20**, 5195.

Sevior, K.D. (1972). *Low Energy Electron Spectroscopy* (Wiley-Interscience, New York).

Shah, M.B., Elliott, D.S. and Gilbody, H.B. (1987). *J. Phys. B* **20**, 3501.

Slevin, J. (1984). *Rep. Prog. Phys.* **47**, 461.

Smith, W.H. and Liszt, H.S. (1971). *J. Opt. Soc. Am.* **61**, 938.

Smith, W.W. and Gallagher, A. (1966). *Phys. Rev. A* **145**, 26.

Sohn, M. and Hanne, G.F. (1992). *J. Phys. B* **25**, 4627.

Soper, J. (1989). *CERN Program Library* (CERN, Geneva).

Srivastava, S.K. and Vušković, L. (1980). *J. Phys. B* **13**, 2633.

Stefani, G., Camilloni, R. and Giardini-Guidoni, A. (1978). *Phys. Lett.* **6A**, 364.

Stelbovics, A.T. (1990). *Phys. Rev. A* **41**, 2536.

Stelbovics, A.T. (1991). *Aust. J. Phys.* **44**, 241.

Stelbovics, A.T. and Bransden, B.H. (1989). *J. Phys. B* **22**, L451.

Storer, P., Caprari, R.S., Clark, S.A.C. and Weigold, E. (1994). *Rev. Sci. Instr.* **65**, 2214.

Tate, J.T. and Smith, P.T. (1932). *Phys. Rev.* **39**, 270.

Temkin, A. (1962). *Phys. Rev* **126**, 130.

Teubner, P.J.O., Buckman, S.J. and Noble, C.J. (1978). *J. Phys. B* **11**, 2345.

Teubner, P.J.O., Furst, J.E., Tonkin, M.C. and Buckman, S.J. (1981). *Phys. Rev. Lett.* **46**, 1569.

Teubner, P.J.O., Riley, J.L., Brunger, M.J. and Buckman, S.J. (1986). *J. Phys. B* **19**, 3313.

Toffoletto, F., Leckey, R.C.G. and Riley, J. (1985). *Nuc. Inst. Meth. B* **12**, 282.

Townsend, J.S. and Bailey, V.A. (1922). *Phil. Mag* **43**, 593.

Trajmar, S. and Register, D.F. (1984). In *Electron—Molecule Collisions,* ed. I. Shimamura and K. Takayanagi (Plenum Press, New York).

Unguris, J., Pierce, D.T. and Celotta, R.J. (1986). *Rev. Sci. Instr.* **57**, 1314.

van den Brink, J.P., Nienhuis, G., van Eck, J. and Heideman, H.G.M. (1989). *J. Phys. B* **22**, 3501.

van Wingerden, B., Kimman, J.T.N., van Tilburg, M. and de Heer, F.J. (1981). *J. Phys. B* **14**, 2475.

van Wingerden, B., Kimman, J.T.N., van Tilburg, M., Weigold, E., Joachain, C.J., Piraux, B. and de Heer, F.J. (1979). *J. Phys. B* **12**, L627.

van Wingerden, B., Weigold, E., de Heer, F.J. and Nygaard, K.J. (1977). *J. Phys. B* **10**, 1345.

van Wyngaarden, W.L. and Walters, H.R.J. (1986). *J. Phys. B* **19**, 929.

Wagenaar, R.W. and de Heer, F.J. (1985). *J. Phys. B* **18**, 2021.

Wallace, S.J. (1973). *Ann. Phys. (N.Y.)* **78**, 190.

Wannier, G.H. (1953). *Phys. Rev.* **90**, 817.

Warner, C.D., King, G.C., Hammond, P. and Slevin, J. (1986). *J. Phys. B* **19**, 3297.

Weigold, E. (1993). *J. de Physique IV* **3C6**, 187.

Weigold, E., Frost, L. and Nygaard, K.J. (1979). *Phys. Rev. A* **21**, 1950.

Weigold, E., Hood, S.T. and Teubner, P.J.O. (1973). *Phys. Rev. Lett.* **30**, 475.

Weigold, E. and McCarthy, I.E. (1978). *Adv. At. Mol. Phys.* **14**, 127.

Weigold, E., Zheng, Y. and von Niessen, W. (1991). *Chem. Phys.* **150**, 405.

Weiss, A.W. (1967). *J. Chem. Phys.* **47**, 3573.

Weiss, A.W. (1974). *Phys. Rev. A* **9**, 1524.

Weyhreter, M., Barzick, B., Mann, A. and Linder, F. (1988). *Z. Phys. D* **7**, 333.

Whitehead, A., Watt, B.J., Cole, B.J. and Morrison, I. (1977). *Adv. Nucl. Phys.* **9**, 123.

Wigner, E.P. and Eisenbud, L. (1947). *Phys. Rev.* **72**, 29.

Williams, J.F. and Willis, B.A. (1975). *J. Phys. B.* **8**, 1641.

Williams, J.F. (1975). *J. Phys. B* **8**, 1683; *ibid.*, 2191.

Williams, J.F. (1976a). *J. Phys. B* **9**, 1518.

Williams, J.F. (1976b). In *Electron and Photon Interactions with Atoms,* ed. H. Kleinpoppen and M.R.C. McDowell (Plenum, New York).

Williams, J.F. (1981). *J. Phys. B* **14**, 1197.

Williams, J.F. (1986). *Aust. J. Phys.* **39**, 621.

Williams, J.F. (1988). *J. Phys. B* **21**, 2107.

Williams, J.F. and Crowe, A. (1975). *J. Phys. B* **8**, 2233.

Williams, J.F. and Heck, L. (1988). *J. Phys. B* **21**, 1627.

Williams, W. and Trajmar, S. (1978). *J. Phys. B* **11**, 2021.

Wolcke, A., Goeke, J., Hanne, G.F., Kessler, J., Vollmer, W., Bartschat, K. and Blum, K. (1984). *Phys. Rev. Lett.* **52**, 1108.

Wollnik, H. (1987). *Optics of Charged Particles* (Academic, New York).

Wübker, W., Möllenkamp, R. and Kessler, J. (1982). *Phys. Rev. Lett.* **49**, 272.

Yamani, H.A. and Reinhardt, W.P. (1975). *Phys. Rev. A* **11**, 1144.

Zetner, P.W., Trajmar, S. and Csanak, G. (1990). *Phys. Rev. A* **41**, 5980.

Zhang, X., Whelan, C.T., Walters, H.R.J., Allan, R.J., Bickert, P., Hink, W. and Schönberger, S. (1992). *J. Phys. B* **25**, 4325.

Zheng, Y., McCarthy, I.E., Weigold, E. and Zhang, D. (1990). *Phys. Rev. Lett.* **64**, 1358.

Zhou, H.L., Whitten, B.L., Snitchler, G. and Norcross, D.W. (1990). *Phys. Rev. A* **42**, 3907.

Zhou, Y. (1992). Unpublished.

Index

alignment 200, 206, 212
alignment parameters 208
alkali-metal atoms
  frozen-core approximation 125
  optical pumping 44
  scattering experiments 226, 248
amplitude
  coordinate space 59
  momentum space 59
  transition 59
analysers electron energy 15
analysing power 237
analytic orbitals 85, 87, 123, 129
angular correlation, electron–photon
  measurements 211
  parameters 206, 208, 210
angular momentum
  coefficients 65–7
  orbital 62, 82, 91, 116
  spin 64, 164, 166
  total 64, 116, 127, 164, 166
  transferred in collision 200, 203,
    206, 212, 240, 248
  vector addition 64
antisymmetrization 72, 157, 263
argon
  target for ionisation 267, 273, 276
  valence structure of ion 295–302
asymmetry (spin) 237

closed-shell targets 255
  exchange 238
  fine-structure 240, 251
  function in scattering 243–6, 251
  lithium 247
  mercury 257
  sodium 248–50
  total ionisation 288
asymmetry parameter 237, 243, 246,
  248
asymptotic scattering region 93
atomic units: definition 2
atoms
  sources 39
  detectors 43
autoionising resonances 279–83

basis
  configuration interaction 128, 131,
    136, 178, 230
  Hartree–Fock 129, 130, 292
  Laguerre 129, 179
  magnesium 136
  magnetic substates 203
  multiconfiguration 127
  one-body Schrödinger equation 86
  orbitals for Hartree–Fock 123
  pseudostates 179, 195–6
  representation 52

Slater-type orbitals 87, 195
weak coupling 292
binary-encounter approximation 290, 294
Born approximation 12, 20, 101, 103, 152
boundary condition
two-body 89, 91, 100
three-body 264, 266
box
normalisation of collision states 142–50
wave functions 82
Breit–Rabi hyperfine energy level diagram 41
Breit–Wigner resonance formula 106

cascades 11, 229, 233
central potentials 56, 62, 82, 87
channel
continuum 140–1
discrete 140–1
Hamiltonian 140, 263
state 140, 164, 263
charge-cloud distribution 200, 203, 254, 256, 259
charged target 96, 152
Clebsch–Gordan coefficients 65
closure approximation 193
closure theorem 52
coherence 201–2, 207
coincidence measurements
electron–electron 22–31
electron–photon 45
collision
box normalisation 144, 146
Hamiltonian 140, 263
state 141, 264
T-matrix element 145, 151, 264
wave packet 143
collisional
alignment 200
orientation 200, 203, 239, 249

collision system (frame) 203, 241
commutation rules
angular momentum operators 62, 63
canonically-conjugate observables 51
creation and annihilation operators 75, 118, 160, 175–7
commuting observables 51
configuration interaction 128–33, 196
argon ion 299
basis for collision theory 178
basis for helium 231
basis for ionisation 292
basis for magnesium 136, 232
ionisation 292
m-scheme 133
practical calculations 130
symmetry basis 132
xenon ion 303
configurations independent particle 72–3, 116, 292
convergent close coupling 178–9, 214, 285, 288
coordinate representation 55–6, 98, 195
correlations of bound electrons 292–4, 299, 303, 304
Coulomb
functions 92
multipole expansion 69, 170
phase shift 93, 97
potential 56
scattering amplitude 96, 153
T-matrix element 269–71
wave 95–6, 104
coupled-channels method 178, 195, 197, 212, 215, 227
coupled-channels-optical method 182, 212–34, 247–50
equivalent local 194, 230–4
total ionisation asymmetry 288
total ionisation cross section 286

creation and annihilation operators 73–5, 169
crossed beam technique 9, 17–20
cross section
  absolute determination 19, 30
  definition of 5, 8
  differential 14, 25, 148, 149
  double differential 22
  integral 6, 8, 11
  ionisation 24–32
  momentum transfer 6, 12
  normalization procedures 20, 31
  phase shift analysis of 20
  total 7, 8
  total ionisation 24
  triple differential 25

delta function 53, 56
density matrix
  magnetic substates spin-dependent 241
  magnetic substates spin-independent 202, 205
  m-scheme 161, 169, 171, 173, 174–7
  orbital space 131
  reduced 170, 171–3
  relating orbitals to channels 161, 169, 174–7
  special cases 174–7
  spin 242
density of final states 146, 147, 149
differential cross section 5, 14, 23–5
  absolute (e, 2e) 30
  double 22, 283
  helium 230
  hydrogen 212
  ionisation 25, 30, 149, 263, 269, 291
  magnesium 231
  polarised electrons 242
  potential scattering 88–90
  scattering 148, 205
  single 23
  sodium 225

spin-dependent scattering 238
time-dependent scattering 111
triple 25, 30
unpolarised electrons 242
diffusion coefficients 12
Dirac equation
  central potential 78
  free electron 77
Dirac–Fock
  energy for xenon ion 303
  multiconfiguration 127, 305
  orbitals in ionisation 294, 302, 306
  problem 122, 125
Dirac matrices 78
direct product space 51, 65
distorted wave 89, 97, 114
  Born approximation 154, 191–3, 231, 266, 275, 276, 286
  impulse approximation 272, 290, 296, 299, 305
  representation 152–5, 174, 263
  second Born 192, 196, 249
  transformation 152
  unitarised DW Born approximation 191
distorting potential 152, 154, 263
double-differential cross section 284

eigenvectors 51
electron
  coincidence spectrometer 26
  detectors 15, 27
  optics 15, 26
  photon correlations 45–9
  polarisation 32–9
  sources 15
  spectrometers 16, 25
  swarm technique 12
electron impact coherence parameters 212, 231
electron–photon correlations 45–9
  with polarised electrons 253
energy-loss spectra 17

energy of electron beam
  resolution 16
  selection 15
equivalent local potential 194
exchange
  amplitudes 159, 238, 246, 263, 290
  asymmetry 239
  polarisation mechanism 238
  potential 159, 161–3, 172
  scattering from spin 1/2 targets 246
Feshbach projection operators 180
final-state interaction 273–4
fine-structure effect 239, 251
first-order many-body theory 231
Fourier transformation 56
frame of reference
  atomic collision natural 203
  relativistic electron 78
frozen-core Hartree–Fock 125, 136,
    225, 247
function theorem 54, 98

Gallium arsenide
  photo-electron source 33
geometry effects scattering 17
golden rule (Fermi) 146
Green's function 98, 150, 181, 184,
    185, 192
ground-state correlations 299, 301,
    304

Hartree–Fock
  basis 129, 130, 292
  configurations for argon 296
  configurations for magnesium 232
  energy for argon ion 297
  energy for xenon ion 303
  equation 119–21
  frozen core 125, 136, 225, 247
  multiconfiguration 126
  nonrelativistic 121
  orbitals for argon 296, 301
  orbitals for lead 307

orbitals for sodium 308
  orbitals for xenon 302
  potential 119, 160
  problem 116–20
  relativistic 122 *see also* Dirac–Fock
  unperturbed states 133
Hartree–Fock–Slater 296
Heisenberg equation of motion 79
helium
  autoionisation 280–3
  configuration interaction 231, 299
  double-differential cross section 284
  ground-state correlations 299
  target for ionisation 231, 280–3,
    271, 274, 284, 299
  target for scattering 230–1
  total ionisation cross section 287
hydrogen
  Laguerre representation 129
  observation of 1s orbital 59
  orbitals 55, 85, 87, 129
  target for ionisation 59, 274, 286,
    288
  target for scattering 212–24
  total ionisation asymmetry 288
  total ionisation cross section 286
Hyperfine interactions 40

Impulse approximation 268–72, 290,
    291, 297
  see also distorted wave
integral equations
  box-normalised collision 144
  charged target 103, 153–4
  collision 144, 264
  coordinate space 99
  momentum space 100, 151
  numerical solution 102
  potential scattering 98, 101
  P-projected 182
  reduced 102, 165, 167
  relativistic 114
integrated channel cross sections

hydrogen 215–18
magnesium 232
sodium 228–9

jj coupling 66, 67, 127, 164

kinetic energy 56, 77
K-matrix element 94

Laguerre
  basis 129, 179
  polynomials 85
laser radiation state selection 42
lead 304
Legendre polynomials 63
limits in scattering theory 142–3,
    149–51
Lippmann–Schwinger equation 100,
    151, 154, 165, 167, 264
lithium 247
LS coupling 67, 127, 166, 172, 173

magnesium
  spectroscopy 136, 231
  target for scattering 231–4
magnetic fields
  inhomogeneous for states selection
    40
  shielding 14, 25
magnetic substate scattering
    parameters 205, 206, 207, 209–12
  helium 230
  hydrogen 218–20
magnetic substates in ionisation 307
magnets
  dipole 40
  hexapole 40
mercury 254
metastable states selection 44
momentum
  amplitude 59
  probability distribution 59
  representation 55, 56, 59
  transfer 57, 263

momentum profile 294, 296–8, 305,
    306, 309
momentum representation 3, 55, 56,
    59
  bound states 290
  potential scattering 100
monochromators 15
Mott scattering 33
  Mott analyser 34–9
multichannel calculation of ionisation
    266
multichannel expansion 151, 157, 161
multiconfiguration Dirac–Fock 127,
    305
multiconfiguration Hartree–Fock 126,
    232
multipole parameters 207

natural orbitals 130, 133, 232
nonlocal potential 56, 125, 159, 180
normalisation 58
  collision states 142, 145
  continuum states 89
  cross section measurements 20
  scattering cross sections 216, 226,
    234
  spectroscopic factors 293

observable 51, 52
occupation number representation 73,
    117
off-shell amplitudes 101
one-hole manifold 293
one-hole state 292
operators 51
  time development 58, 79, 142
optical excitation functions 11
optical potential
  equivalent-local 194
  formal 180–2
  pseudostates 196
  weak-coupling 183–6
optical pumping 42–5, 201

optical theorem 97, 146, 215
orbital energy 262, 293, 297, 303
orbital manifold 293, 297
orbitals 72, 116
  analytic 87, 123, 128
  experimental 293
  Hartree–Fock 120, 123, 292, 307
  hydrogenic 85
  Laguerre 129, 179
  natural 130, 133, 232
  numerical 120
  relativistic 120, 122, 302, 307
  Slater-type 87, 179, 195
  Sturmians 129, 179, 195
orientation 200, 203, 206, 211, 248
oscillator strength 115, 138, 229, 232

parity 63, 112, 116, 128, 164, 166, 202
partial-wave expansion
  Coulomb wave 96, 104
  integral equation 102, 165, 167
  plane wave 91
  potential scattering function 91
  relativistic scattering function 114
  scattering amplitude 20, 96
  T-matrix element 102, 165, 167
  V-matrix element 102, 165, 167
Pauli exclusion principle 71
Pauli matrices 64, 79
perturbation theory for bound states
  133, 298, 303
perturbation theory (time-dependent)
  146
phase of a scattering function 95
phase shift 20, 94, 97, 100, 105
photon helicities 254
photon spectroscopy 115, 137
polarisation potential
  bound state core 136, 225, 250
  channel excitation 180, 252
  long-range dipole 186, 189
  reduced matrix elements 184–6
  T-matrix element 189

polarisation (spin)
  atoms 39-45
  combined mechanisms 241
  correlation measurements 211
  electrons 32, 35, 38, 140, 148, 236–40
  exchange 238
  fine-structure effect 239
  function 237, 243
  photons 45, 200, 210
  spin–orbit mechanism 236
  S, T, U parameters 38, 245–6, 252
Polarised atoms
  analysis 43
  sources 39
polarised electrons
  analysis 34
  sources 32
  S, T, U parameters 38, 245–6, 252
position sensitive detectors 27
potential matrix element
  evaluation 170, 172
  jj coupling 164, 170, 173
  LS coupling 166, 172, 173
  partial-wave expansion 101, 164,
    166
  reduced 101, 164, 166
  special cases 175–77
  spin–orbit 177
potential scattering 20, 87, 98–104,
  113, 139, 141
probability 58, 142
probability amplitude 58, 142
probe amplitude 290
pseudoresonances 179, 197
pseudoscalar 112
pseudostate method 195, 214, 218–20,
  222–4, 286
pseudostates 180, 195

quantum beats 47, 111

Ramsauer 4
Ramsauer–Kollath 14

Ramsauer–Townsend effect 4
recoil technique 9
reduced matrix elements 69
  collision potentials 164, 166
  polarisation potential 187
  potential scattering 101
relative flow technique 21
relativistic
  distorted wave 113
  orbitals 120, 302, 306
  see also Dirac–Fock
  potential 80
  potential scattering 113, 252
representation 52
  coordinate 55, 56, 195
  distorted wave 152
  momentum 3, 55, 56, 290
  occupation number 73
  spin 64
  theorem 52
  weak coupling 292
representation theorem 52, 65
resolvent 98, 150, 153
resonance 21, 104, 220–4, 279
Ricatti–Bessel functions 83
R-matrix method 196–8, 212, 222–4, 251, 257
  intermediate-energy 197, 214, 216, 287

scattering amplitude 20, 90, 96, 97, 100, 106, 202, 204
scattering matrix 202
scattering state
  antisymmetric 157
  many-body 141, 149, 157
  one-body 91
Schrödinger equation 54
  collision 141, 152, 157
  Coulomb potential 85
  cubic box 81
  local central potential 82, 91
  matrix solution 86

potential scattering 98
radial 91
short-range potential 84
spherical box 83
Schrödinger equation of motion 57, 140, 141
screening 121
second Born approximation 191
second quantisation 73, 117
separation energy 262, 291, 293, 295
Sherman function 237, 243, 245
Slater-type orbitals 87, 124, 195
S-matrix element 94, 96, 100, 191
sodium
  spectroscopy 136
  target for ionisation 307
  target for scattering 225–9, 248–51
spectral function 291, 293
spectroscopic factor 293, 298–9, 301, 303, 307
spectroscopic sum rule 293, 296
spherical Bessel functions 83
spherical harmonics 63
  renormalised 63
spherical tensor 68, 169, 207
spin of an electron 64, 79
spin–orbit potential 35, 80, 120, 177, 236
spinors 64, 79, 113
S-polarisation parameter 38, 245, 252
state of a system 50
state multipoles 48, 207, 219, 255, 257
  normalised 255, 257
state parameters 204
stationary state 58, 141
statistical uncertainties 29
Stern–Gerlach magnets 40, 41
Stokes parameters 209–12, 255, 257
structure amplitude 290, 299, 304
STU parameters polarised electrons 38, 245–6, 252
Sturmians 129, 179, 195
sum rule spectroscopic 293, 296

superelastic scattering 48, 248
swarms, electron 12
symmetry configurations 128, 130,
    132, 292
symmetry manifold 116, 128, 292

Temkin–Poet problem 179
tensor operators 68, 169, 207
three-body model of ionisation
    262–74
threshold ionisation 275
time-dependent scattering 107, 140
time-development operator 58, 79, 142
time evolution of excited states 47
time-of-flight technique 10
time reversal 70, 103, 143, 150, 155
T-matrix element
  charged target 103, 152
  collision 145, 151
  ionisation 149, 264–9
  partial-wave expansion 101, 165,
    167
  polarisation 187–90
  potential scattering 100
  reduced 101, 165, 167
  relativistic 114
  resonance 105
  scattering 100, 148
total cross section
  helium 231
  hydrogen 217
  potential scattering 90, 97
  sodium 229
total elastic cross section 96
total ionisation asymmetry 288
total ionisation cross section
  helium 287
  hydrogen 217, 286
  magnesium 234
  threshold 278
total reaction cross section 97
T-polarisation parameter 38, 245, 252
transition rate 146

transport coefficients 12
triple differential cross section 25, 30

uncertainty principle 107, 111
unitarised distorted-wave Born
    approximation 152, 191–2
unitarity 94, 191
units 2
U-polarisation parameter 38, 245, 252

vacuum for scattering experiment 14
variation method 116, 124, 126, 191
vectors
  bra and ket 50
  brackets 50
  scalar products 50
  state 50

Wannier theory of ionisation 276
wave packet 61, 107, 142
weak coupling
  approximation for spectroscopy 292,
    296
  distorted-wave Born 154
  ion states 292
  Q-space 183
  representation for spectroscopy 292,
    296
  verification ion states 295–8, 303,
    307
  verification Q-space 186
Wien filter 37
Wigner–Eckart theorem 69, 170
Wigner symbols
  3-j 66
  6-j 67
  9-j 67

xenon
  configuration interaction for ion 303
  Dirac–Fock orbitals 302
  target for ionisation 302
  target for scattering 252
  valence structure of ion 302